BUILDING BLOCKS OF

CIVIL ENGINEERING

BUILDING BLOCKS OF

CIVIL ENGINEERING

Multiple Choice Questions, True & False Type Questions and Fill-in-the Blanks For all Competitive Examinations

PROF. (DR.) SANTOSH KUMAR

KHANNA BOOK PUBLISHING CO. (P) LTD.

Publisher of Engineering and Computer Books

4C/4344, Ansari Road, Darya Ganj, New Delhi-110002

Phone : 011-23244447-48 **Mobile:** +91-9910909320

E-mail : contact@khannabooks.com

Website : www.khannabooks.com

Dear Readers,

To prevent the piracy, this book is secured with HIGH SECURITY HOLOGRAM on the front title cover. In case you don't find the hologram on the front cover title, please write us to at contact@khannabooks.com or whatsapp us at +91-99109 09320 and avail special gift voucher for yourself.

Specimen of Hologram on front Cover title:

Moreover, there is a SPECIAL DISCOUNT COUPON for you with EVERY HOLOGRAM.

How to avail this SPECIAL DISCOUNT:

Step 1: Scratch the hologram
Step 2: Under the scratch area, your "coupon code" is available
Step 3: Logon to www.khannabooks.com
Step 4: Use your "coupon code" in the shopping cart and get your copy at a special discount
Step 5: Enjoy your reading!

Thanks,
Khanna Publishing House

Price : ₹ 498.00

ISBN: 978-93-55383-37-2

BUILDING BLOCKS OF CIVIL ENGINEERING
by **Prof. (Dr.) Santosh Kumar**

First Edition: 2025

Published by:
Khanna Book Publishing Co. (P) Ltd.
CIN: U22110DL1998PTC095547

Visit us at: **www.khannabooks.com**
Write us at: **contact@khannabooks.com**

To view complete list of books,
Please scan the QR Code:

Printed in Bharat.

FOREWORD

It gives me immense pleasure to introduce this comprehensive compilation of Multiple Choice Questions (MCQ), Fill in the Blanks and True and False questions in Civil Engineering, meticulously designed to cater to the needs of students, aspirants and professionals. In an era where competitive examinations demand not only technical expertise but also sharp analytical and problem-solving skills. This book, I am sure, serves an invaluable resource to bridge the gap between theoretical knowledge and its application.

Civil Engineering, often regarded as the backbone of infrastructure development, encompasses a wide array of disciplines, from Water Resources Engineering, Structural Engineering, Geotechnics, Environmental Science to Transportation Engineering. Mastering such a diverse field requires an organised and thorough approach. I find that this book surely guide learners towards achieving that mastery. The carefully designed MCQs, True and False type questions and Fill-in-the Blank type topics providing clarity, enhancing understanding, offer a systematic pathway to self-assessment.

Besides serving as a preparation tool for competitive examinations like ESE, GATE or Other state level civil engineering services examinations, the book may help in promoting intellectual growth and confidence.

The author of this book Professor Santosh Kumar had been my colleague and a friend for the last almost five decades. In him I find a person of high degree talent, as a teacher, a researcher and a social activist. During his tenure as a teacher in Bihar College of Engineering (Now NIT Patna) he had been at the centre of various activities.

I am sure the present book will prove to be a unique work among other similar books in the market.

Prof. (Dr.) Anil Kumar Sinha
Former Principal, Bihar College of Engineering
(Now NIT Patna)

PREFACE

The discipline of Civil Engineering is vast and diverse, encompassing essential knowledge and skills, that form the backbone of infrastructure development. Preparing for various competitive examination in this field demands a strong conceptual foundation, technical expertise and above all question-answering ability.

This book has been meticulously compiled to serve as comprehensive resource for aspirants preparing for various civil engineering examinations. The book focuses on three key formats of questions-Multiple Choice Questions (MCQ), True/False Statements and Fill-in-the Blanks, ensuring a well rounded approach to knowledge assessment.

The content of this book, spread over 23 sections, has been structured to cover topics to Civil Engineering including Structural Analysis, Soil Mechanics, Fluid Mechanics, Transportation Engineering, Environmental Science, Surveying, CPM, Geology and Engineering Mathematics. At the beginning of each section some important definitions are also provided to supplement main script.

There are many publications mostly providing MCQs. Addition of True/False and Fill-in-the Blanks provide opportunity to the users of this book of self-assessment.

The book is the result of dedicated effort, research and practical insights into the needs of Civil Engineering aspirants. It is designed not only to help Engineering graduates excel in their competitive examinations but also to foster deeper understanding of the subject, which is essential for professional growth.

Users of this book are advised to keep consulting their text books in case some answers are not clear or precise. In this context constructive feedback and suggestions for improvement are always welcome.

Any book of this magnitude cannot see the sumrise without physical and moral support of my own people. My wife Professor Bharti S. Kumar, Former Dean, Faculty of Social Science, Patna University, son Gautam a software engineer in Singapore, daughter-in-law Sungdhi, an economist Daughter Dr. Gogi, son-in-law Ajay Agrawal, Grand Children Yash, Suhani and Tara all acted like catalysts.

I appreciates the moral support of my friends Prof. Anil Kumar Sinha and Prof. R.P. Kashyap. Finally, I concede that without the support of my computer expert Mr. Pankaj Choudhary of Essen Computer this book might not have seen the morning sun.

Thanks are also due to publishers of others MCQ books whom I consulted before starting this project.

Santosh Kumar

ABOUT THE AUTHOR

Prof. (Dr.) Santosh Kumar had his technical education in Civil Engineering from R.I.T. Jamshedpur (Now NIT/J), Roorkee University (Now IIT/R) and Delft University (Netherlands). He had been University Professor at Bihar College of Engineering a constituent of Patna University, where he taught for almost forty years. After retirement he was World Banks Projects Consultant for four years. Before this he was Principal of Apex Institute of Technology (Jaipur), S.J. Engineering College (Dausa) and Gurukul Institute of Engineering & Technology (Kota).

During his teaching years he guided number of Ph.D. and a large number of Master of Engineering dissertations. He has also published more than fifty National and International Research Papers.

He was also consultant to World Bank Projects in Bihar for four years.

He has authored several technical books on subject of Irrigation Engineering, River Engineering and a book for Competitive Examinations and co-authored a book on Environmental Science.

He had been a visiting Professor in Patna University, Magadh University and Patliputra University.

Prof. Santosh Kumar is a well known personality in Social Circles of Patna. Every evening he provides free tuition to poor kids of class VI to X. He had been a noted theater activist having directed more than ten full length stage place and several corner plays of social importance.

CONTENTS

SECTION 1

Airport Engineering

Definitions – 50

Multiple Choice Questions – 101

True & False – 51

Fill in the Blanks – 54

DEFINITIONS

Airport Engineering

Aircraft Capacity: The aircraft capacity determines the number of passengers, baggage, cargo and fuel that can be accommodated in the aircraft. The terminal facilities are planned to receive the aircraft of highest capacity likely to land.

Aircraft Rescuer and Fire fighting (ARFF): These are services dedicated to responding to aircraft emergencies, equipped with specialized vehicles and trained personnel.

Aircraft Speed: It can be referred in different ways. However, **Air speed** and **Ground speed** are worth mentioning. The term **air speed** is used to mean the speed of aircraft relative to the medium in which it is flying. The **ground speed** (or **Cruising speed**) is the speed relative to ground. If the wind is blowing in opposite direction of aircrafts movement, the air speed will be less than the ground speed and vice versa.

Airport Capacity: The landing and takeoff are considered as two main movements of each aircraft. The landing is given priority over take-off. The term **airport capacity** is used to indicate the aircraft movements which an airport can permit within a specified period of time by allowing acceptable time of delay to the departing aircraft. The most important factor affecting airport capacity is the spacing between successive aircrafts.

Apron: It is a defined area of the airport to accommodate aircrafts for loading and unloading of cargo and passengers, parking, refueling etc. It is usually paved and is located in front of the building or adjacent to hangers.

Cargo: The term **Cargo** is used to indicate the freight, other than passengers, baggage and mail which is carried by a transport aircraft.

Control Tower: A facility that keeps an eye on aircraft operations on the ground and in airspace around the airport, ensuring safety and efficient traffic flow.

Correction for Elevation to basic Runway Length: As per ICAO, the basic runway length should be increased at the rate of 7% per 300 m rise in elevation above Mean Sea Level (MSL). This correction is required because air density reduces as the elevation increases.

Drainage: For safe and uninterrupted operation of air traffic the airport and adjoining areas should be completely free from inundation. For this an efficient drainage system has to be designed.

Environmental Impact Assessment (EIA): It is the procedure for finding out the potential effects on the environmental system due to the construction of a new airport or the expansion of an existing one. The EIA is particularly true for large developments where serious environmental issues may come up like- Ecological changes, pollution, socio-economic factors etc. It is now mandatory.

Flaps: It is a hinged section of an aeroplane wing, used in landing or takeoff. Flaps when projected in air, produce an immediate reduction in speed of aircraft and thus, they are intended to serve as air brakes.

Flight Time: The total time from the moment an aircraft moves under its own power for the purpose of taking off to the moment it comes to rest at the end of the flight is known as **flight time.**

Flight Visibility: This term is used to indicate the average range of visibility in forward direction from the cockpit of an aircraft in flight.

Fuselage: It describes the shape of the Central-tube shaped part of airplane. Wings, tails, engines are all parts of an airplane that are attached to fuselage.

Gate Position: The space allotted to an aircraft parking at a loading apron is known as **gate position.**

Gate: The term gate is used to denote an aircraft parking space adjacent to a terminal building and used by a single aircraft for loading and unloading of passengers, baggage and cargo.

Grading Plan: This plan shows the original and final contours of the proposed runway, apron and taxiway. The ultimate development is to be shown in dotted lines together with typical sections required.

Hanger: It is a large shed erected at the airport for housing the aircrafts particularly for the purpose of servicing and repairing of aircrafts.

Head Wind: The runway is usually oriented in the direction of prevailing winds. The head wind indicates the wind from the opposite direction of the head or nose of the aircraft while it is landing or taking off. During landing it provides a braking effect and during takeoff it provides greater lift on the wings of aircraft.

Height Zoning: The height or hazard zoning is mainly aimed to protect the approaches to the airport from obstructions. It regulates the height of structures on land surrounding the airport. The height is fixed depending upon the size of airport and type of aircrafts involved.

Holding Apron: It is the area placed adjacent to the ends of runways to allow checking of aircraft engine and equipments prior to takeoff and also to wait till clearance for takeoff is given.

Horizontal Surface: The imaginary horizontal surface which is circular in plan and which is located at a level of 45 m above the airport established elevation is known as **horizontal surface.**

Instrument Runway: The runway which is adequately equipped with the radio beam facilities and on which landing can be made according to ILS rules known as instrument runway.

International Airport: It is an airport which handles International air traffic and functions according to International aviation rules made by ICAO. It serves as a place of entry and departure from the country and necessary facilities for customs, immigration and other procedures are also available. The minimum area of an International airport is to be 40 km^2, but 80 km^2 is more reliable figure.

Lighting Plan: Proper lighting of airport is essential for safety purpose. The layout plan of runway, apron and taxiway is prepared and then the location of lights, cable connections, controlling ducts and back up facility etc. are marked on this plan.

Master Plan: Usual practice is to start any airport with minimum essential requirements and to go on developing the airport as the demand for air traffic increases. The main purpose for preparing a **master plan** is to provide specific details for future development to satisfy aviation requirements. Thus the master plan must take into consideration the detailed projections of the level of demand on various facilities associated with the airport.

Nose Hangars: These are provided for large size aircrafts. They enclose only the forward portion of the aircraft and leave the major portion of the aircraft fuselage and tail surfaces exposed. Such design provides easy servicing of aircraft.

Number of Gates: The number of gates is determined in such a way that a pre-determined hourly flow of aircraft can be easily and conveniently accommodated. Thus, the number of gates required will depend on - Peak hour volume, Gate occupancy time, Restricted or Unrestricted use of gates etc.

Pay Load: The term **pay load** is used to mean the total revenue-producing load which includes the weight of passengers and their baggage, mail and cargo.

Pressure Altitude: The altitude at which the pressure corresponding to the standard atmosphere is obtainable is called **pressure altitude.**

Range: The distance that an aircraft can fly without refueling is known as its **range.** The range is influenced by the pay load, usually as the range is decreased, the pay load is increased and vice versa.

Rudder: When aircraft is in flight it is one of the major controls. It helps the pilot to turn the nose of the aeroplane in any required direction. It can move to and fro about a vertical axis through about 30°. It is usually hinged to the fin at the tail.

Runway: It is a long and comparatively narrow strip of land which is selected and prepared for the landing and takeoff of the aircraft along its length. It is usually paved.

Runway End Safety Area (RESA): It is an area provided to reduce accidents of aircrafts undershooting or overrunning the runway. Its shape is rectangular and symmetrical about runway lines. It includes runway structural pavements, shoulders and stopways.

Standard Atmosphere: It is an imaginary atmosphere of **dry air** and it has been defined by the ICAO to have the following features.

The air is a perfect dry gas, the temperature at sea level is 15°C and pressure at sea level is 760 mm of Hg and the temperature gradient from sea level to the altitude at which the temperature becomes –15.5°C is –0.0065°C per meter and zero above.

STOL: It is the abbreviation indicating 'Short Take Off and Landing'.

Stopway: It is the rectangular area at the end of runway in the direction of takeoff in which an aircraft can be stopped after an interrupted takeoff. Its width is equal to the width of runway and thickness sufficient to bear the weight of the aircraft.

Surveillance Radar: It provides an overall picture of the surrounding atmosphere within a radius of 50 km to 100 km. It moves through 360° and the information about any aircraft within the range is received and positions of the aircraft are displayed on a screen.

Taxiway: It is the defined path on a land aerodrome, selected or paved for the use of taxiing aircraft to and from the runway and loading apron is known as **taxiway.**

Taxiway on Busy Airports: It is desirable to locate the taxiway at various points along the runway on busy airports. It is then possible for the landing aircrafts to leave the runway as early as possible making room for other aircrafts. Such taxiways are called – **exit taxiways** or **turn-offs.**

Taxiway Turning Radius: A horizontal curve is to be provided whenever there is a change in direction of the taxiway. For comfortable turning large radius is required. The radius should be, $R = \frac{V^2}{125\,f}$ (m) where V = speed in KMPH and f = coefficient of friction (approximately 0.13).

Terminal Area: It is the portion of the airport other than the landing area is called terminal area. It includes terminal building, aircraft apron, cargo storage building, hangars, automobile parking area etc.

Threshold: The beginning of the runway portion to be used for landing is known as threshold.

Traffic Survey: In this, investigations are carried out to predict the probable amount of traffic including expected future traffic. The survey should include– Facilities available, types of aircrafts, volume of traffic, capacity of runway, passengers' facilities etc.

Type A-Airport: Type-A airports can accommodate aircrafts with wingspan of upto 15m, such as small propeller planes.

Type B-Airport: These can accommodate aircraft with wingspan of upto 24 m; such as regional jets and turboprops – Effective Grade – 1.5%; Transverse Grade – 2%; Longitudinal Grad – 2%.

Type C-Airport: Airports that can accommodate aircrafts with a wingspan of 36 m, such as narrow body airliners. All Grades – 2%.

Visibility: The greatest distance to which a prominent object of certain specified dimension is perceivable to the eye, the object being observed under proper light and under the existing atmospheric conditions is known as **visibility**.

Wind Rose: The diagram showing direction, duration and intensity of wind over a certain period in a specified region is known as wind rose. It resembles a rose.

Zoning: It refers to the enactment of legislation for a restricted development of the area surrounding the airport so that no structure protrudes above the obstruction clearance line and thus cause hazard to safe air navigation, especially in the approach and turning areas.

MULTIPLE CHOICE QUESTIONS

1. In which area of airport loading and unloading of cargo and passengers, parking refuelling of aircraft is carried out.

(a) Apron (b) Taxiways (c) Runway (d) Hanger

2. All markings on runway of an airport are

(a) Yellow (b) White (c) Black (d) Red

3. Place where aircraft maintenance is done?

(a) Apron (b) Stockyard (c) Hanger (d) Terminal building

4. The runway Orientation is made so that landing and takeoff are

(a) Against wind direction (b) along the wind direction

(c) Normal to the wind direction (d) Transverse to wind direction

5. The runway length is increased on every rise of 300 m from Mean Sea Level for airport by

(a) 4% (b) 6% (c) 10% (d) 7%

6. At a station mean of average temperature is 25°C and mean of maximum daily temperature is 40°C, what is airport reference temperature?

(a) 20.6° C (b) 25° C (c) 30° C (d) 38.6° C

7. As per ICAO, for A and B type airports maximum effective grade is
   (a) 1.75% (b) 1.5% (c) 1.25% (d) 1.0%
8. In deciding runway length in Airport which case is **not** considered?
   (a) Normal landing (b) Emergency landing
   (c) Normal takeoff (d) Engine failure
9. On a taxiway the colour of Clearance Bar Lights at holding position is
   (a) Yellow (b) Blue (c) Red (d) Green
10. A gradient of + 0.08 is followed by a gradient of – 0.07. If permissible rate of change of grade is 0.003 per 30 m, the length of the Transition Curve is
    (a) 150 m (b) 140 m (c) 160 m (d) 175 m
11. According to International Civil Aviation Organisation (ICAO) the runway length of aerodromes, have been coded by
    (a) Seven English Alphabets (b) Last Seven English Alphabets
    (c) First Seven English Alphabets (d) First Seven Natural Numbers
12. As per ICAO the minimum basic runway length of A and E type of airports will be
    (a) 1500 m and 600 m (b) 2100 m and 750 m
    (c) 1500 m and 750 m (d) 2100 m and 600 m
13. As per ICAO, minimum width of safety area for instrumental runway should be
    (a) 78 m (b) 150 m (c) 300 m (d) 450 m
14. For high temperature of 196° C, the taxi ways and aprons are made with
    (a) Asphaltic concrete (b) Rubberized tar concrete
    (c) Plain concrete (d) All these
15. Choose the correct answer
    (a) Landing speed is directly proportional to the wing loading
    (b) For constant wing loading, take off distance is directly proportional to the powder loading
    (c) Neither (a) nor (b)
    (d) Both (a) and (b)
16. The lift off distance is the distance along the centre of the runway between the starting point and
    (a) End of runway (b) Point at which craft is airborne
    (c) End of stopway (d) Point where craft attains 10.7 m height
17. An aircraft is flying in an atmosphere of 30° C and speed of 1260 km per hour, its speed is known as
    (a) Subsonic (b) Sonic (c) Supersonic (d) Mach
18. Airport elevation is the Reduced Level (RL) above Mean Sea Level (MSL) of the
    (a) Control tower (b) Highest point of landing area
    (c) Lowest point of the landing area (d) None of these
19. The thickness of the pavement is decided on load carried by
    (a) Main gears (b) Nose-wheel (c) Tail wheel (d) All these

20. The meteorological condition which influences the size and location of an airport is?
(a) Atmospheric pressure (b) Air density
(c) Reduced Level RL (d) All these

21. The distance between main gears = 10 m.
Distance of nose gear from centre of main gear = 30 m
Angle of turning = 60°, then the distance of centre of rotation from the nearer main Gear is
(a) 12.30 m (b) 11.30 m (c) 10.30 m (d) 9.30 m

22. The height of the pilots eye above the runway surface is assumed as
(a) 1 m (b) 3 m (c) 4 m (d) 5 m

23. The wind intensity during calm period in runway should be
(a) Below $\frac{4.6 \text{ km}}{\text{H}}$ (b) Above $\frac{5 \text{ km}}{\text{H}}$
(c) Between 5 to $\frac{10 \text{ km}}{\text{H}}$ (d) Below $\frac{6.4 \text{ km}}{\text{H}}$

24. Wing loading of an aircraft is
(a) Load of wings
(b) Gross total weight of the aircraft/load of the wings
(c) Gross total weight of the aircraft/Wing area
(d) Gross total weight of the aircraft/Total H.P. of engines

25. The bearing of the runway at threshold is 290°, the runway number is
(a) N 70° W (b) 290° (c) 29° (d) W 20° N

26. International Civil Aviation Organisation (I.C.A.O.) was set up in Montreal (Canada) in
(a) 1929 (b) 1939 (c) 1947 (d) 1950

27. If lift off distance of an aircraft is 2000 m, the clear way at the end of runway should not be less than
(a) 145 m (b) 152.5 m (c) 162.5 m (d) 172.5 m

28. Total correction for elevation, temperature and gradient for runway should not be more than
(a) 15% (b) 20% (c) 25% (d) 35%

29. If the width of approach area near the runway end is 150 m, the width of the approach area at a distance of 3 km from runway end will be
(a) 1500 m (b) 1200 m (c) 1000 m (d) 800 m

30. Maximum gross takeoff weight of an aircraft is
(a) Equal to maximum structural landing weight
(b) Less than maximum structural landing weight
(c) More than maximum structural landing weight
(d) Equal to the empty operating weight plus payload

31. Zero fuel weight of an aircraft is
(a) Equal to empty operating weight
(b) Equal to maximum landing weight

(c) Less than empty operating weight
(d) Equal to sum of empty operating weight and maximum pay load

32. The landing and takeoff of the aircraft is made against the wind direction. In no case the centre line of runway should make an angle with the wind direction exceeding.
(a) 10° (b) 20° (c) 30° (d) 40°

33. The engine failure case for determining the basic runway length may require
(a) Only clearway (b) Either a clear way or stop-way
(c) Only stop-way (d) Either a clearway or a stop-way or both

34. The capacity of parallel runway pattern depends upon
(a) Weather conditions and navigational aids available
(b) Lateral spacing between two runways and weather condition
(c) Lateral spacing and navigational aids available
(d) Lateral spacing, weather conditions and navigational aids available

35. The width and interval of transverse centre line bars along the extended centre line of runway in approach lighting system are
(a) 3 m and 30 m (b) 4.2 m and 30 m (c) 4.2 m and 50 m (d) 3 m and 45 m

36. The slope of the obstructions clearance line from the boundary of the Heliport should be
(a) 1 : 2 (b) 1 : 5 (c) 1 : 8 (d) 1 : 40

37. Which of the following is an example of failure in flexible pavements ?
(a) Aligator cracking (b) Mud pumping (c) Warping cracks (d) Shrinkage cracks

38. Beaufort scale is used to determine
(a) Strength of Winds (b) Direction of Winds
(c) Height of Aircraft (d) None of these

39. For a class A airport the difference of Reduced Level (RL) of higher and lower edges of the conical surface is
(a) 25 m (b) 50 m (c) 75 m (d) 100 m

40. An airport has 4 gates. If the weighted average gate occupancy time is 30 minutes and gate utilisation factor is 0.5, then the capacity of gate will be
(a) 1 aircraft per hour (b) 2 aircrafts per hour
(c) 4 aircrafts per hour (d) 16 aircrafts per hour

41. The length of runway is increased per 300 m rise above Mean Sea Level (MSL) by
(a) 3% (b) 4% (c) 5% (d) 7%

42. For a supersonic transport aircraft, the minimum turning radius of taxiway is
(a) 60 m (b) 120 m (c) 180 m (d) 240 m

43. The minimum width of the clearway is
(a) 50 m (b) 100 m (c) 150 m (d) 250 m

44. Two single runways may be arranged so has to have
(a) L-shape (b) T-shape (c) X-shape (d) All these

45. The maximum value of the angle turning of nose gear large jet aircrafts is limited to

(a) 20° (b) 30° (c) 45° (d) 60°

46. As per ICAOA, for airports serving big aircrafts, the cross-wind component should not exceed

(a) 15 kmph (b) 25 kmph (c) 35 kmph (d) 45 kmph

47. To determine basic runway length, the landing case requires that aircraft should come to a stop p% of landing distance, the value of p is

(a) 40 % (b) 50 % (c) 60 % (d) 75 %

48. For night landing thresholds are lighted

(a) Green (b) Red (c) White (d) Yellow

49. Runway threshold is indicated by a series of parallel lines starting from a distance from runway end

(a) 3 m (b) 6 m (c) 10 m (d) 15 m

50. The FAA classification of airport is based on

(a) Function (b) Geometry design

(c) Airport approach speed (d) Length of Runway

51. Which of the following is **not** a characteristic of centralised system of terminal area

(a) Passengers, cargo routed centrally (b) Passengers facilities in small units

(c) Walking distance to aircraft (d) Common facilities for different gate positions

52. Which of the following does not affect the site-selection of an airport site?

(a) Adequate access (b) Air traffic potential

(c) Sufficient air space (d) Number of ground staff

53. The length of a runway under standard atmospheric conditions is 1800 m. If the actual Reduced Level (RL) of the site is 1200 m, the design length of runway is

(a) 2360 m (b) 2460 m (c) 2560 m (d) 2660 m

54. The cruising of the aircraft is 500 KMPH. If there is a head wind of 50 KMPH, then the air speed and ground speed of the aircraft respectively will be

(a) 450 KMPH and 500 KMPH (b) 500 KMPH and 450 KMPH

(c) 450 KMPH and 450 KMPH (d) 500 KMPH and 500 KMPH

55. Castor angle is defined as the angle

(a) Formed by longitudinal axis of aircraft and the direction of movement of the nose gear

(b) Between the direction of wind and longitudinal axis of runway

(c) Between the true speed of aircraft and the crosswind component

(d) Between the horizontal and the fuelage axis

56. The **main** disadvantage of angle nose out parking configuration of aircraft is that, the

(a) Aircraft rear loading door is far away from terminal building

(b) Hot blast is directed towards the terminal building

(c) Overall apron area required is more

(d) All these

57. The air blowing at a speed of 77 knots, in the direction of aircraft flying at a ground speed of 775 knots, the air speed of aircraft, is

(a) 775 knots (b) 75 knots (c) 850 knots (d) 675 knots

58. Calm period is the percentage of time during which wind intensity is less than.

(a) 4.8 KMPH (b) 6.4 KMPH (c) 8.0 KMPH (d) 9.6 KMPH

59. For a proposed airport, the survey project provides

(a) Master plan (b) Topographic plan
(c) Grading plan (d) All these

60. The application of which diagram is used to find the orientation of the runway to get the desired wind coverage

(a) Wind Butterfly (b) Wind Cycle (c) Wind Star (d) Wind Rose

61. Which is the largest airport in India

(a) Delhi (b) Mumbai (c) Bangluru (d) Kolkata

62. The colour of Clearance Bar Lights in pavement installed at holding position on taxi ways is

(a) Red (b) Blue (c) Yellow (d) Green

63. Airports are classified on how many basis?

(a) 5 (b) 4 (c) 3 (d) 2

64. How many fly rules are there?

(a) 2 (b) 5 (c) 4 (d) 3

65. The co-ordinates of the ends of runway are (5000, 5000) and (8000, 7000), the co-ordinates of another runway are (4600, 5100) and (7000, 5300), then the co-ordinates of Airport Reference Point (ARP) are

(a) (6500, 6000) (b) (5800, 5200) (c) (6500, 5600) (d) (8000, 7000)

66. As per UK design criteria, if Load Classification Number (LCN) of aircraft is between 1.25 and 1.5 times the LCN of pavement, then number of movements allowed are

(a) Zero (b) 300 (c) 3000 (d) Unrestricted

67. Effective length of a runway is the distance between

(a) Ends of runway
(b) Point of intersection of the obstruction clearance line and the extended plane of the runway surface and the other end of runway
(c) Point of intersection of the glide path and extended plane of runway surface and the other end of the runway
(d) Ends of clear way on either side

68. The minimum width of clearway is

(a) 50 m (b) 100 m (c) 150 m (d) 250 m

69. The bearing of the runway at threshold is 290°, the runway number is

(a) N 70° W (b) 290° (c) 29° (d) 20° N

70. Which one is **not** the advantage of air transport?
(a) Save time (b) Reach inaccessible areas
(c) Relief operation in emergencies (d) None of these

71. According to ICAO the recommended length of air ports is decided on
(a) Sea level elevation (b) Standard sea level temperature (15° C)
(c) Effective gradient percentage (d) All these

72. Which of the following factors are taken into account for estimating the runway length required for aircraft landing?
1. Normal maximum temperature 2. Airport elevation
3. Maximum landing weight 4. Effective runway gradient

Select the correct combination
(a) 1, 2, 3, 4 (b) 1, 3, 4 (c) 2, 3 (d) 1, 2, 4

73. The Reduced Level (RL) of the proposed site of an airport is 2500 m above MSL. If the recommended length by ICAO for runway at sea level is 2500 . The required length of the runway is
(a) 2500 m (b) 3750 m (c) 3000 m (d) 3250 m

74. Conical surface of the approach area rise outwards
(a) 1 in 10 (b) 1 in 15 (c) 1 in 20 (d) 1 in 25

75. The slope of the obstruction clearance line from the boundary of the heliport should be
(a) 1 : 2 (b) 1 : 5 (c) 1 : 8 (d) 1 : 40

76. The centre to centre spacing of heliport lighting along the periphery of landing and takeoff area should be
(a) 2.5 m (b) 5.0 m (c) 7.5 m (d) 10 m

77. The threshold marking are
(a) 4 m wide
(b) 1 m clear space between adjacent
(c) Placed symmetrically on either side of the runway centerline
(d) All these

78. In approach areas of runways equipped with instrumental landing facilities any object within 4.5 km distance from runway end shall be considered as an obstruction if its height is more than
(a) 20 m (b) 30 m (c) 40 m (d) 51 m

79. In instrument landing system
(a) Locator Outer Marker (LOM) and Locator Middle Marker (LMM) are installed on one side and Localizer antenna on the other side
(b) L.O.M. and Localizer are on one side and L.M.M on other
(c) Localizer and L.M.M on one side and L.O.M on other
(d) None of these

80. Pick up the component not applicable to aeroplanes
(a) Wings (b) Engines (c) Air Scrents (d) None of these

81. According to I.C.A.O the slope of transitional surface at right angles to the centre line of runway is kept

(a) 1 in 4 (b) 1 in 5 (c) 1 in 6 (d) 1 in 7

82. Castor angle is defined as the angle

(a) Formed by longitudinal axis of aircraft and direction of movement of the nose gear
(b) Between wind direction and longitudinal axis of runway
(c) Between true speed of aircraft and crosswind component
(d) Between horizontal and fuselage axis

83. Pickup the correct statement from the following

(a) F.I.R stands for Flight Information Regions
(b) Radius of control area is 160 km
(c) Radius of control zone is 40 km
(d) All these

84. Consider the following of ICAO recommendations and choose the correct ones

1. The basic runway length should be increased at the rate of 7% per 300 m rise in elevation above Mean Sea Level.
2. The basic runway length after having been corrected for elevation should be further increased at 1% for every 1° C rise in airport reference temperature above the atmospheric temperature at that elevation.
3. The runway length after having been corrected for elevation and temperature should be further increased at the rate of 20% every 1% of effective gradient.

(a) 1 and 2 are correct (b) 2 and 3 are correct
(c) 1 and 3 are correct (d) All are correct

85. Pick up the correct statement from the following

(a) Air Screw converts the energy given by engine into speed
(b) The propeller which are driven by turbine engines are technically called Turboprops
(c) The aircraft which obtain thrust directly from turbine engine, are called Turbojets
(d) All these

86. Pick up the incorrect statement from the following

(a) Channelization of pavement is caused by constant use of tri-cycle gears of aircraft
(b) Blast pads are used over cohesion-less soils to resist erosion due to tremendous speed of the jet blast
(c) Over-run areas of least 300 m length on either side of the runway are provided
(d) None of these

87. Pick up the incorrect statement from the following

(a) In a single engine aeroplanes, the engine is provided in the nose of the aircraft
(b) In double engine aeroplanes, one engine on either wing is placed symmetrically
(c) In three engine aeroplanes, two engines are placed on both wings and one engine is placed in the tail
(d) None of these

88. Pick up the correct statement from the following
   (a) Landing speed is directly proportional to wins loading
   (b) Wing loading remains constant, the take-off distance is directly proportional to the power loading
   (c) Both (a) and (b)
   (d) Neither (a) nor (b)

89. Pick up the correct statement from the following
   (a) Runway edge from the landing side, is called threshold
   (b) Localizer indicates to the pilot his position with respect to the proposed alignment
   (c) The glide slope indicates the correct angle of descent
   (d) All these

90. Pick up the correct statement from the following
   (a) The concrete pavements are suitable if fuel spillage occurs
   (b) Rubberized tar concrete hot blast as well as spillage
   (c) Epoxy asphalt concrete sets in very small time
   (d) All the above

91. Pick up the correct statement from the following
   (a) Approach Zone Survey is carried out to determine the elevation of protruding obstructions above horizontal, conical and transitional surfaces
   (b) The wind data of airport is depicted in the form of a chart known as wind rose
   (c) The landing and takeoff of the aircraft is made against the wind direction
   (d) All these

92. Pick up the correct statement from the following
   (a) Approximate geometric centre of the landing area is called Airport reference point
   (b) The boundaries of horizontal surface and conical surface are marked with reference to airport reference point
   (c) The location of the airport on a map is the position of airport reference point
   (d) All these

93. Pick up the correct abbreviation from the following
   (a) L.M.M (Low-Powered Middle Marker) (b) V.H.F. (Very High Frequency)
   (c) L.O.M (Low-Powered Outer Marker) (d) All these

94. Match the designated components in List 1 with its purpose at airports in List 2.
   List 1
   (A) Basic Runway length (B) Runway capacity
   (C) Runway Geometric Design (D) Hanger
   List 2
   1. Width and Length of safety area of airport 2. Housing servicing of airport
   3. location of exit taxiway 4. Engine failure class

Options

(a) A–1, B–2, C–4, D–3 (b) A–4, B–3, C–1, D–2
(c) A–4, B–2, C–3, D–1 (d) A–4, B–1, C–3, D–2

95. As per ICAO, for A, B and C type of Airports, maximum effective, transverse and longitudinal grades in percentage should respectively are
(a) 1.0, 1.5 and 1.5 (b) 1.0, 1.5 and 2.0 (c) 1.5, 1.5 and 2.0 (d) 2.0, 2.0 and 2.0

96. The bearing of the longest line of a wind rose is 845° E, the bearing of the runway will be numbered
(a) 135° (b) 31° (c) 13° (d) Both (b) and (c)

97. Select the correct statement of the following
(a) Tar concrete pavements are suitable if fuel spillage occurs
(b) Rubberized tar concrete hot blast as well as spillage
(c) Epoxy asphalt concrete sets in very small time
(d) All these

98. The slope of the transitional surface for A, B and C type of runway shall be
(a) 1 : 5 (b) 1 : 7 (c) 1 : 10 (d) 1 : 12

99. As per ICAO, the minimum basic runway length for A and E type of airport will be
(a) 1500 m and 600 m (b) 2100 m and 750 m
(c) 1500 m and 750 m (d) 2100 m and 600 m

100. The centre to centre spacing of heliport lighting along the periphery of landing and takeoff area should be
(a) 2.5 m (b) 5.0 m (c) 7.5 m (d) 10.0 m

101. Maximum certified capacity of Air Bus 380 introduced 2009 is
(a) 853 passengers (b) 900 passengers (c) 1005 passengers (d) 690 passengers

TRUE OR FALSE QUESTIONS

1. Elevation of airport site above MSL is a factor that controls airport size. (T or F)
2. Runways are oriented in a direction against the prevailing wind. (T or F)
3. There are two types of fly rules. (T or F)
4. The Visual Flight Rule (VFR) allows the aircraft to be operated within reasonable conditions by oneself. (T or F)
5. Main cause of flight delays is weather condition. (T or F)
6. Asphalt has been used in construction of apron pavements. (T or F)
7. Indira Gandhi Rashtriy Uran Academy is at Amethi. (T or F)
8. Rajiv Gandhi National Aviation University is at Amethi. (T or F)
9. Pawan Hans is Helicopter support service. (T or F)

10. All markings on Runways are Red. (T or F)
11. Aligator cracking is example of failure of flexible pavements. (T or F)
12. For supersonic transport aircraft the minimum turning radius of taxiway is 100 m. (T or F)
13. The minimum width of clearway is 100 m. (T or F)
14. If the monthly mean minimum and maximum temperatures are 25° C and 46° C then the airport reference temperature is 35.5° C. (T or F)
15. Zero fuel weight of an aircraft is equal to maximum landing weight. (T or F)
16. For airports serving big aircrafts, the crosswind component should not be more than 35 KMPH. (T or F)
17. If a is deceleration in $\frac{m}{s^2}$ and V speed in KMPH, stopping distance should be $\frac{V^2}{51a}$ m. (T or F)
18. General aviation airports are classified according to usage into-basic utility airports and general utility airports. (T or F)
19. Basic utility general aviation airports accommodate smaller aircrafts. (T or F)
20. Basic utility airports serve airplanes with wingspans of less than 59 feet. (T or F)
21. In conventional air transport system, aircraft and airports are dependent on each other in providing service to passengers. (T or F)
22. Maximum certified capacity of airbus-380 is 380. (T or F)
23. The balanced field length for Piston-Engine Aircraft is 20 m. (T or F)
24. For a level flight Lift is equal to weight. (T or F)
25. Aircrafts usually are **rigid** in the makeup of their maximum weight. (T or F)
26. The layout plan scale for Airport terminal area should be 1 : 5000 to 1 : 10, 000. (T or F)
27. For airport surveillance Radio based equipments are used. (T or F)
28. There is mention of aircraft in Ramayan. (T or F)
29. The first aeroplane was flown by Wright Brothers. (T or F)
30. Air India is the largest airline in the world. (T or F)
31. The speed of aircraft on taxiway is lower than its speed on runway. (T or F)
32. For A and B types of airports the taxiwidth should be 15.0 m. (T or F)
33. As per ICAO, for A, B and C types of airports maximum effective grad is 1 % (T or F)
34. As per ICAO, for D and E type of airports maximum effective gradient is 2 %. (T or F)
35. Transverse gradient as per ICAO is for A , B type airports is between 0.5% to 1.5%. (T or F)
36. The height of the pilots eye above runway surface is assumed to be 5 m. (T or F)
37. The length of Runway is increased per 300 m rise above Mean Sea Level by 5%. (T or F)
38. The equation to find out the stopping distance of the aircraft is $S.D = \frac{V^2}{25.5\ d}$ where (d) is deceleration rate. (T or F)
39. Airports have fuel farms consisting of giant storage tanks. (T or F)
40. According to ICAO Runway should be provided 5 m wide drains on each side. (T or F)

41. The Runway Centreline width should be > 1 m. (T or F)
42. On the taxiway the centreline is 15 cm yellow strip. (T or F)
43. In airports wind direction is indicated by wind vane. (T or F)
44. The wind cone should be clearly visible from a height of 50 m. (T or F)
45. A strong beam of light to indicate geographical location of airport is called approach lighting. (T or F)
46. The visibility at night is the distance from which a human can see a 25 candela light. (T or F)
47. Blind landing system enables the pilot to be guided with safety on the optimum approach path for safe landing. (T or F)
48. The climbing speed of helicopter must be 6 m per minute. (T or F)
49. The diameter of landing area of helicopter should not be less than 50 m. (T or F)
50. A STOL aircrafts requires not more than 150 m of runway. (T or F)
51. Maximum certified capacity of Airbus-380 introduced in 2009 is 853 passengers. (T or F)

FILL IN THE BLANKS

1. A non-powered and lighter-than-air aircraft is known as ....
2. The freight other than passengers baggage and mail is called ....
3. The velocity of wind at 10 m height $\leq$ 1.6 KMPH is knows as .... period.
4. The main body of an aircraft to which wings and other parts are attached is called .... ?
5. The Ratio of the speed to speed of sound is known as .... Number.
6. An umbrella like device made of silk sheet with chords used to jump out of aircraft is called ....
7. The portion of the airport other than the landing area is called .... area.
8. The diagram showing direction, duration and intensity of wind is known as ....
9. The distance that an aircraft can fly without refuelling is called the ....
10. For A-type airport minimum basic runway length should not be less than .... m.
11. A wheeled staircase rolled up in airplane for use in getting on or of is called ....
12. In airports Approach Surface is .... in shape.
13. In Airports conical is .... in shape.
14. Airports are involved in two types of zoning, one is Height zoning and other is .... zoning.
15. The minimum area required for an International Airport is .... km^2.
16. The major Environmental impact of any airport is .... pollution.
17. Percentage of space in airports required for Airline operation is .... %.

18. The building provided for housing and maintenance of aircrafts is called ....
19. For large aircrafts .... hangars are provided.
20. For small aircrafts .... Hangars are provided.
21. The paved area used for lading and unloading of aeroplanes is called ....
22. Full form of AAI ....
23. The GAGAN project has technological collaboration with ....
24. According to BEAUFORT Scale Velocity of wind at 10 m above flat ground of 110 Km/Hour is termed as ....
25. Full form of STOL ....
26. The greatest distance to which a prominent object of specified dimension is perceivable by eye is called ....
27. The wings have the shape of an ....
28. The total revenue-producing load is called .... load.
29. Overall Sound pressure level or noise is measured in the units ....
30. Runway threshold is indicated by a series of parallel lines starting from a distance of .... m from runway end.
31. In approach area of runway any object within 4.5 km distance from runway end is considered obstruction if the height is more than .... m.
32. As per ICAO minimum width of safety area for instrumental runway should be .... m.
33. An airport has 4 gates with average gate occupancy 30 minutes and gate utilization factor 0.5, the capacity of gate per hour will be .... aircrafts.
34. As per ICAO the minimum basic runway length for E-type airport will be .... m.
35. Air speed is the speed of aircraft relative to ....
36. The cruising speed of the aircraft is 500 KMPH. If there is a head wind of 50 KMPH then the air speed will be .... KMPH.
37. Aircraft should come to rest on landing within .... % of the landing distance.
38. The interval of transverse centre line bars along the extended centreline of runway in approaching lighting system are .... m.
39. The Runway length after elevation and temperature corrections should be further increased @20% for every .... % of effective gradient.
40. If standard atmospheric temperature at Sea Level is 15° C, then standard temperature at an altitude of 800 m will be ....
41. According to ICAO all marking on the runway are of .... colour.
42. For supersonic transport aircraft the minimum turning radius of taxiway is ....
43. Airports are classified on how many basis ....
44. The FAA classification of Airports is based on Airport .... speed.
45. Runways are oriented in the .... of the prevailing wind.

46. The application of .... diagram is used to find the orientation of the runway to get the desired wind coverage.
47. The International Civil Aviation Organisation (ICAO) was created in the year ....
48. Term used by FAA to identify very busy commercial service airports as measured by passengers emplanement is ....
49. The taxiway provided to minimize the Runway occupancy time by landing aircraft is called .... Taxiway.
50. Approximately there are about .... airports in the world.
51. India has .... number of airports.
52. U.S.A has more than .... airports.
53. Largest airport in terms of area is in ....
54. The busiest airport in the world is in .... USA.

ANSWERS

Multiple Choice Questions

1. a	**2.** b	**3.** c	**4.** a	**5.** d	**6.** c	**7.** d	**8.** b	**9.** a
10. a	**11.** c	**12.** d	**13.** c	**14.** d	**15.** d	**16.** a	**17.** b	**18.** b
19. a	**20.** d	**21.** a	**22.** b	**23.** d	**24.** c	**25.** c	**26.** c	**27.** d
28. d	**29.** b	**30.** c	**31.** d	**32.** c	**33.** d	**34.** d	**35.** b	**36.** c
37. a	**38.** a	**39.** d	**40.** c	**41.** d	**42.** c	**43.** c	**44.** d	**45.** d
46. c	**47.** c	**48.** a	**49.** b	**50.** c	**51.** b	**52.** d	**53.** c	**54.** a
55. a	**56.** b	**57.** d	**58.** b	**59.** d	**60.** d	**61.** a	**62.** c	**63.** b
64. a	**65.** c	**66.** b	**67.** b	**68.** c	**69.** c	**70.** d	**71.** d	**72.** d
73. b	**74.** c	**75.** c	**76.** c	**77.** d	**78.** b	**79.** a	**80.** d	**81.** d
82. a	**83.** d	**84.** a	**85.** d	**86.** d	**87.** c	**88.** c	**89.** d	**90.** d
91. d	**92.** d	**93.** d	**94.** b	**95.** a	**96.** b	**97.** d	**98.** b	**99.** d
100. c	**101.** a							

True & False

1. T	**2.** F	**3.** T	**4.** T
5. T	**6.** F	**7.** T	**8.** F Rai-Braeli
9. T	**10.** F White	**11.** T	**12.** F 180 m
13. F 180 m	**14.** F 32°	**15.** F Sum of Empty Operating weight + Maximum pay load	
16. T	**17.** F $\frac{V^2}{25.5a}$	**18.** T	**19.** T
20. F 49 feet	**21.** T	**22.** F 853	**23.** F $\frac{11m}{35'}$
24. T	**25.** F Flexible	**26.** T	**27.** T
28. T	**29.** T	**30.** F 16th	**31.** T
32. F 22.5 m	**33.** T	**34.** T	**35.** T
36. F 3 m	**37.** F	**38.** T 7%	**39.** T
40. F 10 m	**41.** F 90 cm	**42.** T	**43.** F Wind cone
44. F 30 m	**45.** F Beacon	**46.** T	**47.** T
48. T	**49.** F 30 m	**50.** T	**51.** T

Fill in the Blanks

1. Bolloon	**2.** Cargo	**3.** Calm	**4.** Fuselage
5. Mach	**6.** Parachute	**7.** Terminal	**8.** Wind rose
9. Range	**10.** 2100 m	**11.** Ramp	**12.** Trapezoidal

13. Circular **14.** Land use **15.** 40 Km^2 **16.** Noise

17. 38% **18.** Hangar **19.** Nose **20.** T

21. Apron **22.** Airport Authority of India **23.** ISRO

24. Storm **25.** Short Takeoff and Landing **26.** Visibility

27. Aerofoil **28.** Pay **29.** Decibel (dB) **30.** 6 m

31. 30 m **32.** 300 m **33.** 4 (Four) **34.** 600 m

35. Wind **36.** 450 KMPH **37.** 60 % **38.** 30 m

39. 1% **40.** 9.8° C **41.** Yellow **42.** 180 m

43. 4 (Four) **44.** Approach **45.** Direction **46.** Wind rose

47. 1947 **48.** Hub **49.** Exit **50.** 4000

51. 48 **52.** 14,000 **53.** Saudi-Arabia (776 Km2)

54. Atlanta (USA)

SECTION 2

Construction-Planning and Management

Definitions – 25

Multiple Choice Questions – 66

True & False – 26

Fill in the Blanks – 34

DEFINITIONS

Construction-Planning and Management

Bar Chart Scheduling: It is the conventional method of scheduling used in construction industry. It is also known as **Gantt Chart**. It consists of two co-ordinate axes, one showing the time and the other showing jobs or activities to be performed. Each job is depicted on the horizontal axis. Length of the bar shows duration of activity.

Construction Planning and Management (CPM): CPM involves the systematic process of organising, scheduling, and overseeing construction projects to ensure they are completed on time, within budget, and according to specified quality standards.

Construction Safety Management: It focuses on ensuring that workers and the site are safe throughout the project. It includes strictly following the safety regulations, provide training, using safety equipments, and monitoring on-site injuries or deaths.

Critical Path: The critical path is a sequence of project tasks that determines the minimum time required to complete the project. Any delay in critical path task will cause delay of the overall project. Tasks on critical path have zero slack time.

Critical Path Method (CPM): It is a project management tool used to determine the longest sequence of activities in a project, called the **critical path**. It helps identify tasks that are dependent on others and critical to meeting project deadlines. Delays in tasks on the critical path will directly affect the overall project completion time.

Earliest Finish Time (EF): The earliest point of time that a task can be completed. It can be calculated by adding the task's duration to its earliest start time. (EF = ES + Duration).

Earliest Start Time (ES): The earliest point of time that a task can start, given the project's start date and the duration of preceding tasks.

Economic Perspective of CPM: This refers to the optimization of resources such as labour, materials, and equipment to reduce costs while ensuring quality and timely delivery of construction projects.

Indirect Project's Cost: It consists of expenditures which cannot be clearly allocated to individual activities of the project e.g. establishment charges, insurance money, administration charges, bonus etc. It normally increases with time.

Latest Finish Time (LF): It is the latest point of time a task can be completed without delaying the project's overall completion.

Latest Start Time (LS): It is the latest point of time that a task can start without delaying the project's overall completion.

Management Perspective of CPM: It is the discipline of defining project goals, estimating resources, scheduling tasks, managing project risks, and ensuring seamless communication among stakeholders to achieve objectives.

Network Planning: The main objective of network planning is to compete the work within stipulated time and at minimum overall cost. Sometimes it becomes necessary to accelerate project completion which can be achieved by reducing the duration of critical activities through deployment of additional manpower and supply of material and equipment.

Planning: It involves formation of a number of alternative realistic work plans for achieving specific objectives and finally selecting one plan which is best suited from the stand-point of available resources and constraints imposed upon the project.

Project Duration: It is the total time required to complete the project. It is determined by the duration of **critical path**.

Project Network Diagram: It is the visual representation of the project's tasks, their durations and their inter-dependence. This diagram helps identify the critical path and other important sequences of activities.

Project's Direct Cost: It consists of expenditure which are directly chargeable to and can be identified specifically with activities of the project e.g. material cost and workers cost. It increases if duration increases beyond normal time.

Quality Control (QC): It is the process of monitoring construction work to ensure that it meets the specified standards, regulations and clients expectations. It includes inspection, testing of materials, and continual assessment to identify and correct any defects and deviations from the required quality standards.

Resource Allocation: It is the process of assigning and scheduling available resources in the most effective and economical way possible. The resources which allocated can be- financial, physical, manpower or technological resources.

Risk Management: Risk management in construction planning involves identifying, assessing, and mitigating potential risks that could impact the project, such as delays, cost overruns, safety hazards, or supply chain disruptions. It involves creating strategies to minimise the likelihood and impact of **risks.**

Scheduling: It is the fitting of the final work plan to a time scale. It shows the duration and order of various construction activities.

Slack (or Float): It refers to the amount of time that a task can be delayed without affecting the completion date. Tasks not on critical path often have slack time, which allows some flexibility in scheduling.

Stakeholder Management: It involves identifying all parties who have interest in the construction project, including clients, contractors, regulatory agencies, and the local community. Effective communication and collaboration with stakeholders ensure the project runs smoothly and meets everyone's expectations.

Technical Perspective of CPM: It is the application of project management principles, methods, and tools in planning, co-ordination and controlling construction projects, focussing on resource allocation, risk assessment, and efficient execution.

Tender: It is an offer or invitation to bid for a project or to accept a formal offer. This term usually refers to the process through which the Government and financial institutions put forward invitation bids for large projects. These bids are to be submitted within a given deadline.

MULTIPLE CHOICE QUESTIONS

1. In a bar chart vertical (y-axis) represents
(a) Time (b) Activities of project
(c) Number of workers (d) Types of activities

2. Inter-dependencies between various activities is generally observed in
(a) Bar charts (b) Milestone charts (c) Network analysis (d) Job layouts

3. Gantt Charts show
(a) Comparison of actual to scheduled progress
(b) Balance of work to be completed
(c) Progressive cost of project
(d) Inventory costs

4. Gantt Chart Provides information about
(a) Break even point analysis (b) Selling price
(c) Production schedule (d) Material handling

5. Which of the following does not represent an activity ?
(a) Foundation is dug (b) Area cleaned
(c) Invitation sent (d) Site located

6. In resource levelling
(a) Total duration of project is reduced (b) Total duration of project is increased
(c) Uniform demand of resources is achieved
(d) Cost of project is controlled

7. Sensitivity analysis is the study of
(a) Comparison of profit and loss (b) Comparison of assets and liabilities
(c) Change in output due to change in input (d) Economics of costs and benefits of project

8. In PERT analysis the time estimates of activities and probability of their occurrence follow
(a) Normal distribution curve (b) Poisson's distribution
(c) Beta distribution (d) None of these

9. Free Float mainly used to
(a) Identify the activities which can be delayed without affecting the total float of preceding activities
(b) Identify the activity which can be delayed without affecting the total float of succeeding activities
(c) Identify the activities which can be delayed without affecting either preceding or succeeding activities
(d) Establishing priorities

10. .... float is defined as the time through which an activity can be delayed without affecting the succeeding or preceding activities?
    (a) Total (b) Free (c) Independent (d) Interfering
11. In PERT analysis, a critical activity has
    (a) Maximum float (b) Zero float (c) Maximum cost (d) Minimum cost
12. A critical path has
    (a) Zero slack (b) Minimum slack (c) Maximum slack (d) Infinite slack
13. Interfering float is the difference between
    (a) Total float and free float
    (b) Total float and independent float
    (c) Free float and independent float
    (d) None of these
14. The independent float affects only
    (a) Preceding activity
    (b) Succeeding activity
    (c) Particular activity involved
    (d) None of these
15. In PERT chart, the activity time distribution is
    (a) Normal (b) Binomial (c) Poisson (d) Beta
16. Critical Path Method (CPM) is good for
    (a) Small projects only
    (b) Large projects only
    (c) Any project
    (d) None of the project
17. PERT is used in preparation of
    (a) Budgeting
    (b) Scheduling
    (c) Evaluating
    (d) Finalizing
18. Playing Float concept is used in
    (a) Cash Management
    (b) Inventory Management
    (c) Receivable Management
    (d) All of these
19. If the value of variance is more
    (a) Certainty is more
    (b) Probability of certainty is more
    (c) Uncertainty is more
    (d) Probability distribution curve has unsymmetrical shape
20. The variance of an activity when pessimistic time is 10 days and optimistic time is 4 days is
    (a) 1 (b) 6 (c) 12 (d) 24
21. Probabilistic time estimates for project management is used by
    (a) CPM (b) Gantt chart (c) OC curve (d) PERT
22. In construction project planning, free float can affect which of these?
    (a) Only that particular activity
    (b) Overall completion
    (c) Succeeding activity
    (d) Preceding activity
23. From the network, the earliest start time in days for activity 5-6 will be (Number on each arrow is activity duration in days)

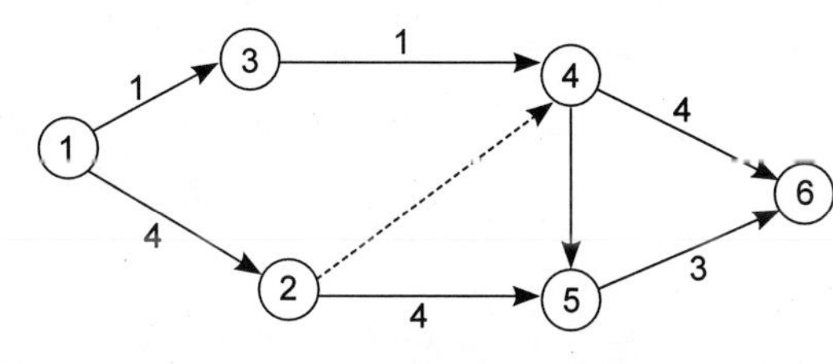

(a) 8 (b) 7 (c) 9 (d) 11

24. A merge event is represented by

(a) 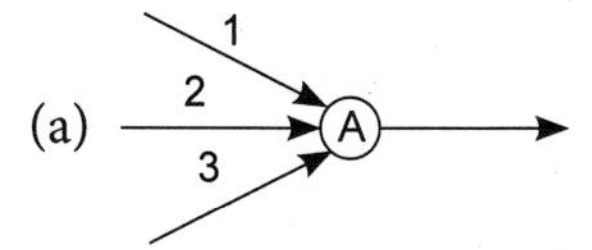

(b) 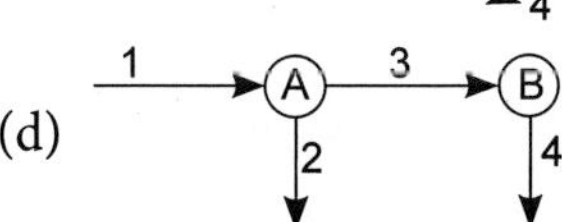

(c) —1→(A)—2→(B)—3→(C)—4→

(d) —1→(A)—3→(B), (A)—2↓, (B)—4↓

25. PERT analysis, event means

(a) Start or finish of a task (b) Time taken for a task

(c) End of an activity (d) Work involved in project

26. A PERT network has 9 activities on its critical path. The standard deviation is of each activity is on critical path is 3. The standard deviation of critical path is

(a) 3 (b) 9 (c) 27 (d) 81

27. The basis of PERT analysis is

(a) An optimistic time (b) A pessimistic time period

(c) The most likely date (d) All of these

28. Define critical path

(a) Most direct route (b) Combination of various paths

(c) Shortest route (d) Connecting start and end nodes

29. Actual progress in a bar chart is illustrated by which colour?

(a) Green (b) Blue (c) Red (d) Yellow

30. First person to use bar charts?

(a) H. Hoseph (b) Henry Gantt (c) W. Playfair (d) Rene Descarte

31. PERT requires

(a) Single time estimate (b) Double time estimate

(c) Triple time estimate (d) One time estimate

32. The total slack or float of an activity in a project network is estimated as

(a) Earliest finish time + Latest finish time (b) Latest finish time – Earliest finish time

(c) Latest start time – Earliest start time (d) Earliest start time – Latest start time

33. In PERT an event is represented by

(a) A milestone in project time line (b) A sequence of activities

(c) A resource collection

(d) A specific point in time when activity starts or finishes

34. Latest finish time in CPM technique is indicated by a

(a) Circle (b) Square (c) Line (d) Triangle

35. If t_o is optimistic time, t_p is pessimistic time and t_m most likely time then probabilistic time is

(a) $t_o + t_p + \frac{t_m}{3}$ (b) $t_o + 2t_p + \frac{t_m}{4}$ (c) $t_o + 4t_p + \frac{t_m}{6}$ (d) $t_o + t_p + \frac{4t_m}{6}$

36. Actual performance of a task is called

(a) An event (b) An activity (c) A duration (d) Any of these

37. For planning of projects, the earlier method used was

(a) Bar charts (b) PERT (c) CPM (d) Milestone charts

38. Milestone chart is an improvement over

(a) Bar chart (b) PERT (c) CPM (d) All these

39. The first stage of a large construction work is

(a) Contract (b) Design (c) Conception (d) Study and Evaluation

40. A critical activity has

(a) Maximum float (b) Minimum float (c) Zero float (d) Average float

41. The occurrence of the starting of any activity is called its

(a) Head event (b) Tail event (c) Dual role event (d) None of these

42. The occurrence of the completion of an activity is called its

(a) Head event (b) Tail event (c) Dual role event (d) None of these

43. A dummy activity in a network

(a) Is represented by a dotted line (b) Is an artificial activity

(c) Does not consume time or resources (d) All these

44. Dummy activities are used to

(a) Determine critical path (b) Find project completion time

(c) Maintain the required network (d) None of these

45. When the activity A precedes B but succeeds C then the network is

(a) (6) —A→ (7) —B→ (8) —C→ (9) (b) (6) —B→ (7) —C→ (8) —A→ (9)

(c) (6) —C→ (7) —A→ (8) —B→ (9) (d) (6) —C→ (7) —B→ (8) —A→ (9)

46. The flow diagram below depicts

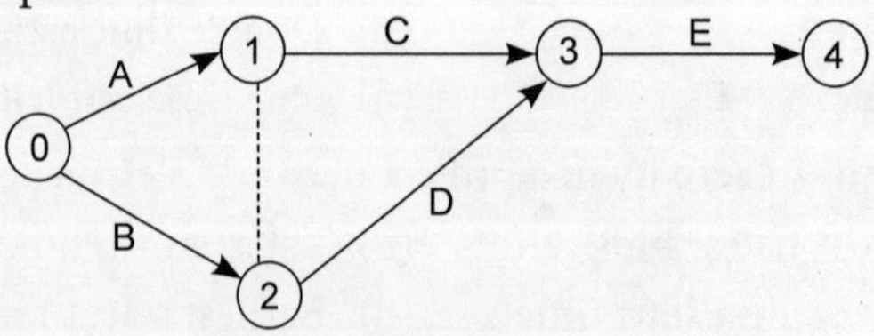

(a) Activity C can be started only after completion of A and B

(b) Activity D can be started only when B is completed

(c) Activity E can be started only when A, B, C, D are completed

(d) All of the above

47. Which of the following is an activity?
(a) Construction of foundation (b) Construction of roof
(c) Construction of super-structure (d) All these

48. The time with which direct cost does not reduce with increase in time is known as
(a) Crash time (b) Normal time (c) Optimistic time (d) Standard time

49. The system of organization introduced by F.W. Taylor is known as, ... organisation
(a) Effective (b) Functional (c) Line (d) Line and staff

50. In the network shown here the critical path is

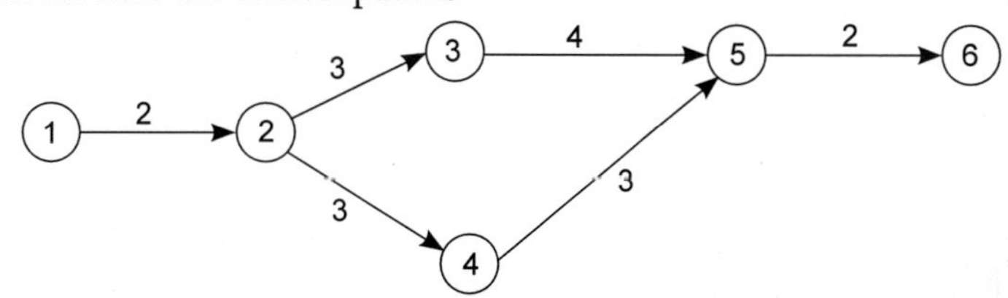

(a) 1 – 2 – 3 – 4 – 5 – 6 (b) 1 – 2 – 4 – 5 – 6
(c) 1 – 2 – 3 – 5 – 6 (d) 1 – 2 – 4 – 3 – 5 – 6

51. Which of the following network is correctly drawn?

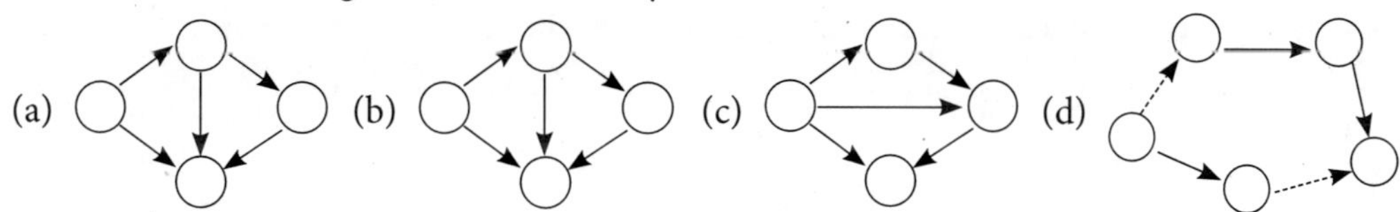

52. A PERT net has 3 activities on critical path with mean times 3,8 and 6 and standard Deviations 1, 2 and 2 respectively. The probability of project being completed in 20 days is
(a) 0.5 (b) 0.66 (c) 0.84 (d) 0.95

53. For a given size of bucket, the ideal output of a dragline will be least in
(a) Sand and Gravel (b) Common earth (c) Moist loam (d) Wet sticky clay

54. Given that,
t = Duration of various jobs
t_m = Mean time of different durations
n = Number of observations
Formula for Standard Deviation (SD) is
(a) $\frac{\Sigma t}{n}$ (b) $t - t_m$ (c) $\frac{\Sigma(t - t_m)^2}{n}$ (d) $\sqrt{\frac{\Sigma(t - t_m)^2}{n}}$

55. In the time-cost optimization using CPM method, the crashing of activities along the critical path is done starting with the activity having
(a) Longest duration (b) Highest cost slope (c) Least cost slope (d) Shortest duration

56. Which of the following is not a PERT event?
(a) Start of site investigation (b) Sessional works completed
(c) Bus starts from Patna (d) Class is being attended

57. Playing Float is a concept used in management of

(a) Cash (b) Receivable (c) Inventory (d) All these

58. What is the rule for earliest start time?

(a) It compares the activity's time to that of successor activity
(b) It compares the activities end time to that of previous activity
(c) It specifies when a project may begin
(d) It establishes the start date of project

59. The goal of network analysis is to

(a) Reduce the overall project duration
(b) Reduce entire project cost
(c) Keep delays due to interruption and conflicts to a minimum
(d) Extend the project's entire time

60. The activity which does not require any resources or time?

(a) When there is no activity (b) A dummy action
(c) Unique action (d) The preceding activity

61. The activity-on-arrow network of activities for a construction project is shown here. The duration (expressed is days) of the activities are mentioned below the arrows

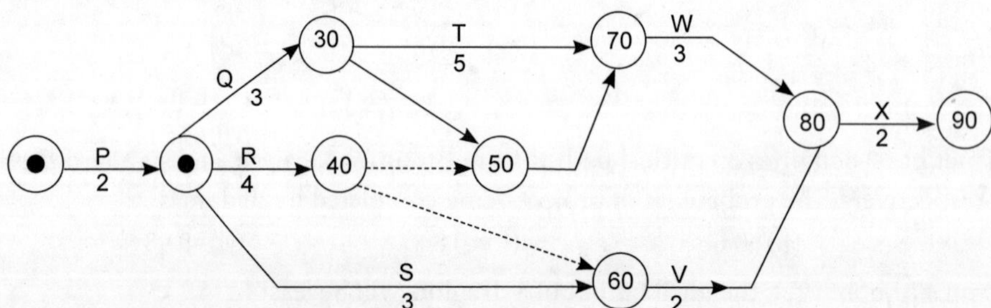

The critical duration for this construction project in days is

(a) 13 (b) 14 (c) 15 (d) 16

62. If the excavation of Earth is done manually then the cost is ₹ 10 per m^3. A machine has ₹ 4000 as fixed cost plus ₹ 2 per m^3 as variable cost. The quantity of earth for which both the costs will be same?

(a) 500 m^3 (b) 1000 m^3 (c) 1500 m^3 (d) 200 m^3

63. A bulldozer operates on a 30 m stretch and moves at 2.4 KMPH when pushing earth and returns back at 6 KMPH. The time lost in loading and shifting gears is 18 seconds per trip. Its operating factor is nearly 50 minutes per hour. What is the number of trips made per hour?

(a) 31 (b) 33 (c) 35 (d) 37

64. Given that the scope of construction work is well-defined with all drawings, specifications, quantities and estimates, which type of contract will be most preferred?

(a) EPC contract (b) % age rate contract
(c) Item rate contract (d) Lump

65. For a given activity the optimum time, pessimistic time and most probable estimates are 5, 17 and 8 days respectively. The expected time in days is
(a) 8 (b) 9 (c) 10 (d) 15

66. Which of the following is **not** an excavating and moving type equipment?
(a) Bulldozer (b) Clam shell (c) Scraper (d) Dump truck

TRUE OR FALSE QUESTIONS

1. While CPM network focuses an Events PERT's focus is on Activities. (T or F)
2. Generally PERT is preferable over CPM ? (T or F)
3. An event is a definite position of an activity. (T or F)
4. An event is a function of only one activity. (T or F)
5. The graphical representation of activities and events of any project is known as network. (T or F)
6. PERT is applied for long-time-range planning. (T or F)
7. In PERT analysis the time-estimates of activities and probability of their occurrence follow Poisson's distribution curve. (T or F)
8. The independent float affects only the particular activity involved. (T or F)
9. Total float affects preceding activities only. (T or F)
10. Free float affects succeeding activities only. (T or F)
11. The basic action involved in sheep foot rolling is kneading. (T or F)
12. Light construction is work involving light structural members on massive foundation. (T or F)
13. Proper planning, design and execution of any work is called Modular co-ordination. (T or F)
14. The process of calculating resource requirement of project is known as Scheduling. (T or F)
15. Milestone chart shows the events-chronological order, but not - logical sequence. (T or F)
16. A Gantt chart indicates balance of work to be done. (T or F)
17. When float of an activity is zero the activity is critical and any delay in its performance will delay in completion of project. (T or F)
18. A dummy activity becomes critical when its earliest start time is same as its latest finishing time. (T or F)
19. An activity of project is graphically represented by a circle. (T or F)
20. A negative float represents that programme which falls behind schedule due to shortfall of resources. (T or F)
21. CPM analysis presumes that times are related to cost. (T or F)
22. The most likely time (m) is mode of the Normal distribution. (T or F)
23. Fulkerson's rule is connected with numbering of event in PERT/CPM. (T or F)

24. Reduction in duration is represented by variance. (T or F)
25. In resource levelling uniform demand of resources is achieved. (T or F)
26. CPM and PERT use same approach to calculate critical path in a project. (T or F)

FILL IN THE BLANKS

1. Various activities of any project are represented on the .... axis of the bar chart.
2. Probability of finishing a project earlier than expected is less than .... %.
3. The process of inspiring the subordinates to work for achieving objectives is called as ....
4. The direct cost of a project with respect to normal time is ....
5. If expected time for completing of project is 10 days with S.D. of 2 days, the expected time of completion of project with 99% probability is .... days.
6. Here is a arrow diagram for a project the arrow A′ is known as .... activity.

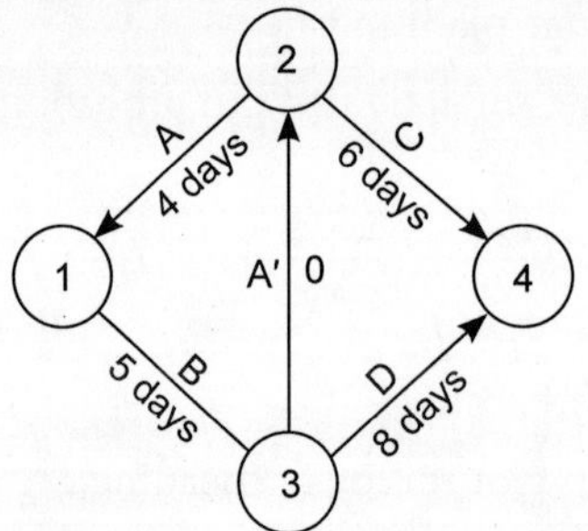

7. With respect to normal time the direct cost of a project is ....
8. The early finish time of an activity is the .... of early start time and duration of activity.
9. The difference between the Total Float and Free float is called .... float.
10. The estimated time required to complete an activity is known as ....
11. The resource in a construction project is made up of ....
12. Whenever there is difference in the planned and actual performance CPM network is to be ....
13. The working range of a crane is limited horizontally for maximum lift only by .... length.
14. Two centrifugal pumps in parallel have following characteristic

Delivery Head	Delivery	Efficiency
18 m	90 lps	60%
18 m	60 lps	50%

The overall efficiency of the system is ....

15. Total project cost Vs Item curve is a/an .... shaped curve.
16. The process of incorporating changes and rescheduling or replanning is called ....

17. Expected time of completion of a project is 10 days with standard deviation of 2 days, the expected time of the project with 99.9% probability (if Normal Deviate Z for 99.0% probability is 3) is ....

18. If the scheduled completion time of a project is more than earliest expected time for completion of the project, then the probability of completion of the project within scheduled completion time will be more than .... %.

19. Controlling the quantity of each material into each batch is called ....

20. A machine costs ₹ 16,000. By constant rate of depreciation, its salvage value after 3 years is ₹ 2000. The rate of depreciation is .... %.

21. PERT technique of network analysis is particularly useful for Research and .... projects.

22. The constraint in resource smoothening operation is project .... time.

23. A machine cost ₹ 20,000 and its useful live is 8 years, the money is borrowed at 8% interest per annum. The capital recovery factor is at 8% interest per annum of 8 years is 0.174. The annual equipment cost will be ₹ ....

24. The starts and completion of task is known as ....

25. Any activity requires time and ....

26. CPM requires .... time estimate.

27. PERT is a project planning and .... technique.

28. Military type of organisation can be termed as .... organisation.

29. F.W. Taylor's system of organisation is known as .... organisation.

30. CPM allows .... time estimates of project activities.

31. PERT uses .... time estimates of project activities.

32. PERT originated for managing large defense project in ....

33. CPM and PERT both are .... techniques used in Project management.

34. Amount of time a task can be delayed from the early start without changing the completion date is known as ....

ANSWERS

Multiple Choice Questions

1. b	**2.** a	**3.** a	**4.** c	**5.** d	**6.** c	**7.** c	**8.** c	**9.** b
10. c	**11.** b	**12.** b	**13.** a	**14.** c	**15.** d	**16.** b	**17.** b	**18.** d
19. c	**20.** a	**21.** d	**22.** d	**23.** c	**24.** a	**25.** a	**26.** b	**27.** d
28. c	**29.** a	**30.** c	**31.** c	**32.** b	**33.** d	**34.** d	**35.** d	**36.** b
37. a	**38.** a	**39.** c	**40.** c	**41.** b	**42.** a	**43.** d	**44.** c	**45.** c
46. d	**47.** d	**48.** b	**49.** b	**50.** a	**51.** a	**52.** b	**53.** d	**54.** d
55. c	**56.** d	**57.** d	**58.** b	**59.** a	**60.** b	**61.** c	**62.** a	**63.** d
64. c	**65.** b	**66.** d						

True & False

1. F CPM on Activities		**2.** T	**3.** T
4. F $\geq$ 2 activities	**5.** T	**6.** T	**7.** F β-Distribution curve
8. T	**9.** F All the activities	**10.** F All the activities	**11.** T
12. F	**13.** T	**14.** F Resource aggregation	
15. T	**16.** F Compares between actual and scheduled progress		
17. T	**18.** T	**19.** F By an Arrow	**20.** T
21. T	**22.** F β-distribution	**23.** T	**24.** F Crashing
25. T	**26.** F		

Fill in the Blanks

1. 'Y' or Vertical Axis	**2.** Less than 50%	**3.** Motivation	**4.** Minimum
5. 16 (Sixteen) days	**6.** Dummy	**7.** Minimum	**8.** Sum of
9. Interfering	**10.** Duration	**11.** Manpower	**12.** Updated
13. Boom	**14.** 55.5%	**15.** U-shape	**16.** Updating
17. 16 days	**18.** 50%	**19.** Botching	**20.** 0.5 %
21. Development	**22.** Duration	**23.** ₹ 3480	**24.** An Activity
25. Resources	**26.** Single	**27.** Control	**28.** As Line
29. Functional	**30.** Deterministic	**31.** Probabilistic	**32.** U.S.A.
33. Control	**34.** Float		

SECTION 3

Dams

Definitions – 17

Multiple Choice Questions – 55

True & False – 25

Fill in the Blanks – 21

DEFINITIONS

Arch Dam: An arch dam made mostly of concrete is curved in plan and carries a major part of its water pressure horizontally to the abutments by arch action. The site for such dam should be narrow valley (V type) with strong abutments. For its construction skilled labour and elaborate design is required.

Buttress Dam: It is a 'hollow' dam with solid upstream panels (slabs, arches or massive head) supported at intervals on the downstream side by buttresses at regular intervals. This type is preferred in wide valleys where solid rock base is not available.

Crest Gates: In controlled spillways the flow over spillway is controlled by crest gates. They are designed to prevent the water level from exceeding safe limits, ensuring the structural integrity of the dam and avoiding uncontrolled overflow. During low flows the gates are in place to provide additional storage, and during the floods they are removed to increase the spillway capacity.

Dam: It is a hydraulic structure built across a river, a stream or an estuary to retain water on its upstream side. Dams provide water for human consumption, for irrigation, for industrial and for domestic use. They are also used for generating hydro-electricity. Dams are also useful in checking floods.

Earth Dam: These are made of locally available soil and gravel by compacting successive layers of earth (soil), using the impervious material as a **core** and placing more permeable substance on the upstream and downstream portion of the dam. These can be made on relatively weaker foundation. Upstream face is protected by laying stone (Riprap) membrane against erosion by waves or wind. A downstream toe is provided to keep the downstream portion dry.

Elementary Profile: For starting the design of gravity dam only three forces are considered- weight of the dam, water pressure on the upstream face and uplift force at the bottom of the dam. For this a triangular section (like hydrostatic pressure on upstream side) is assumed and base width is designed. Proper amendments are imposed and stability checked.

Gravity Dam: They are made either of masonry (for low height dams) or concrete (for medium and high dams). Gravity dams usually run in straight line across a broad (U-type) valley and resist the horizontal thrust of the retained water entirely by their own weight. The three main forces acting on a gravity dam are Dam's weight (stabilizing Force), water pressure and uplift force (Dislodging Force).

Middle-third Rule: For no tension to develop in the gravity dam, the resultant of all external forces should always lie within the middle third portion of the dam's base.

Non-Rigid Dams: They are constructed of non-rigid material such as earth (soil) or broken rocks. Earth dams, rockfill dam and combine earth and rockfill dam are under this category.

Phreatic Line: The top flow line of a saturated soil mass (like an earth dam) below which seepage takes place, is called the Phreatic Line. Hydrostatic pressure acts below the phreatic line whereas atmospheric pressure above the phreatic line. This line is the boundary between saturated and 'dry' soil mass. In case of Earth Dams phreatic line must remain well within the dam section to avoid seepage failure.

Pore Water Pressure: It is the pressure experienced by water trapped in the voids in a saturated soil mass. In earth dams increase in pore water pressure reduces the shear strength of the soil. The variation of pore water pressure in earthen dams plays an important role in maintaining its stability. The pore water pressure within the dam is altered by the external loading like rapid drawdown, earthquake and rising water table. an increase in pore water pressure may cause earth dam's failure.

Rigid Dams: These are constructed of rigid materials such as masonry, concrete, steel or timber. Gravity dam, arch dam, buttress dam, steel and timber dam come under this category.

Rockfill Dam: A rockfill dam is an embankment which uses various size rocks to provide stability and an impervious membrane to provide water tightness. The impermeable layer is provided on the upstream face.

Spillway: It is the overflow portion of the dam over which surplus discharge flows from the reservoir to the downstream. Spillways are necessary part of the dam as a safety measure against overtopping and consequent damages and failure. Spillway is thus a safety valve. These may be provided inside the body of the dam or in the periphery of the reservoir. Among various types of spillway Ogee profile is most popular.

Stilling Basin: It is a structure provided as a means to absorb or dissipate the energy of falling water from the spillway and protect the spillway area from erosion and undermining, IS-4997-1968 lays down the criteria for stilling basin design. A hydraulic jump is formed inside the stilling basin through which energy dissipation is accomplished. Stilling basin also has chute blocks, basin blocks and end sill etc. to support energy loss.

Swedish Circle Method: It is the most popular method of stability analysis of earth dams. In this method potential failure surface is assumed to be cylindrical and factor of safety (Ratio of the average shearing strength to the average shear stress) determined by statics on potential sliding surface.

Upstream Impervious Blanket: Also known as riverside blankets, overlaying a pervious foundation. They help in reducing the quantity of seepage. To some extent they also reduce uplift pressure and **escape gradient** downstream or land side.

MULTIPLE CHOICE QUESTIONS

1. In a concrete dam main stabilizing Force is,
(a) Friction (b) Water Pressure (c) Weight (d) Uplift

2. Minimum concrete is used in which type of Dam?
(a) Gravity (b) Buttress
(c) Constant Angle Arch (d) Constant Radius Arch

3. In which seismic zone the intensity is severe?
(a) III (b) II (c) IV (d) V

4. The hydrodynamic Pressure distribution due to Earthquake is
(a) Parabolic (b) Triangular
(c) Ellipse cum parabola (d) Ellipse

5. The limiting height of a low gravity dam is about
(a) 90 m (b) 110 m (c) 125 m (d) 135 m

6. For no-tension condition the Resultant of all forces should pass through the base of Gravity dam at,

(a) Toe (b) Heel (c) Lower $\frac{1}{3}$ (d) Middle $\frac{1}{3}$

7. An elementary profile of a gravity dam considers only,

(a) Water Pressure & Weight (b) Weight & uplift
(c) Weight only (d) Weight, water pressure and uplift

8. According to (IS 6512-1984), the free board in gravity dam should be, wind set plus.

(a) $\frac{4}{3}$hw (b) hw (c) $\frac{3}{4}$hw (d) $\frac{3}{2}$hw

9. In an arch dam the central angle for minimum concrete is, about

(a) 120° (b) 133° (c) 66° (d) 140°

10. In an earth dam, the phreatic line is at,

(a) Atmospheric pressure (b) Below atmospheric pressure
(c) Above atmospheric pressure (d Depends on Height of dam

11. When Pore water pressure increases factor of safety,

(a) Increases (b) Does not depend on
(c) Decreases (d) Depends on toe filters

12. When in an Earth dam if there is high degree of stratification the drain used is

(a) Toe drain (b) Chimney drain (c) Blanket drain (d) Impervious cutoff

13. The downstream crest profile of an Ogee spillway is,

(a) $x = 2\ H_d^{0.85}\ y$ (b) $x = H_d^{0.85}\ y$ (c) $x = 2\ H_d^{0.85}\ y$ (d) $x^{1.85} = 2\ H_d^{0.85}\ y$

14. For Hydraulic Jump Upstream Froude Number is,

(a) $F = 1$ (b) $F > 1$ (c) $F < 1$ (d) $F > 1.5$

15. The stilling basin below spillway for Froude Number less than 4.5 is, of type?

(a) I (b) II (c) III (d) IV

16. The discharge over an Ogee spillway in terms of total head He is proportional to,

(a) He (b) He $\frac{1}{2}$ (c) He $\frac{2}{3}$ (d) He $\frac{3}{2}$

17. Seepage through foundation in an earth dam can be controlled by

(a) Rock Toe (b) Cut-off
(c) Horizontal blanket (d) Chimney drain

18. The major stabilizing force in a gravity dam is

(a) Water pressure (b) Self weight (c) Friction (d) Pore pressure

19. Bhakra-Nangal River Valley Project is on the rivers

(a) Ganga (b) Yamuna (c) Sutlej-Beas (d) Rabi

20. Hirakund Dam is in which state of India?

(a) Bihar (b) Odisa (c) West Bengal (d) Punjab

21. Hirakund Dam is in which river?
(a) Son (b) Ram Ganga (c) Mahanadi (d) Chenab

22. The height of Tehri Dam, the tallest in India is
(a) 301.5 m (b) 280.5 m (c) 205.5 m (d) 260.5 m

23. Damodar Valley Corporation (DAV) is in which states,
(a) Bihar-Jharkhand (b) Bihar-W. Bengal
(c) Jharkhand-Odisa (d) Jharkhand-W. Bengal

24. Sardar Sarovar Dam on Narmada River is what type
(a) Gravity (b) Earth (c) Rock (d) Arch Dam

25. Who described Dams in India as 'Temple of Modern India'
(a) Mahatma Gandhi (b) Jawahar Lal Nehru
(c) Ambedkar (d) Sardar Patel

26. Which one of the following is a non-rigid dam
(a) Gravity Dam (b) Steel Dam (c) Rockfil Dam (d) Buttress Dam

27. Arch dams are best suited for the site.
(a) Flat land (b) Plateaus (c) Narrow valleys (d) Broad valleys

28. Which of the following zone in zoned earth dam prevents piping through cracks?
(a) Central core (b) Core wall (c) Outer zone (d) Transition zone

29. Which of the following is the earliest known dam?
(a) Sada-el-kafara dam (b) Dam of Marib
(c) Jawa dam (Jordan) (d) Ha-ilar dam

30. Which of the following is the largest Hydroelectric project in Maharashtra ?
(a) Khopoli (b) Bhaira (c) Vaitarna (d) Koyna

31. Which of the following Dams is located in South India
(a) Rihand (b) Mettur (c) Gandhi Sagar (d) Tilaiya

32. Salal Dam is built on which river
(a) Chenab (b) Mahanadi (c) Krishna (d) Sutluj

33. Which of the following dams is in Karnataka ?
(a) Tung-Bhadra (b) Gandhi Sagar (c) Srisailam (d) Mnuaiperiyar

34. Ban Sagar Dam is on the river
(a) Mahanadi (b) Krishna (c) Sone (d) Sharavati

35. Tehri Dam is located at the confluence of rivers
(a) Bhagirathi & Bhilangna (b) Bhagirathi & Alaknanda
(c) Alaknanda & Vanganga (d) Banas & Khari

36. No reinforcement is required in
(a) Multiple arch buttress dam (b) Mushroom head buttress dam
(c) Massive head buttress dam (d) Free head buttress dam

37. For gravity dam for seismic zone V the zone factor is

(a) 0.24 (b) 0.36 (c) 0.12 (d) 0.16

38. The wave height (hw) for the Fetch > 32 km is given by

(a) $\sqrt{VF}$ (b) $0.5\sqrt{VF}$ (c) VF (d) $0.03\sqrt{VF}$

* (where V is wind velocity in KMPH)

39. Gallery is provided in dams to

(a) Reduce uplift pressure (b) Operate Equipments
(c) To inspect stress condition (d) All these

40. Which of the following gate is called flush boards?

(a) Dropping shutters (b) Tainter gates
(c) Drum gates (d) Vertical lift gates

41. Near the top of dam for passage of flood, structural arrangement is called

(a) Sluice way (b) Spillway (c) Gallery (d) Abutment

42. The temporary dam constructed to exclude water from specific area is

(a) Debris dam (b) Diversion dam (c) Coffer dam (d) Storage dam

43. The largest dam in the world by volume of concrete used is

(a) Three Gorges-China (b) Hoover-USA
(c) Itaipu-Brazil/Paraguay (d) Grand Coulee-USA

44. Type of dam suitable for mountainous regions is

(a) Embankment (b) Gravity (c) Arch (d) Buttress dam

45. Which dam in the world has created largest artificial lake by area?

(a) Aswan (b) Kariba (c) Bhakra Nangal (d) Hirakud dam

46. The negative impacts associated with dam construction

(a) Submergence of large land area (b) Loss of bio-diversity
(c) Earthquake possibility (d) All these

47. The three Gorges Dam is built on river

(a) Yangtze (b) Yellow (c) Mecong (d) Brahmputra

48. Which country has maximum number of large dams

(a) USA (b) Brazil (c) India (d) China

49. Which is the primary purpose of dam construction?

(a) Flood control (b) Generate energy
(c) Water supply for irrigation etc. (d) All these

50. The discharge per m length over an Ogee spillway with coefficient of discharge $C_d = 2.5$ at a head of 4 m, neglecting velocity of approach head is

(a) 200 cunces (b) 20 cumces (c) 10 cumces (d) 2 cumces

51. A medium head hydroelectric scheme is for head between

(a) 15 m to 60 m (b) 60 m to 150 m (c) 150 m to 300 m (d) 60 m to 150 m

52. A medium head turbine is

(a) Pelton (b) Francis (c) Kaplan (d) None of these

53. Draft Tube Shape can be

(a) Conical (b) Elbow (c) Moody (d) Any of these

54. Plant capacity Factor is the ratio of Average load output and ($\alpha \times$ Peak Load), where α is

(a) 1.5 (b) 1.3 (c) 1.1 (d) 0.9

55. Which of the following is longest dam in India

(a) Bhakra (b) Tehri (c) Hirakund (d) Tilaiya

TRUE OR FALSE QUESTIONS

1. In a barrage the crest level is low with large gates. (T or F)

2. Embankment dams are generally of trapezoidal shape. (T or F)

3. On weak foundations gravity dam is preferred. (T or F)

4. In earth dams compacted core is made of shale. (T or F)

5. Arch-gravity dams are thinner and less expensive. (T or F)

6. Dams are constructed for flood control only. (T or F)

7. Dams can cause ecological disruption. (T or F)

8. Dams provide positive environmental effects with no negative effects. (T or F)

9. All large dams are made only for generating power. (T or F)

10. Hydropower is a source of renewable energy. (T or F)

11. Only earth dams can be made on weak foundation. (T or F)

12. The hydrodynamic water pressure is assumed to be straight line. (T or F)

13. Silt pressure is a function of angle of internal friction (ϕ). (T or F)

14. The Factor of Safety against overturning should not be less than 1.2. (T or F)

15. The factor of safety against sliding should not be less than 1. (T or F)

16. In an elementary profile of a Gravity dam only three forces, weight of dam, uplift force and water pressure are considered. (T or F)

17. An elementary profile is of trapezoidal shape. (T or F)

18. Pre-cooling of concrete helps in control of cracks. (T or F)

19. Constant radius arch dam has vertical face. (T or F)

20. Ambersen type dam is multiple dome type. (T or F)

21. Upstream slope in earth dam fails due to sudden flood. (T or F)

22. As per USBR the minimum free board in an earth dam of over 6 m high, should be 3 m above the top of gate. (T or F)

23. Darcy's Law of seepage is analogous to Ohm's law of electric flow. (T or F)
24. If pore water pressure decreases factor of safety also decreases. (T or F)
25. In general the upstream slope of earth dams is kept as 3 : 1. (T or F)

FILL IN THE BLANKS

1. Pressure at Phereatic surface is ....
2. In a dam volume of water below minimum pool level is known as ....
3. An Overflow dam is known as ....
4. Highest gravity dam in the World is ....
5. Highest gravity dam in India ....
6. Highest dam in India ....
7. Highest dam is the World ....
8. Longest Earthen dam in the World ....
9. Who said, "Irrigation Projects are the modern temples of our Society ?" ....
10. Biggest Arch Dam in India ....
11. How many Earthquake Zones exist in India, ...
12. In a gravity dam upstream water depth of 100 m will exerts pressure force equal to ....
13. Limiting height of a **low** gravity dam is about ....
14. For no tension in a gravity dam resultant of all the forces should be within ....
15. Besides inspection facility etc, galleries in dam also help in reducing ....
16. Joints in concrete dams are provided to prevent ....
17. According to IS 6512-1984 the Free Board in a gravity dam should be wind set-up plus ....
18. The factor of safety against overturning in a Gravity dam should not be less than ....
19. The elementary profile of a gravity dam is of shape ....
20. In a gravity dam the wave pressure acts above still water at a distance of ....
21. In an arch dam major part of water load is carried to ....
22. In Arch dams the Trial Load Analysis theory is given by ....
23. Which is perhaps only buttress dam in India ....
24. Inside an earth dam top seepage line is called ....
25. The diaphram in an Embankment dam is made of ....
26. Maximum Earth Dam failures are due to ....
27. Seepage of water through the body or foundation in an earth dam leads to ....
28. During sudden drawdown the .... slope may fail in Earth dam.

29. In earth dams slopes are protected by ....
30. According to Indian standards no earth dam should have top-width less than ....
31. In stratified formation of earth dam a .... drain is given.
32. Graphical method of determining phreatic line is given by ....
33. Swedish circle method is used to check stability of .... dams.
34. If pore water pressure increases the Factor of safety ....
35. Factor of safety in earth dam is given by, $FS = \dfrac{CL + \tan\phi\,\Sigma\,(N - U)}{\Sigma T}$, Here (U) represents ....
36. If in an earth dam the foundation is pervious to large depth, to reduce loss of water an .... is provided.
37. To get maximum dry density, the .... water content is used.
38. In a rock fill dam the impervious membrane is provided at ....
39. A spillway is called the .... for a dam.
40. For downstream profile of a spillway the equation is $x^n = k H_d^{n-1}\, y$ here (n) is equal to ....
41. Morning glory spillway is the ....
42. Baby siphon in a siphon spillway helps in ....
43. In a hydraulic jump the initial and Final depths are related as $\dfrac{D_2}{D_1} = \dfrac{1}{2}\left[\sqrt{(1+F_1^2)} - 1\right]$, where F_1 is ....
44. The most commonly used spillway type is ....
45. The Water Power Development with no pondage is called ....
46. Hydro-electric plant having Turbine-cum-pump installation is called .... plant.
47. A high head hydroelectric scheme is for head more than .... m.
48. Surge tank is provided to reduce water .... pressure.
49. For heads more than 300 m the .... turbine (wheel) is most suitable
50. A draft tube helps in .... efficiency of the Turbine.
51. Load Factor is the ratio of Average Load over a certain period to .... load occurring during the same period.

ANSWERS

Multiple Choice Questions

1. c	**2.** b	**3.** d	**4.** c	**5.** a	**6.** d	**7.** d	**8.** a	**9.** b
10. a	**11.** c	**12.** b	**13.** d	**14.** b	**15.** a	**16.** d	**17.** b	**18.** b
19. c	**20.** b	**21.** c	**22.** d	**23.** d	**24.** a	**25.** b	**26.** c	**27.** c
28. d	**29.** c	**30.** d	**31.** b	**32.** a	**33.** a	**34.** c	**35.** a	**36.** d
37. b	**38.** d	**39.** d	**40.** a	**41.** b	**42.** c	**43.** a	**44.** c	**45.** b
46. d	**47.** a	**48.** d	**49.** d	**50.** d	**51.** a	**52.** b	**53.** d	**54.** c
55. c								

True & False

1. T	**2.** T	**3.** F Earth Dam	**4.** F Clay
5. T	**6.** F Multipurpose	**7.** T	**8.** F
9. F	**10.** T	**11.** T	**12.** F Parabolic
13. T	**14.** F Not less than 1.5	**15.** T	**16.** T
17. F Triangular	**18.** T	**19.** T	**20.** F Flat slab
21. F Due to Sudden drawdown		**22.** T	**23.** T
24. F Increases	**25.** T		

Fill in the Blanks

1. Atmospheric	**2.** Dead storage	**3.** Spillway
4. Grand Dixen (285 m)	**5.** Bhakra (226m)	**6.** Tehri Dam (260.5 m)
7. Nurek Dam (300 m)	**8.** Hirakud Dam (15 km)	**9.** Pandit Jawahar Lal Nehru
10. Idduki 168.91 m	**11.** Four	**12.** 49000 KN
13. 88 m	**14.** Middle Third of base	**15.** Uplift Force
16. Cracking	**17.** $\frac{4}{3}$hw	**18.** 1.5
19. Right Angled Triangle	**20.** $\frac{3}{8}$hw	**21.** Abutments
22. USBR	**23.** Mir Alam Dam (46 m high, 1298 m long)	
24. Phreatic Line	**25.** Impervious Marterial	**26.** Hydraulic Failure
27. Piping	**28.** Upstream	**29.** Riprap
30. (Six) 6 m	**31.** Chimney	**32.** Casagrande
33. Earth dam	**34.** Decreases	**35.** Pore Water Pressure
36. Upstream Impervious Blanket	**37.** Optimum	**38.** Upstream face
39. Safety valve	**40.** 1.85	**41.** Shaft Spillway

42. Priming

43. $\frac{V}{\sqrt{g^{D}}}$

44. Ogee Type

45. Run-off River Plant

46. Pumped Storage Plant

47. 60 m

48. Water Hammer

49. Pelton/Wheel (4)

50. Increasing

51. Peak

SECTION 4

Docks and Harbours

Definitions – 20

Multiple Choice Questions – 33

True & False – 15

Fill in the Blanks – 15

DEFINITIONS

Anchorage: These are where anchors are lowered and utilized. The locations usually have condition for safe anchorage in Protection from weather condition or other hazards. The purpose is of resting a ship securely at sea, for waiting to enter the port or taking on cargo or passengers where port facility is inadequate.

Ballast: It is any heavy load that is added to a ship to make it stable. When Cargo is unloaded ship becomes unstable, then ballast is added temporarily.

Beacon: It is a conspicuous device designed to attract attention to a specific location. A common example is a **lighthouse**, which draws attention that is used to navigate around obstacles or into port.

Beam: It is the measure of the width of the boat, ship at its widest point. Generally, larger the beam, the more stable is the boat/ship.

Breakwater (Jetty): It is a structure built offshore to protect a harbour, port, or coastline from the effects of waves, storms, and erosion. It creates a calm, sheltered area for ships to dock. Actually it reduces wave energy and thus prevents damage to vessels, docks, piers etc.

Cargo: It is the goods or merchandise conveyed specifically in ship or airplane. **Freight** refers to goods conveyed by air, water or **land**.

Deadweight Tonnage (or Deadweight) (DWT): It is the measurement of total contents of a ship including cargo, fuel, crew, passengers, food, and water aside from boiler water.

Dry Dock: It is a type of dock meant for the maintenance and repair of ships and boats. Water can be drained out of the basin to allow cleaning and repair of underwater components of the ship.

Estuary: It is a partially enclosed coastal body of brackish water with one or more rivers or streams flowing into it, and with a free connection to the open sea. Estuaries form a transition zone between river environment and maritime environments and are an example of **ecotone**.

Keel: It is a long piece of wood or metal on the bottom of the boat that stops it from falling over sideways in the water. It runs longitudinally along the centre of the bottom of the hull. It provides stability to the boat.

Lock: It is a mechanism used in harbour or canal to raise and lower ships between different water levels. It consists of a chamber where water can be controlled to allow vessels to pass from one level to another, often between a **wet dock** and an open sea or rivers.

Marina: It is a dock or basin provided to secure moorings for **pleasure boats** and often offering supply, repair and other facilities.

Mooring: It is a procedure to anchor the ship to a fixed or floating element and keep it connected during loading or unloading operations. Safe mooring must be able to withstand several forces like-wind, current, waves and tides.

Pier: It is a raised structure extending from shore into the water, typically used for docking ships, boats and other vessels. Piers often provide direct access from land to the vessel. They may also serve as loading and unloading point. Some piers are also used for shopping and entertainment.

Quay: It is a long, usually rigid structure beside water, where boats can be tied up and their goods can be loaded or unloaded.

Reefer Ship: It is a refrigerated ship typically used to transport perishable cargo-such as fruits, milk products meat etc. Typically, reefer containers have temperature range of –30°C to 30°C.

Riprap: These are loose stones used to form a foundation for a breakwater or other structures. Main purpose of riprap is to protect shoreline from damaging effects of waves and currents. Rock riprap involves laying large stones or concrete blocks along the shoreline to provide a barrier against erosion.

Turning Basin: It is a wider body of water, either located at the end of a ship canal or in a port to allow cargo ships to turn and reverse their direction of travel or, to enable long narrow barges in a canal to turn a sharp corner. It is also called a **winding basin** or **swinging basin.**

Warf: It is a structure built along the shoreline for ships to dock and load and unload cargo. It is similar to a **quay**, but a Warf is larger with additional facilities for loading and unloading like cranes or warehouses.

Wet Dock: This type of dock is flooded with water to allow ships to anchor. A constant water level is maintained in wet docks. They are provided with lock system. Primarily they are used for loading, unloading and as storage of vessels.

MULTIPLE CHOICE QUESTIONS

1. For designing the dock, the proportion of ship load assumed to be borne by keel block is
   (a) $\frac{5}{8}$ (b) $\frac{3}{8}$ (c) $\frac{3}{16}$ (d) $\frac{5}{16}$
2. As compared to wall type breakwater, mound type breakwater
   (a) Requires skilled labour (b) Requires low maintenance
   (c) Requires less material (d) Has less damage due to gradual failure
3. Which of the following structures protects the shore by trapping of littoral drift?
   (a) Groynes (b) Sea walls (c) Revetments (d) Moles
4. Minimum width of ship clearance shall be (where B is beam length of the design vessel)
   (a) B or 30 m (b) 1.5 B or 50 m (c) 1.5 B (d) 50 m
5. When a ship floats at its designed water line, the vertical distance from water line to bottom of the ship is known as?
   (a) Beam (b) Depth (c) Draft (d) Free Board
6. The difference in height between Highest high water and lowest low water is called
   (a) Mean Range (b) Maximum Range (c) Maximum Rise (d) Mean Rise
7. It maximum spring rise is 2 m and height of wave expected is 4 m, then the breakwater height above the datum will be
   (a) 2.5 m (b) 4 m (c) 5 m (d) 7 m
8. Which of the following is a fixed type mooring accessory?
   (a) Bollard (b) Buoys (c) Cables (d) Anchors

9. The significant wave height (h_w) is defined as the average height of the
   (a) $\frac{1}{3}$ Highest wave (b) $\frac{1}{4}$ Highest wave (c) $\frac{1}{5}$ Highest wave (d) $\frac{1}{10}$ Highest wave
10. At a given port, fetch is 400 Nautical Miles, the maximum height of storm wave will be
    (a) 20 feet (b) 25 feet (c) 30 feet (d) 35 feet
11. Dead weight tonnage of a ship
    (i) Varies with latitude and season (ii) Is more than displacement tonnage
    (iii) Is difference between displacement load and displacement light
    (a) (i) and (ii) are correct (b) (ii) and (iii) are correct
    (c) (i) and (iii) are correct (d) Only (iii) is correct
12. The minimum diameter of turning basin, where ships turn by going ahead and without tug assistance should be? (L is length of the ship)
    (a) L (b) 1.5 L (c) 2.0 L (d) 4.0 L
13. Basin subjected to strong winds and tide, the length of berthing area should not be less (Length of Vessel = L)
    (a) L (b) L + 10% clearance between adjacent vessels
    (c) L+20% clearance between adjacent Vessels (d) 2 L
14. A ship is berthed in a chamber and lifted by buoyancy principle. Such a chamber is called
    (a) Dry dock (b) Wet dock (c) Floating dock (d) Refuge dock
15. The anchorage area required by a 181 m long ship anchored by a single anchor in a harbour of 15 m depth
    (a) 12 Ha (b) 16 Ha (c) 18 Ha (d) 20 Ha
16. The lowest tide which occurs in half lunar month is called?
    (a) Spring tide (b) Neap tide (c) Lunar tide (d) Tidal bore
17. Which of the following type of sea walls results in greater protection of shore structures?
    (a) Vertical sea wall (b) Sea wall with batter
    (c) Stepped sea wall (d) Sea wall having concave face
18. For height of the wave, if Fetch (F) is known, formula used is
    (a) $0.34\sqrt{F}$ (b) $0.5\sqrt{F}$ (c) $1.5\sqrt{F}$ (d) $3.4\sqrt{F}$
19. A ship strikes the berth generally at an angle, with the face of the dock
    (a) 90° (b) 45° (c) 30° (d) 10°
20. By increasing the rise of lock-gates
    (i) The length of the lock gate will increase
    (ii) Transverse stress due to water pressure on the gate will increase
    (iii) Compressive force on gate will increase
    (a) (i) and (ii) are correct (b) (i) and (iii) are correct
    (c) Only (ii) is correct (d) Only (iii) is correct

21. The weight added to improve the stability of ship when it has discharged its Cargo is known as
(a) Ballast (b) Dead weight (c) Gross tonnage (d) Cargo tonnage

22. The main purpose of breakwater is to
(a) Provide docking facility for vessels (b) Control sedimentation within harbour
(c) Protect vessels from waves and currents (d) Improve water circulation in the harbour

23. The main function of a quay wall is
(a) Anchoring floating docks (b) Protecting shore from erosion
(c) Loading and unloading of cargo facility (d) Giving berthing space for smaller boats

24. For improving navigation by directing vessels into desired channel, the harbour structure designed
(a) Break water (b) Quay wall (c) Pier (d) Beacon

25. For site selection for harbour which of these factors is not considered?
(a) Water depth (b) Wave climate
(c) Air pollution level (d) Sedimentation rate

26. Most common material used for construction of revetment and backwaters?
(a) Concrete (b) Riprap (c) Steel (d) Hard Timber

27. If H is the expected height of wave, then height of breakwater is generally adopted is
(a) 1.2 to 1.25 H above datum (b) 1.2 to 1.25 H above low water level
(c) 1.2 to 1.25 H above high water level (d) 1.2 to 1.25 H above mean sea level

28. As per Berlin's formula, the length of wave in meters is given by
(a) $1.56\ t^2$ (b) $1.34\ t^2$ (c) $1.74\ t^2$ (d) $1.94\ t^2$
(where t is period in seconds for two successive waves to pass same section)

29. Littoral drift as bed load is proportional to
(a) $\frac{V^2}{(G-1)^2}$ (b) $\frac{V^4}{(G-1)^4}$ (c) $\frac{V^3}{(G-1)}$ (d) $\frac{V^2}{G-1}$
(where V is velocity and G specific gravity of transported material)

30. The most popular method of construction of wall breakwater is
(a) Barge method (b) Staging method (c) Low level method (d) None of these

31. Which of the following conditions of loading imposes greatest load on the foundation in case of dry docks?
(a) When the dock is empty (b) When dock is full of water
(c) When the dock is empty with the ship of maximum tonnage
(d) When the dock is empty and under construction

32. What is the main purpose of a dock?
(a) To store cargo containers (b) To anchor ships during bad weather
(c) To store fuel for vessels
(d) To facilitate passenger embarkation and disembarkation

33. Which of the following land-marks on the coast line must be depicted on hydrographic maps?
(a) Shore line (b) Light houses (c) Church spires (d) All of these

TRUE OR FALSE QUESTIONS

1. Wet dock is used for carrying maintenance work of vessels. (T or F)
2. Beam of a vessel means the width of the vessel. (T or F)
3. The height of ship's side between waterline and the deck is known as free board. (T or F)
4. The width of entrance channel of medium side harbous is 120 m to 200 m. (T or F)
5. The clearance under the keel at the lowest water level when ship is fully loaded has to be 50 cm to 120 cm. (T or F)
6. Mooring is the process of inspecting ship's hull for any damage. (T or F)
7. An important navigational aid at the entrance is the lighthouse. (T or F)
8. Filling of water into ships tank to stabilize it during loading unloading is called Ballast. (T or F)
9. Marina is facility for mooring recreational boats, yats and offering fuel, water, electricity. (T or F)
10. Dry dock is basin that can be flooded to allow a ship to float for repair work. (T or F)
11. Docks are open areas for berthing ships and are classified as wet docks for loading-unloading and dry docks for repairs. (T or F)
12. Every port is a harbour. (T or F)
13. Basin subjected to strong winds and tide, the length of berthing area should not be less than the length of the design vessel (L). (T or F)
14. The low wall built into sea about perpendicular to coast line, to resist travel of sand etc. along a beach is called Groin. (T or F)
15. For designing of dock, the proportion of ship load assumed to be borne by keel block is $\frac{5}{6}$. (T or F)

FILL IN THE BLANKS

1. Refrigerated cargo ship carrying perishable goods is called .... ship.
2. Mean Sea Level (MSL) adopted by survey of India is at ....
3. Elevations of bottom of water bodies is found from .... charts.
4. The weight of a ship built in shipyard including all machines, frames etc. is called .... tonnage.
5. Plimsol Mark or the load Line is marked on the .... of the ship.
6. The process of unloading cargo from a ship at docks is called ....
7. The structure built along the shoreline where ships can load/unload cargo is called ....

8. .... is the depth of water alongside a dock.
9. For guiding the ship into the harbour entrance, the structure is called ....
10. Guiding the ship-into a specific dock (or berth) is called ....
11. For securing a ship to the dock device used is known as ....
12. Loading a ship with cargo is known as ....
13. The floating platform attached to the shore at which vessels can dock is called ....
14. While waiting for their next job the ships are assigned an area where ships anchor is called an ....?
15. The place where goods before loading are safely kept is called ....

ANSWERS

Multiple Choice Questions

1. a 2. d 3. a 4. a 5. c 6. b 7. d 8. a 9. a
10. c 11. c 12. d 13. c 14. c 15. d 16. a 17. d 18. a
19. d 20. a 21. a 22. c 23. c 24. d 25. c 26. b 27. a
28. a 29. a 30. b 31. b 32. d 33. d

True & False

1. F 2. T 3. T 4. F 100 to 160 m 5. T 6. F Secures Ship to Docks
7. T 8. F Only during voyages 9. T 10. T 11. F Closed Area
12. T 13. F Not less than 14. T 15. F Is 5/8

Fill in the Blanks

1. Reefer ship 2. Mumbai 3. Hydrographic 4. Light Weight
5. Hull 6. Discharging 7. Warf 8. Draft
9. Lighthouse or Buoy 10. Berthing 11. Mooring or Bollard
12. Embarkation 13. Pier or Jetty 14. Anchorage 15. Warehouse or Storage Yard

SECTION 5

Engineering Geology

Definitions – 25

Multiple Choice Questions – 56

True & False – 55

Fill in the Blanks – 33

DEFINITIONS

Asbestos: It is an excellent thermal and electric insulator, and highly fire resistant. It was commonly used as a building material, until its adverse effects on human health were discovered. Inhalation of asbestos fibres can lead to dangerous lung condition and even cancer.

Basalt: It is an extrusive igneous (volcanic) rock that is low in silica content, dark in colour and relatively rich in iron and magnesium. Many basalts are very fine-grained and some are glassy.

Dolomite: It is a type of limestone. Dolomite is made of 60% calcium carbonate and 40% magnesium carbonate and traces of heavy metals such as lead. Its chemical formula is $CaMg\,(CO_3)_2$. Sometimes it is also referred as dolostone. It is used as an ornamental stone and also a source of magnesium.

Earth's Crust: It is the outside layer of Earth and is made of solid rock, mostly basalt and granite. It has two divisions-continental crust (under the land) and other is oceanic crust (under the ocean). The continental crust is thicker (30 km to 50 km) mostly granite. The oceanic crust is thin (5 km to 10 km) mostly made of basalt.

Felspar: It belongs to any groups of hard crystalline minerals that consist of aluminium silicates with varying amounts of Potassium, Sodium, Calcium and Sometimes barium. Several verities of felspar minerals are used as gemstones. They exhibit unique optical phenomena.

Geomorphology: It is the study of landforms and landform evolution. The topic traditionally has been studied both qualitatively, which describes landforms, and quantitatively, which is process-based study describes forces acting on Earth's surface to produce landforms and landform changes.

Geophysics: It is the study of physical, processes and physical properties of Earth and its surrounding space environment. It is used to help locate hydrocarbons, minerals, aggregate and other natural resources. The geophysical data is used for geologic mapping. It uses gravity, magnetic, electrical and seismic methods for data.

Gypsum: It is a common mineral found in layers that were formed under salt water millions of years ago. When water evaporated, it left the mineral behind. Gypsum is composed of calcium sulphate ($CaSO_4$) and water. Its chemical name is calcium sulphate dihydrate ($Ca\,SO_4.\,2\,H_2O$). Gypsum can by milled and mixed with water resumes its original rock-like state. It can be endlessly recycled.

Hardness Number: Hardness testing involves pushing a hard indenter into a sample with a known force and measuring the size of the residual indent. A **hardness** number is defined as the force divided by the area of contact between indenter and specimen. A typical hardness test is that of Vickers. It has a diamond indenter.

Isomorphism: In isomorphism, two or more compounds show identical morphologies, whereas in **polymorphism** the same compound shows different morphology. Examples of such compounds are Ferrous sulphate ($Fe\,SO_4$) and Zinc Sulphate ($Z_4\,SO_4$). Isomorphous substances have an equal number of ions in their unit cells.

Kaolin (China Clay): Is a hydrated aluminum silicate crystalline mineral formed over millions of years by hydro-thermal decomposition of granite rocks. It is characterised by its fine particle size. It is chemically inert substance. It is widely used in ceramics. Kaolin is also used in medicines, toothpaste and coated paper.

Lignite: Also known as **Brown Coal**, is the lowest grade coal with least concentration of carbon. It has low calorific value and high moisture content. Mostly used in thermal plants.

Lithosphere: It is the rigid outermost shell of Earth, which includes both the continental and oceanic crust as well as uppermost mantle. It is strong and elastic layer that sits above the weaker and ductile asthenophere. It has 3 layers-rock, mantle and core.

Magma: It is the molten or semi-molten natural material from which all igneous rocks are formed. Magma is inside the earth while the molten rock that breaks through the Earth's surface is called **Lava**. The temperature of Magma is between 700°C and 1400°C.

Marble : It is a metomorphic rock consisting of carbonate minerals, mostly calcite ($Ca\ CO_3$) and dolomite ($Ca\ Mg\ (CO_3)_2$) that have recrystalized under the influence of heat and pressure. It is found mostly as White or Pink colour. Its hardness is very high. It is used in building monuments and sculptures.

Mica: If is in the group of silicate minerals whose outstanding characteristic is that individual mica crystals can easily be split into fragile elastic sheets. It has perfect cleavage. Found in many colours. Mica's best use is in electronics.

Palaeontology: It is the study of fossils and the evolution of life on Earth. It is scientific study of life of the geologic past which involves analysis of plants and animal fossils including those of microscopic size, preserved in rocks.

Pegmatite: It is an igneous rock showing a very coarse texture, with large interlocking crystals. Most pegmitites are composed of quartz, felspar and mica having similar sillicic composition as granite. Its size varies from 1 cm to even 1 m.

Petrology: It is the branch of geology that studies rocks, their mineralogy, composition, texture, structure and conditions under which they are formed. Petrology has three sub-divisions- Igneous, Metamorphic and Sedimentary Petrology.

Plutonic Rocks: These rocks are igneous rocks formed from magma that cools and solidifies within the crust. surrounded by pre-existing rocks.

Quartz: It is made of two most abundant chemical elements on Earth-Oxygen and Silicon. The crystals are of tetrahedrons shape. Even a small crystal has billions of tetrahedrons. An important use is in clocks and watches; computers and radios, GPS etc.

Sediment: It is a naturally occurring material that is broken down by process of weathering and erosion, and is subsequently transported by the action of wind, water, ice or by the force of gravity acting on the particles. Most often sediment are transported by water (fluvial process) or wind (acolian process) and glaciers.

Serpentine: It is any group of hydrous magnesium-rich silicate minerals. The composition of these common rocks forming minerals is approximately $Mg_3\ Si_2\ O_5\ (OH)_4$. Used as a source of magnesium and asbestos.

Stratigraphy: It is concerned with the study of rock layers (strata) and stratification (layering). Primarily the study includes of sedimentary and layered volcanic rocks.

Weathering: It is breaking down or dissolving of rocks and minerals on Earth's surface. Once a rock disintegrated, **erosion** transports the rock pieces and minerals away. Water, acids, salt, plants, animals and most importantly changing temperature all are agents of weathering.

MULTIPLE CHOICE QUESTIONS

1. Lithosphere is the
   (a) Upper atmospheric zone (b) Lower atmospheric zone
   (c) Solid part of earth (d) Seas and oceans
2. The branch of geology dealing with ancient life is called
   (a) Palaeontology (b) Stratigraphy (c) Lithology (d) Anthology
3. On Moh's scale hardness of granite is
   (a) 8 (b) 6 (c) 4 (d) 2
4. Asbestos has .... structure
   (a) Foliated (b) Fibrous (c) Granular (d) Bladed
5. Petrology confines itself to the study of
   (a) Petroleum products (b) Earth's features
   (c) Mineral deposits (d) Composition of rocks
6. The rocks having alumina or clay as major constituents are known as
   (a) Argillaceous (b) Siliceous (c) Calcarious (d) Sedimentary, rocks
7. The most suitable road material out of the following would be
   (a) Coarse grained igneous rocks (b) Volcanic rocks
   (c) Plutonic rocks (d) Hypabyssal rocks
8. Rocks formed from molten magma are known as
   (a) Sedimentary (b) Metamorphic (c) Igneous (d) Plutonic, rocks
9. Kaolin is chemically classified as
   (a) Metamorphic (b) Argillaceous (c) Calcerious (d) Silicious, rock
10. Shale, Slate, Marble, Limestone and Sand stones, are
    (a) Stratified (b) Unstratified (c) Foliated (d) Fibrous, rocks
11. Gypsum has got a .... structure
    (a) Bladed (b) Fibrous (c) Reniform (d) Granular
12. Volcanic rocks are suitable for use as crushed stone because they are
    (a) Hard and Tough (b) Have cementing property
    (c) Resistant to abrasion (d) All these
13. Dolomite is a limestone containing carbonate of magnesia upto
    (a) 15% (b) 20% (c) 25% (d) 45%
14. Pegmatite is a
    (a) Intrusive igneous rock (b) Extrusive igneous rock
    (c) Sedimentary rock (d) Metamorphic rock

15. Basalt is a rock type

(a) Sedimentary (b) Metamorphic
(c) Extrusive igneous (d) Intrusive igneous

16. Pegmatite has .... cementing property

(a) Exceptional (b) Excellent (c) Good (d) Poor

17. The branch of Geology which deals with occurrence, origin and history of rocks is known as

(a) Hydrogeology (b) Pedology (c) Petrology (d) Geomorphology

18. Gneiss is resulted from metamorphic transformation of

(a) Dolomite (b) Shale (c) Granite (d) None of these

19. Stones used for ornamental work must be

(a) Soft (b) Hard (c) Light (d) Heavy

20. Shingle is

(a) Decomposed laterite (b) Water bound pebbles
(c) Crushed granite (d) Air weathered rock

21. Laterite is found in

(a) UP (b) Punjab (c) West Bengal (d) Kerala

22. Stones used for rubble masonry must be

(a) Soft (b) Hard (c) Light (d) Heavy

23. Black marble is found in district of

(a) Jodhpur (b) Jaipur (c) Jabalpur (d) Jaisalmer

24. The tendency of stone to split along certain parallel planes is known as

(a) Texture (b) Fracture (c) Cleavage (d) Structure

25. The property of mineral by virtue of which it can be cut with knife is

(a) Parting (b) Sectile (c) Malleable (d) Ductile

26. Mica is

(a) Flexible (b) Rigid (c) Elastic (d) Both a & c

27. Colour of a mineral depends upon, its

(a) Atomic structure (b) Outer surface (c) Composition (d) Both a & c

28. Which mineral group is abundantly found in Earth's crust?

(a) Mica group (b) Felspar group (c) Oxide group (d) Silicate group

29. Felspar is abundantly found in

(a) Igneous (b) Sedimentary (c) Metamorphic (d) Fossil, rocks

30. Felspar minerals are free from

(a) Iron (b) Iron and Manganese
(c) Magnesium (d) Iron and Magnesium

31. What is the chemical composition of Quartz?

(a) SiO_4 (b) Si_2O_3 (c) SiO_2 (d) Al_2O_3

32. Which quartz mineral shows blue colour?
(a) Rose (b) Smoky (c) Rock crystal (d) None of these

33. Rock crystal belongs to which mineral group?
(a) Quartz (b) Felspar
(c) Carbonate (d) Ferro-magnesium group

34. What is the hardness of quartz mineral?
(a) 4 (b) 2 (c) 7 (d) 5

35. The maximum angle of slope of a bed (or layer of rock) with horizontal is known as
(a) Dip (b) Strike (c) Bedding (d) Decline

36. The mineral which could be scratched by finger nail has the hardness between
(a) 3 and 4 (b) 2 and 3 (c) 5 and 6 (d) 7

37. The molten rock under the earth's surface is called
(a) Lava (b) Magma (c) Lava + Magma (d) None of these

38. Igneous rocks are formed by
(a) Weathering & Erosion (b) Heat and Pressure
(c) Compaction & Cementation (d) Cooling of magma

39. Marble is resulted from metamorphic transformation of
(a) Dolomite (b) Sand stone (c) Felsite & Marl (d) Any of those

40. Singbhum belt is known for presence of
(a) Manganese (b) Iron, copper (c) Coal (d) All these

41. Lignite (brown coal) contains carbon upto
(a) 30% (b) 70% (c) 95% (d) 50%

42. Stones used for construction of Retaining walls must be
(a) Soft (b) Hard (c) Light (d) Heavy

43. The rocks used for roofing is
(a) Basalt (b) Granite (c) Slate (d) Pumice

44. Isomorphism is a character exhibited by some minerals which have analogous chemical composition and crystallize in closely related crystal forms, which of the following are isomorphous?
(a) Calcite group (b) Barite group (c) Argaonite group (d) All these

45. Which of the following statement is correct in the context of Gondwana system?
(a) Middle Gondwana deposits are completely devoid of coal seams
(b) Lower Gondwana deposits contain the most important coal bearing strata in the Indian stratigraphy
(c) Upper Gondwana deposits contain a few coal seams
(d) All the above

46. Small pieces of rocks are
(a) Pebbles (b) Particles (c) Sediment (d) Chips

47. Igneous rocks change to metamorphic rocks by process of?
(a) Magma cooling (b) Heat and pressure (c) Compaction (d) Weathering and Erosion

48. Method of Rock forming is called
(a) Weathering cycle (b) Rock cycle (c) Magma cycle (d) Rock formation

49. Which classification of rock is not found in rock cycle?
(a) Cemented rock (b) Metamorphic rock
(c) Sedimentary rock (d) Igneous rock

50. How thick is Earth's crust about
(a) 6.5 km (b) 4.0 km (c) 40 km (d) 400 km

51. Which is the parental rock for metamorphic rock?
(a) Sedimentary (b) Metamorphic (c) Igneous (d) All these

52. The lithosphere is approximately thick
(a) 50 to 100 km (b) 100 to 200 km (c) 5 to 10 km (d) 1 to 2 km

53. The hardness number of Diamond is
(a) 5 (b) 6 (c) 7 (d) 10

54. Calcite shows number of sets of cleavage
(a) Two (b) Three (c) One (d) Four

55. Cleavage means
(a) Development of crystal faces during mineral growth
(b) Splitting a mineral along planner surfaces
(c) Development of irregular fractures when mineral breaks
(d) Density and specific gravity of minerals

56. Study of rock deformation and geometry is known as
(a) Geophysics (b) Engineering Geology
(c) Structural Geology (d) Petrology

TRUE OR FALSE QUESTIONS

1. Geomorphology is the scientific study of landforms and landform evolution. (T or F)
2. Ionosphere is the layer of Earth's atmosphere between 80 to 100 km above Earth's surface. (T or F)
3. Core is the innermost structural shell of Earth. It starts at a depth of 5100 km. (T or F)
4. The specific gravity of earth's mass is about 2.75. (T or F)
5. Hematite has a granular structure. (T or F)
6. Kyanite has bladed structure. (T or F)
7. Granite, Basalt and Trap are unstratified rocks. (T or F)
8. Plutonic rocks are Extrusive igneous rocks. (T or F)

9. The volcanic rocks are fine grained rocks. (T or F)
10. Granite belongs to Plutonic group of rocks. (T or F)
11. Talc is the softest and diamond is the hardest rock. (T or F)
12. As per Moh's scale hardness of Topaz is 10. (T or F)
13. Slate is metamorphic transformation of sandstone. (T or F)
14. The property of mineral by which it can be cut with a knife is called Ductile. (T or F)
15. When mineral occurs in flattened or square form it is called tabular. (T or F)
16. The rose colour of rose quartz is due to iron. (T or F)
17. Yellow variety of quartz is known as citrine. (T or F)
18. Monomineralic rock are made of only one type of mineral. (T or F)
19. The earth crust constitute maximum volume percent is silicon. (T or F)
20. Average geothermal gradient of earth's crust is 15° C/km. (T or F)
21. Petrography is the study of rocks in the laboratory to analyse mineral composition, chemical composition, texture etc. (T or F)
22. The physical property of minerals by which they are identical on account of their colour of powder form is called lusture. (T or F)
23. Talc has pearly lusture. (T or F)
24. The layer that separates crust from core is known as Mantle. (T or F)
25. The internal heat engine of Earth is derived by Radioactivity. (T or F)

FILL IN THE BLANKS

1. Chemical substances having same chemical formula but different crystal structure are called ....
2. Very dark coloured igneous rocks are called hyper ....
3. Crystals which are better developed in two directions than the third are known as ....?
4. Light coloured minerals have smaller specific gravity than .... coloured minerals.
5. Splitting of mineral along the planner surface is known as ....
6. The most suitable materials for road construction are .... rocks.
7. Sedimentary rocks resulting from accumulation of plant and animal remains are ....
8. Pegmatite is an intrusive .... rock.
9. Decomposed laterite is called ....
10. Himalayas are .... mountains.
11. The maximum permissible water absorption of Limestone is .... %.
12. Bituminous coal, common or cooking coal contains carbon upto .... %.

13. Boghead coal is a variety of .... coal.
14. Maximum permissible water absorption of granite is .... %.
15. The dark coloured Basalts, Granites and Dolerites are all .... rocks.
16. Sandstone and Limestone are normally .... rocks.
17. Slate, marble and Serpentine are normally .... rocks.
18. Calcium carbonate is found in two forms (Calcite and Aragonite); former in hexagonal and later in orthorhomic systems. The phenomena is described as ....
19. Chalcopyrite is an ore of ....
20. The centre of most layers of earth is .... core.
21. The layer of earth which is both liquid and solid is ....?
22. The only liquid layer of the earth is .... core.
23. As we go deeper into earth density ....
24. As we go deeper into the earth temperature ....
25. The layer of earth in which we live is ....
26. Movement of continents moving over the earth's surface is called continental ....
27. The rock which is stable even in vertical slopes is ....?
28. When the layers are horizontal the slope is stable at all angles upto ....
29. While cutting stratified rocks, consideration should be made of .... factor.
30. Most potential surfaces for rock slips are .... zones.
31. For slope failure removal of .... is the most important factor.
32. The rock mass capable of Yielding contained water economically are ....
33. The rocks formation absolutely impermeable is ....

ANSWERS

Multiple Choice Questions

1. c	**2.** a	**3.** b	**4.** b	**5.** d	**6.** a	**7.** b	**8.** c	**9.** c
10. a	**11.** d	**12.** d	**13.** b	**14.** a	**15.** c	**16.** d	**17.** c	**18.** c
19. a	**20.** a	**21.** a	**22.** b	**23.** b	**24.** c	**25.** b	**26.** d	**27.** d
28. b	**29.** a	**30.** d	**31.** c	**32.** d	**33.** a	**34.** c	**35.** a	**36.** b
37. b	**38.** d	**39.** a	**40.** a	**41.** b	**42.** d	**43.** a	**44.** d	**45.** d
46. c	**47.** d	**48.** b	**49.** a	**50.** c	**51.** d	**52.** a	**53.** d	**54.** b
55. b	**56.** c							

True & False

1. T	**2.** T	**3.** F at 6378 km	**4.** F 5.52	**5.** F
6. T	**7.** T	**8.** F Intrusive	**9.** T	**10.** T
11. T	**12.** F is 8	**13.** F Mud stone	**14.** F Sectile	**15.** T
16. F Titanium	**17.** T	**18.** T	**19.** F Oxygen	
20. F 30° C/km	**21.** T	**22.** F Called streak	**23.** T	
24. F Magma Layer	**25.** T			

Fill in the Blanks

1. Polyorphs	**2.** Melanic	**3.** Tabular	**4.** Dark
5. Cleavage	**6.** Volcanic	**7.** Limestone Chalkbond	
8. Igneous	**9.** Shingle	**10.** Tectonic	**11.** 6%
12. 90%	**13.** Bituminous	**14.** 1.0%	**15.** Igneous
16. Sedimentary	**17.** Metamorphic	**18.** Dimorphism	**19.** Copper
20. Inner	**21.** Mantle	**22.** Outer	**23.** Increases
24. Increases	**25.** Crust	**26.** Drift	**27.** Granite
28. 90°	**29.** Dip	**30.** Shear	**31.** Vegetation
32. Aquifers	**33.** Aquifuge		

Section 6

Engineering Mathematics

Definitions – 14

Multiple Choice Questions – 30

True & False – 10

Fill in the Blanks – 10

DEFINITIONS

Cross-Product: It is a binary operation on two vectors in three-dimensional space. It results in a vector that is perpendicular to both vectors. The vector product of vectors a and b is denoted by a × b. It's formula is, $|\vec{A} \times \vec{B}| = |\vec{A}||\vec{B}||\sin\theta|$.

Curl: A curl is a vector field F = $P_i + Q_j + R_k$ is a vector that describes the rotation or circulation of the field. It is defined as: Curl = ∇ × F.

In Cartesian co-ordinates this is given by Curl

$$F = \begin{vmatrix} i & j & k \\ \frac{\partial}{\partial x} & \frac{\partial}{\partial y} & \frac{\partial}{\partial z} \\ P & Q & R \end{vmatrix}$$

where i, j, k are unit vectors in x, y, z directions.

Determinant: It is a scalar value derived from a square matrix. It gives information about the matrix, such as whether it is invertible (a non-zero determinant indicates an invertible matrix). For example – A 2 × 2 matrix.

$$A = \begin{bmatrix} a & b \\ c & d \end{bmatrix},$$

Determinant of A = ad – bc.

Directional Derivative: The directional derivative of a scalar function f (x, y, z) in direction of a vector *v* is the rate of change of *f* in that direction. It is given by

$$D_v f = \nabla f . \hat{v}$$

where ∇f is the gradient of *f*, and $\hat{v}$ is the unit vector in the direction *v*.

Divergence of Vector: The divergence of a vector field is the extent to which the vector field flux behaves towards or away from a specific point. It does not indicate in which direction the expansion is occurring. Hence (in contrast to the **curl** of a vector field), the divergence is scalar. The formula is, div (F) = ∇.F.

Dot Product: The dot product also called scalar product of two vectors a and b is a scalar value defined as

$$a.b = |a|\,|b| \cos\theta$$

where θ is angle between the two vectors. In component form,

$$a.b = a_1b_1 + a_2b_2 + a_3b_3$$

Euclidean Norm of Vector: It is the square root of the sum of the squares of the vectors co-ordinates. It finds the shortest distance of the route from the vector to the zeroth vector. The **norm** of a vector is simply the square root of the sum of each component squared. It is a geometric object having that has magnitude (Length) and direction.

Fourier Series: It is a way to represent a periodic function as a sum of **sine** and **cosine** functions. It is type of trigonometric series, but not all trigonometric series are Fourier series. It's formula is,

$$f(x) = \frac{1}{2}a_0 + \sum_{n=1}^{\infty} a_n \cos n x + \sum_{n=1}^{\infty} b_n \sin n x$$

Inverse Matrix: The inverse matrix A (denoted as A^{-1}) is another matrix such that $A.A^{-1} = A^{-1}.A = I$, where I is the identity matrix. A matrix is invertible only if it is square and its determinant is non-zero.

Matrix: It is a rectangular array of numbers, symbols, or expressions, arranged in rows and columns. It is denoted as,

$$A = \begin{bmatrix} a_{11} & a_{12} & \cdots & a_{1n} \\ a_{21} & a_{22} & \cdots & a_{2n} \\ \vdots & & & \\ a_{m1} & a_{m2} & \cdots & a_{mn} \end{bmatrix}$$

Matrices are used in solving systems of linear equations, transformations, and various engineering computations,

Maxima: In calculus, a maximum is the highest point of a function in a given interval. Mathematically, if f(x) has maximum at x = c, then. $f'(c) \simeq 0$ and $f''(c) < 0$ for local maxima.

Order of the Differential Equation: The order of a differential equation is defined to be that of the highest order derivative it contains. The **degree** of a differential equation is defined as the power to which highest order derivative is raised. Example, $(f'')^2 + (f'')^4 = x$ is an example of a second degree, third-order differential equation.

Singularity: In mathematics singularity is a point at which a mathematically object is not defined. It is a point where a function, equation, or surface becomes degenerate or diverges to infinity. Singularity is also known as **singular** point.

Taylor Series: In mathematics, the Taylor series or **Taylor Expansion** of a function is an infinite sum of terms that are expressed in terms of function's derivatives at a single point. The Taylor series formula is

$$f(x) = \sum_{n=0}^{\infty} \frac{f^{(n)}a}{n!} \times (n-a)^n$$

where, f(x) = Real or complex-valued function

n = Total number of terms.

MULTIPLE CHOICE QUESTIONS

1. The rank of the matrix $\begin{bmatrix} 1 & 2 & -1 \\ 2 & 4 & 6 \\ 0 & 0 & -8 \end{bmatrix}$ is

(a) 2 (b) 1 (c) 3 (d) 9

2. The inverse $A = \begin{bmatrix} -3 & 5 \\ 2 & 1 \end{bmatrix}$ is of Matrix

(a) $\begin{bmatrix} \frac{5}{13} & \frac{-1}{13} \\ \frac{2}{13} & \frac{3}{13} \end{bmatrix}$ (b) $\begin{bmatrix} \frac{2}{13} & \frac{5}{13} \\ \frac{-1}{13} & \frac{3}{13} \end{bmatrix}$ (c) $\begin{bmatrix} \frac{-1}{13} & \frac{5}{13} \\ \frac{2}{13} & \frac{3}{13} \end{bmatrix}$ (d) $\begin{bmatrix} \frac{1}{13} & \frac{-5}{13} \\ \frac{2}{13} & \frac{3}{13} \end{bmatrix}$

3. If $A^T = A^{-1}$, where A is real matrix, then A is

(a) Normal (b) Symmetrical (c) Hemitian (d) Othogonal

4. The inverse of 2 × 2 matrix $\begin{bmatrix} 1 & 2 \\ 5 & 7 \end{bmatrix}$ is

(a) $\frac{1}{3}\begin{bmatrix} -7 & 2 \\ 5 & -1 \end{bmatrix}$ (b) $\frac{1}{3}\begin{bmatrix} 7 & 2 \\ 5 & 1 \end{bmatrix}$ (c) $\frac{1}{3}\begin{bmatrix} 7 & -2 \\ -5 & 1 \end{bmatrix}$ (d) $\frac{1}{3}\begin{bmatrix} -7 & -2 \\ -5 & -1 \end{bmatrix}$

5. $\underset{\theta \to 0}{\text{Lt}} \sin\frac{\theta}{\theta}$ is

(a) 1 (b) 0 (c) ∞ (d) None of these

6. The series $f(x) = x + \frac{x^3}{3!} + \frac{x^5}{5!} \ldots$ is the expansion of

(a) cos hx (b) sin x (c) sin hx (d) cos x

7. Value of the function $\lim_{x \to a} (x - a)^{x - a}$ is

(a) 1 (b) 0 (c) ∞ (d) a

8. The value of following improper integral $\int_0^1 x \ln x$ is

(a) $\frac{1}{4}$ (b) 0 (c) $\frac{-1}{4}$ (d) 1

9. The function $f(x) = 2x^3 - 3x^2 - 36x + 2$ has its maxima at

(a) $x = -2$ only (b) $x = 0$ only (c) $x = 3$ only (d) Both at $x = -2$ and $x = 3$

10. The angle of intersection of the curves $x^2 = 4y$ and $y^2 = 4x$ at point (0, 0) is

(a) 0° (b) 30° (c) 45° (d) 90°

11. The directional derivative of the following function at (1, 2) in the direction of (4i + 3j) is $f(x, y) = x^2 + y^2$.

(a) $\frac{4}{5}$ (b) 4 (c) $\frac{2}{5}$ (d) 1

12. A velocity vector is given by $\bar{v} = 5xy\,\bar{i} + 2y^2\,\bar{j} + 3yz^2\,\bar{k}$. The divergence of this vector at (1, 1, 1) is

(a) 9 (b) 10 (c) 14 (d) 15

13. The dot product of two vectors $\overline{P}$ and $\overline{Q}$ is zero. The angle between them is

(a) 0° (b) 30° (c) 90° (d) 120°

14. For the parallelogram OPQR as shown here $\overline{OP} = a\bar{i} + b\bar{j}$ and $\overline{OR} = c\bar{i} + d\bar{j}$. The area of parallelogram is

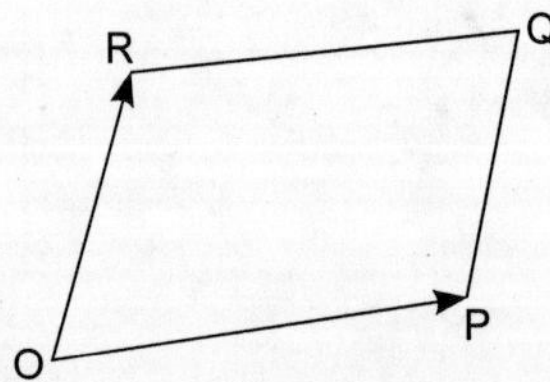

(a) ad – bc (b) ac + bd (c) ad + bc (d) ab – cd

15. The curl of the vector field is $2x^2 yi + 5z^2 j - 4x yz k$

(a) $-14zi - 2x^2k$ (b) $6zi + 2x^2j - 2x^2k$

(c) $-14zi + 6yj + 2x^2k$ (d) $6zi - 8xyj + 2x^2yk$

16. The differential equation, $EI\frac{d^4y}{dx^4} + \rho\frac{d^2y}{dx^2} + ky = 0$ is

(a) Linear of 4th order (b) Non-linear of 4th order

(c) Non-homogeneous (d) Linear and 4th degree

17. If C is a constant, solution of the equation $\frac{dy}{dx} = 1 + y^2$, is

(a) $y = \sin(x + c)$ (b) $y = \cos(x + c)$

(c) $y = \tan(x + c)$ (d) $y = e^x + c$

18. The degree of the differential equation $\frac{d^2x}{dt^2} + 2x^3 = 0$

(a) 0 (b) 1 (c) 2 (d) 3

19. Solution of differential equation $3y\frac{dy}{dx} + 2x = 0$, represents a family of

(a) Ellipses (b) Circles (c) Parabolas (d) Hyperbolas

20. Solution of $\frac{dy}{dx} = -\frac{x}{y}$, at x = 1 and $y = \sqrt{3}$ is

(a) $x - y^2 = -2$ (b) $x + y^2 = 4$ (c) $x^2 - y^2 = -2$ (d) $x^2 + y^2 = 4$

21. The solution of the equation $x\frac{dy}{dx} + y = 0$ passing through the point (1, 1) is

(a) x (b) x^2 (c) x^{-1} (d) x^{-2}

22. The solution of the ordinary differential equation, $\frac{dy}{dx} + 2y = 0$ for the boundary condition y = 5 at x = 1, is

(a) $y = e^{-2x}$ (b) $y = 2e^{-2x}$ (c) $y = 10.95e^{-2x}$ (d) $y = 36.95e^{-2x}$

23. The analytic function, $f(z) = \frac{z - 1}{z^2 - 1}$ has singularity at

(a) 1 and –1 (b) 1 and i (c) 1 and –i (d) i and –i

24. In an experiment positive and negative values are equally likely to occur. The probability of obtaining at most one negative value in five trials is

(a) $\frac{1}{32}$ (b) $\frac{2}{32}$ (c) $\frac{3}{32}$ (d) $\frac{6}{32}$

25. A probability distribution with right skew is shown here. The correct statement for the probability distribution is

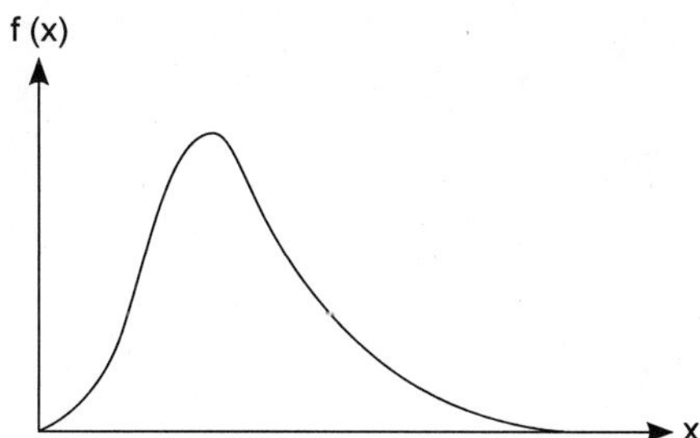

(a) Mean is equal to mode

(b) Mean is greater than median but less than mode

(c) Mean is greater than median and mode

(d) Mode is greater than median

26. Given that one root of the equation, $x^3 - 10x^2 + 31x - 30 = 0$ is 5 the other two roots are

(a) 2, 3 (b) 2, 4 (c) 3, 4 (d) −2, −3

27. The infinite series $1 + \frac{1}{2} + \frac{1}{3} + \ldots$

(a) Converges (b) Diverges (c) Oscillates (d) Unstable

28. The summation of series, $S = 2 + \frac{5}{2} + \frac{8}{2^2} + \frac{11}{2^3} + \ldots \infty$ is

(a) 4.5 (b) 6.0 (c) 6.75 (d) 10.0

29. The number of boundary conditions required to solve the given differential equation is, $\frac{\partial^2\phi}{\partial x^2} + \frac{\partial^2\phi}{\partial y^2} = 0$

(a) 2 (b) 0 (c) 4 (d) 1

30. A continuous real function can be expressed as

(a) Taylor's series and Fourier's series

(b) Taylor's series but not Fourier's series

(c) Neither Taylor's nor Fourier's series

(d) Not by Taylor's but by Fourier's series

TRUE OR FALSE QUESTIONS

1. The value of (p) for the following set of equations is 4.5. (T or F)

   $2x + 3y = y$

   $3x + py = 10$

2. If $A = \begin{bmatrix} 1 & 5 \\ 6 & 2 \end{bmatrix}$ and $B = \begin{bmatrix} 3 & 7 \\ 8 & 4 \end{bmatrix}$ then AB^T is equal to. (T or F)

3. $\begin{bmatrix} 2 & 4 \\ 3 & 6 \end{bmatrix}$ is a singular matrix. (T or F)

4. The function, $f(x) = e^x$ is Even or Odd. (T or F)

5. The Euclidean norm (length) of the vector $(4, -2, -6)^T$ is $\sqrt{12}$. (T or F)

6. In the given second order differential equation $\frac{d^2y}{dx^2} = -12x^2 + 24x = 20$. (T or F)

   The boundary conditions are at x = 0, y = 5 and at x = 2, y = 21. The value of y at x = 1 is 18.

7. $Z = \frac{2 - 3i}{-5 + i}$ can be expressed as, $-0.5 + 0.5\,i$. (T or F)

8. The series, $S = 1 - 1 + 1 - 1 \ldots$ is monotonous. (T or F)

9. The limit of the series $f(x) = x - \frac{x^3}{3!} + \frac{x^5}{5!} - \frac{x^7}{7!} \ldots$ as x approaches $\frac{\pi}{2}$ is π. (T or F)

10. The solution at x = 1 and t = 1 of the partial differential equation subject to initial conditions of u (o) = 3 x and $\frac{\partial u(o)}{\partial t} = 3$ is 6. (T or F)

FILL IN THE BLANKS

1. The rank of the matrix $\begin{bmatrix} 6 & 0 & 4 & 4 \\ -2 & 14 & 8 & 18 \\ 14 & 14 & 0 & -10 \end{bmatrix}$ is ....

2. The determinant of matrix $\begin{bmatrix} 0 & 1 & 2 & 3 \\ 1 & 0 & 3 & 0 \\ 2 & 3 & 0 & 1 \\ 3 & 0 & 1 & 2 \end{bmatrix}$ is ....

3. The $\lim_{x \to \infty} x \sin\left(\frac{1}{x}\right)$ is ....

4. Value of $\int_0^\infty \frac{\sin t}{t} dt$ is ....

5. The divergence of the vector field, $V = x^2i + 2y^2j + z^4k$ at $x = 1$, $y = 2$ and $z = 3$ is ....

6. The directional derivative of function, $f(x, y, z) = x + y$ at the point P (1, 1, 0) along the direction $i + j$ is ....

7. e^x is periodic with period of ....

8. A box contains 10 screws, 3 of which are defective. Two screws are drawn at random with replacement. The probability that one of the two screws is defective will be ....

9. The summation of series $S = 2 + \frac{5}{2} + \frac{8}{2^2} + \frac{11}{2^3} + \ldots \infty$ is ....

10. The type of partial differential equation, $\frac{\partial^2 P}{\partial x^2} + \frac{\partial^2 P}{\partial y^2} + 3\frac{\partial^2 P}{\partial x \partial y} + 3\frac{\partial P}{\partial x} - \frac{\partial P}{\partial y} = 0$, is ....

ANSWERS

Multiple Choice Questions

1. a **2.** c **3.** d **4.** a **5.** a **6.** c **7.** a **8.** c **9.** a

10. d **13.** c **14.** a **19.** a **20.** d **25.** c **26.** a **15.** a

21. c **27.** a **16.** a **22.** d **28.** d **11.** b **17.** c **23.** d

29. a **12.** d **18.** b **24.** d **30.** d

True & False

1. T 2. F $\begin{bmatrix} 38 & 28 \\ 32 & 56 \end{bmatrix}$ 3. T 4. F Neither even nor odd 5. F is $\sqrt{56}$

6. T 7. T 8. T 9. F 1 (One) 10. T

Fill in the Blanks

1. 2 (Two) 2. 88 3. 1 (One) 4. $\frac{\pi}{2}$ 5. 134

6. $\sqrt{2}$ 7. 2 pi 8. 47% 9. 10 10. Hyperbolic

SECTION 7

Environmental Science

DEFINITIONS

Abiotic (Physical) Components: It is the non-living component consist of chemical substances found in soil, water and atmosphere. Abiotic components include physical conditions and non-living resources that affect living organisms in terms of growth, maintenance and growth. They include-water, sunlight, soil, atmosphere, humidity, wind etc.

Acid Rain: It is a condition in which natural precipitation becomes acidic after reacting with pollutants in the air. Acid rain is formed when industrial emissions of gases like sulphur dioxide, Nitrogen oxides and hydrogen chloride combine with water droplets. The ill-effects of acid rain can be found in vegetation, soil, marine resources, monuments as well as on all kinds of organisms.

Adaptation: It is the ability of living organisms to adjust themselves to their surroundings. Each population is fully adapted to its environment and is capable of using the available resources.

Afforestation: It means planting and covering the soil surface by vegetative covers. It helps in reducing soil erosion and increasing infiltration causing enhanced groundwater recharge.

Bio-Diversity: It means the variety of living organisms available on earth including plants and animals. Diversity in simple terms is the 'number of species found in a community'. Bio-diversity can be-Genetic biodiversity, species bio-diversity or ecosystem bio-diversity.

Biomass: It is the standing crop expressed in terms of weight (i.e. organisms mass) of the living matter present.

Biosphere: It is the part of earth where life can exist. Biosphere represents a highly integrated and interacting zone comprising of atmosphere (air), hydrosphere (water) and lithosphere (land).

Biosphere Reserves: Biosphere reserve consists of a **Core area** and a **buffer zone**. The core area is strictly protected in order to maintain its ecological integrity and diversity. In buffer zone besides developmental activities traditional culture and land use are also considered. **Rann of Kachch** and **Gulf of Mannar** are major biosphere reserves in India.

Biotic (Living) Components: These include plants (flora) and animals (fauna) including human beings as an important factor. Micro-organism is the huge part of biotic ecosystem.

Bird's Migration: It refers to the regular, recurrent and cyclical seasonal movements of birds from one place to other. The distance of migration may range from few kilometers to thousands of kilometers. But at the end of period, they go back to their original home. The main reasons being extreme climatic condition, food or water shortage and loss of habitat.

Carbon Credits: A carbon credit is a commercial term used to assign a value to a metric tonne of green house gas emissions. When you purchase a carbon credit from a third party, you are helping to fund a project that will either reduce or isolate a metric tonne of carbon emission from atmosphere in another location. As a result You are effectively 'offsetting' some or all of Your own carbon emissions.

Climate: It can be defined as an average pattern of weather over the long term in a particular area.

Conservation: It means to ensure the preservation of a quality environment for the organisms in terms of their nutrition, recreation etc., and to ensure at the same time through planned manipulations, a continuous Yield of resources.

Consumers: Organisms, mostly animals, that feed either directly or indirectly on producers are called consumers. Consumers may belong to-Herbivores (Plant eaters), carnivores (animal-meat eaters) and Omnivores (Plant and meat eaters).

Decomposers (Detritus Feeders): This category of consumers feed on dead plants and animals. Bacteria, fungi, earthworms, termites etc. are examples of this category.

Deforestation: It is defined as the destruction of forests by cutting trees. This has many negative effects, most important being increased carbon-di-oxide in the atmosphere, main cause of global warming. It also causes increased soil erosion, landslides, loss of habitat to wild life. Also causes desertification.

Disaster: It means an event, which brings sudden great misfortune and grief causing disruption to normal life and causing colossal loss of life and property. Examples are - Floods, earthquake are natural disasters, while war is a man-made disaster.

Ecology: It is the science which deals with various principles which govern relationships between the two components of nature viz., organisms and their environment.

Ecosystem: It is defined as structural and functional unit of biosphere consisting of community of living beings and physical environment, both interacting and exchanging materials between them. It includes plants, trees, animals, fish, birds, microorganisms, water, soil and people.

Endangered Species: It has been realised that for species to survive and flourish, enough numbers must be present in the habitat to make reproduction possible. A threatened or endangered species is one which is vulnerable and whose number is continuously declining.

Environment: It is defined as the surroundings of an object. Environment is a complex of Several Things (light, water, soil, temperature, air etc.) which surround an organism.

Environmental Impact Assessment (EIA): It is the study to predict the effect of a proposed activity on environment. The United Nations Environmental Programme (UNEP) defines EIA as tool used to identify the environmental, social and economic impacts of a project prior to decision-making. It is a means of avoiding any environmental disturbances, due to proposed project, that are much more difficult and expensive to correct after their occurrence than before.

Extinction of Species: A number of species are becoming extinct every year due to human economic activities e.g. extension in agricultural land, increase in agricultural productivity, construction of dams and reservoirs, accelerated soil erosion, industrial development, urbanisation etc. Also due to pollution vulnerable species are getting extinct every year.

Food Chain: It refers to the order of events in an ecosystem where one living organism eats another organism and later that organism is consumed by another stronger organism. The flow of energy and nutrients from one organism to another at different trophic levels forms a food chain.

Food Web: Under natural conditions in the same ecosystem, depending upon the variety of organisms as based on their food habits, a number of linear food chains operate at a time. These chains are interlinked with each other at several points. This inter-locking pattern of number of food chains form a web-like arrangement known as food-web.

Fossil Fuels: Coal, oil and natural gas are known as fossil fuels. These are remains of organisms (flora and fauna) which buried under the layers of sediment millions of years back and converted by heat and overburden (Pressure) into fossil fuels.

Global Warming: It refers to gradual rise in atmospheric and ground surface air temperature and consequent change in global radiation, leading to climate change at different levels (local, regional and global). It is a result of various factors, such as ozone depletion, emission of green house gases, deforestation etc.

Green House Effect: Greenhouse means a building made of glass which allows the short-wave incoming solar radiation to enter, but does not allow the long-wave outgoings radiation to escape. In a similar way the earths atmosphere acts like a greenhouse. The increasing amount of Green house gases (CO_2 56%, CH_4 18%, CFC 13%, O_3 7%, N_2O 6%) are trapping more and more radiation causing global warming.

Habitat: It is the place, where an organism lives, is known as habitat of that organism. The habitat of an organism represents a particular set of environmental conditions suitable for its successful growth.

Heritage Sites: These have been designated by the UNESCO as having outstanding universal value. The idea is to provide framework for international co-operation in preserving and protecting cultural treasures and natural areas all over the world. India has 26 such sites. Globally there are 1223 heritage sites.

Kyoto Protocol: It is an International Treaty to control global warming. It was held is 1997 in Kyoto (Japan) to initiate measures for reducing green house gases in the atmosphere. 84 countries committed to 5.2% reduction in green house effect. India and China did not commit. The Kyoto protocol failed but it gave a positive message.

Mega-Diversity Nations: India is among the 12 mega-biodiversity nations. It is estimated that on earth there are about 10 million biological species existing. About 80% of them live in tropical forests. In India around 1,28,000 species are found, out of these 47,000 are plant species and 81,000 are animal species.

Municipal Solid Waste (MSW) or Refuse: These are all sorts of solid waste from community. MSW is a heterogeneous mixture of various kinds of solid wastes. They include- biodegradable (putrescible) food waste called **garbage** and non-puterscible solid waste called **rubbish**.

National Park: In these no habitation or ownership is allowed. Even traditional activities are prevented like grazing or firewood collection. There are 89 National Parks in India. Common size is 100 to 500 km^2. Important National Parks in India are - Kaziranga, Gir, Kanha, Corbett etc.

Natural Resources: These are provided by nature in limited amounts : they include-fossil fuels, minerals, timber, air and water. These may be renewable or non-renewable resources, biotic or abiotic resources and land resources.

Nitrogen Cycle: In this cycle nitrogen gas in the air serves as a reservoir pool, and this nitrogen is fixed with the help of by certain micro-organisms and natural electrification process, forming plant proteins. The dead bodies and dead plants are decomposed by bacteria to produce ammonia, then nitrites, which by demineralization goes back to atmosphere as nitrogen gas.

Noise Pollution: Noise is defined as 'unwanted sound at a wrong time and at wrong place'. Generally prolonged and loud sound is considered as noise pollution. It can cause physiological as well as Psychological damage.

Ozone Layer: The stratosphere region that contains high concentration of ozone (O_3) is called ozone layer. It exists between 15 km to 35 km above earth. It's thickness over the Earth's surface is about 3 mm (300 Dobson Units). It has the potential to absorb around 97 to 99% of harmful Ultraviolet (UV) radiation from the **Sun**, that can damage the life on earth.

Paris Agreement: It is a legally binding International Treaty on climate-change. It was adopted by 196 parties at the U.N. Climate Change Conference (COP_{21}) in Paris (France) in December 2015. The aim was to limit increase in global temperature well below 2°C above pre-industrial level and pursue effort to fix the limit at 1.5° C.

Poaching: It is the illegal killing of prohibited wild animals. The persons involved are known as **poachers.** Poaching is done to earn illegal money through sale of their hides, tusks, horns, furs, skins etc.

Pollution: It is an undesirable charge in the physical, chemical, or biological characteristics of air, land and water, that may or will harmfully affect human life directly or indirectly.

Producers (Autotrophic) Organism: The green plants are producers, as they have the capacity to manufacture their own food from existing supplies of water and carbon dioxide (CO_2), utilising sunlight as energy source. This process is known as photosynthesis. Oxygen, the much needed gas for survival is released as by-product.

Protected Areas: In order to keep ecosystem intact, and to avoid their being encroached upon or damaged by human beings, it is necessary to keep aside adequate forest areas, representing different types of ecosystems with their peculiar biological life, as **Protected Area**.

Resource Conservation: It is defined as "the planning and management of natural and man-made resources so as to secure their judicious use and continuity of supply while maintaining their quality, value and diversity. Ever increasing population has put much pressure on our natural resources, particularly non-renewable resources.

Sanctuaries: In these some traditional activities are allowed like- grazing, collection of fire wood etc. Ownership of land is not permitted. These are species oriented. For example, Bharatpur Sanctuary (bird), Abhor Sanctuary (Black buck), Hazaribagh Sanctuary (Tiger and Leopard) etc. There are about 500 sanctuaries in India. Common size is 100 to 500 km^2.

Silviculture: It is the science of managing forests on sustainable basis. It can also be defined as that branch of forestry which deals with the establishment, development, care and reproduction of timber.

Species: It is a natural biological unit tied together by the sharing of a common gene pool. Species can also be defined as a uniform interbreeding population spread over time and space.

Sustainable Development: The concept is to meet the needs of the present without compromising the ability of future generation to meet their needs. Sustainability also means, conservation of natural resources and management of bio-sphere for human needs and use for an indefinite period of time.

The Earth Summit (Rio-Summit): The first Earth Summit of United Nations conference on Environment and Development (UNCED) was organised in June 1992 in **Rio De Janerio** (Brazil), for protection of Earth and its Environment, maintenance of ecological balance and to enrich bio-diversity.

Types of Ecosystems: Based on the particular type of habitat ecosystems are classified as, Terrestrial Ecosystem or Biome – on land; Aquatic Ecosystem – in water; Forest Ecosystem; Desert Ecosystem; Wetland Ecosystem; Freshwater and standing water ecosystem; Man-made Eco-systems like – Cropland or agro-ecosystem.

Vegetation: It is the totality of plant growth including large or small populations of each species inter-mixed in a region. In other words, we may say that vegetation is the sum total of plant population covering a region.

Water Conservation: Water conservation is mainly in the context of freshwater. With increasing population, industrialisation and urbanisation water is being used in enormous quantities. Agriculture also consumes a lot of water. To conserve water, we must reduce total water consumption and reduce wastage. Recycling of water is an effective way of conserving water.

MULTIPLE CHOICE QUESTIONS

1. The physical components of the earth are called
   (a) Resources (b) Environment (c) Atmosphere (d) Flora-Fauna
2. Which of the following is a biotic component of Environment
   (a) Plants (b) Animals (c) Rivers (d) Micro-organism
3. Which of the following is not the cause of Environmental
   (a) Deforestation (b) Over population (c) Industry (d) Sports
4. The development that meets the needs of present and also future generation is called
   (a) Sustainable (b) Continuous (c) Dynamic (d) Popular
5. Who said, "Earth provides enough to satisfy every man's need, but not every man's greed".
   (a) Mahatma Gandhi (b) Lord Buddha
   (c) Lord Mahavir (d) Sai Baba
6. The 1972 UN conference on Environment was held at
   (a) New York (b) Stockholm (c) Rio de Janerio (d) Rome
7. The Earth's axis is inclined to vertical at an angle
   (a) 13.5° (b) 66.5° (c) 22.1° (d) 23.5°
8. Which of the following is not a rock form?
   (a) Igneous (b) Metamorphic (c) Soft (d) Sedimentary
9. Which of the following elements occur in maximum quantity in Earth's crust?
   (a) Oxygen (b) Silicon (c) Aluminium (d) Iron
10. The Indian islands entirely built by Coral deposits are
    (a) Andaman (b) Lakshdweep (c) Nicobar (d) None of these
11. 'Geysers' are
    (a) Hot water springs (b) Crater Lakes (c) Caves (d) Glaciers
12. Which of the following is largest desert?
    (a) Kalahari (b) Gobi (c) Kutch (d) Sahara
13. The percentage of Carbon-di-Oxide in atmosphere is
    (a) 1.0% (b) 0.93% (c) 0.03% (d) 1%
14. The Troposphere at the Equator is upto a height of
    (a) 8 km (b) 18 km (c) 50 km (d) 80 km

15. Ozone Layer is in
(a) Troposphere (b) Stratosphere (c) Thermosphere (d) Masosphere

16. Which is the largest Biosphere reserve of India?
(a) Nilgiri (b) Sundarban (c) Gulf of Mannar (d) Narmada Devi

17. Which of the following are producers
(a) Plants (b) Animals (c) Human beings (d) Mircro-organisms

18. The average thickness of Ozone Layer is
(a) 3 mm (b) 3 cm (c) 3 m (d) 3 km

19. Who first defined Ecology
(a) Taylor (b) Odum (c) Birch (d) Haeckelian

20. Which of the following is biotic component?
(a) Soil (b) Non-Green Plants (c) Water (d) Energy

21. Which of the following is not a Natural Hazard?
(a) Forest Fire (b) Tsunamis (c) Auto-Exhaust (d) Tornado

22. In a food chain which is the final player
(a) Hawk (b) Snake (c) Mouse (d) Grass

23. The salt content in Oceans is about
(a) 1.5% (b) 2.5% (c) 3% (d) 3.5%

24. Which of the following is not classified as wetland?
(a) Bog (b) Marsh (c) Swamp (d) Lake

25. Maximum freshwater is as
(a) Groundwater (b) Polar Ice/Glaciers
(c) Rivers/Lakes (d) Atmospheric vapour

26. India's coastline is about
(a) 5000 km (b) 6000 km (c) 7000 km (d) 1000 km

27. If all the ice caps and glaciers were to melt, the sea level may rise by
(a) 10 m (b) 20 m (c) 45 m (d) 65 m

28. Which one of the following has maximum genetic diversity in India?
(a) Mango (b) Teak (c) Wheat (d) Tea

29. The population of Tigers in the world in 2022
(a) 50,500 (b) 40,500 (c) 20,500 (d) 4,500

30. Which of the following is not an endangered species?
(a) Giant Panda (b) Donkey (c) Tiger (d) Blue Whale

31. Which country is Earth's bio-diversity champion
(a) Australia (b) India (c) Brazil (d) Russia

32. Estimated biological species on earth are
(a) 8 to 10 million (b) 1.5 to 2 million (c) 1 billion (d) 80,000

33. Which of the following is man made cause of Biodiversity loss
(a) Flood (b) Earthquake (c) Landslide (d) Poaching

34. The loss of diversity is most prominent in
(a) Agricultural Eco-system (b) Forest Ecosystem
(c) Marine Ecosystems (d) Grass land Ecosystem

35. India's contribution to world's bio-diversity is about
(a) 15% (b) 2.4% (c) 7% (d) 11%

36. Which of the following is not an endangered species?
(a) Python (b) Leopard (c) Sandalwood (d) Pigeon

37. Which of the following is the largest Biosphere Reserve in India?
(a) Rann of Kachch (b) Gulf of Mannar (c) Sundarvans (d) Nilgiri

38. Ghana Bird Sanctuary is situated in which state?
(a) Haryana (b) Rajasthan (c) Punjab (d) Bihar

39. Kaziranga National Park is in which state?
(a) Uttrakhand (b) Assam (c) MP (d) Rajasthan

40. Where is Chilka Lake?
(a) West Bengal (b) Bihar (c) Odisa (d) Manipur

41. How many Tiger reserves are now?
(a) 9 (b) 27 (c) 18 (d) 3

42. According UNESCO criteria India has how many Heritage sites
(a) 5 (b) 10 (c) 19 (d) 26

43. The Natural Heritage site 'Sunderbans' is in
(a) West Bengal (b) Andhra Pradesh
(c) Gujrat (d) Odisa

44. Wildlife (Protection) Act was made in
(a) 1980 (b) 2002 (c) 1972 (d) 2014

45. Which of the following is a non-renewable resource?
(a) Plants (b) Water (c) Air (d) Coal

46. Sustainability is not
(a) Quick Development (b) Living in harmony with nature
(c) Generation equity (d) Preserving nature

47. Which is not included in three ₹ ?
(a) Reducing (b) Recollecting (c) Reusing (d) Recycling

48. Total forest cover in India as percentage of Geographical Area
(a) 9.38% (b) 33.3% (c) 21.67% (d) 76.92%

49. India's mangrove cover area is
(a) 2000 km^2 (b) 3000 km^2 (c) 4000 km^2 (d) 5000 km^2

50. Under the National Forest Policy desired forest cover should be
(a) 25% (b) 50% (c) 37.5% (d) 33%

51. India overtook China's population in the year
(a) 2023 (b) 2020 (c) 2021 (d) 2022

52. Chipko movement was meant to protect
(a) Water bodies (b) Wild life
(c) Agriculture land (d) Trees

53. Science of managing Forests in known as
(a) Forestry (b) Plantation (c) Silviculture (d) Conservation

54. What percentage of irrigation water comes from groundwater
(a) 58% (b) 48% (c) 52% (d) 39%

55. The water stress condition occurs when per capita water availability per year is below
(a) 1700 m^3 (b) 1000 m^3 (c) 700 m^3 (d) 7200 m^3

56. Flood land next to the river is known as
(a) Flood Plain (b) Valley (c) Catchment (d) Basin

57. The embankments constructed parallel to river bank for flood protection are called
(a) Groyens (b) Guide banks (c) Terraces (d) Levees

58. Devastating floods in Eastern India are caused by river
(a) Ganga (b) Hoogli (c) Kosi (d) Brahamputra

59. India has water conflict with
(a) Bangla Desh (b) Sri Lanka (c) Bhutan (d) Myanmar

60. Which of the following human activities can trigger draught?
(a) Over fishing (b) Deforestation (c) Urbanisation (d) Pollution

61. Period of dry weather longer than normal is known as
(a) Flood (b) Disaster (c) Drought (d) Famine

62. Which state is least drought prone?
(a) Odisa (b) Tamil Nadu (c) Rajasthan (d) Assam

63. How can dams cause drought?
(a) By reducing rainfall levels (b) By reducing Water level in downstream
(c) By submerging upstream land (d) Causing more infiltration

64. How climate change likely to affect the occurrence of drought?
(a) Reduce number of droughts (b) No change
(c) Increase in number of droughts (d) Severity of drought will increase

65. Which country produces maximum Coal?
(a) China (b) India (c) Australia (d) US

66. Which state produces maximum coal in India
(a) Jharkhand (b) Odisa (c) Chattisgarh (d) Madhya Pradesh

67. In India Agriculturable land is about

(a) 22% (b) 62% (c) 42% (d) 52%

68. Which of the following is man-made cause of landslide?

(a) Earthquake (b) Steep slope (c) Rainfall (d) Mining

69. Desertification is caused due to

(a) Deforestation (b) Overgrazing (c) Climate change (d) All of these

70. Terracing is an effective method of soil conservation in

(a) Hills (b) Plains (c) Riverine area (d) Coastal belt

71. To avoid soil erosion the slope of cultivable land should not be more than

(a) 25% (b) 15% (c) 10% (d) 5%

72. Which of the following is a Secondary fuel?

(a) Wood (b) Coal (c) Petroleum (d) L.P.G.

73. Which of the following is Primary fuel?

(a) Natural gas (b) Kerosene oil (c) Gobar gas (d) Coke

74. Which of the following is a non-renewable source of energy?

(a) Solar (b) Wind (c) Petroleum (d) Wood

75. Which type of energy is in maximum use in India?

(a) Nuclear (b) Thermal (c) Solar (d) Natural gas

76. The average ash content in Indian Coal is about

(a) 35% (b) 15% (c) 20% (d) 5%

77. Sun's Rays reach earth in

(a) 38 m 20 s (b) 28 m 20 s (c) 18 m 20 s (d) 8 m 20 s

78. The percentage of visible light from Sun is

(a) 44% (b) 34% (c) 24% (d) 14%

79. The Ozone layer absorbs

(a) Visible light (b) U.V. Radiation

(c) Infra-red Radiation (d) γ-ray

80. Which of the following part of Solar radiation has maximum energy?

(a) x-Rays (b) U.V.-Radiation (c) Gamma rays (d) Micro-Waves

81. Maximum Energy content (Calorific value) is in which of the following fuels.

(a) Coal (b) Hydrogen (c) Petrol (d) Methane

82. Natural gas contains maximum of

(a) Methane (b) Oxygen (c) Ethane (d) Hydrogen

83. Which of these is not a part of Acid Rain?

(a) Sulphuric Acid (b) Nitric Acid (c) Carbonic Acid (d) Lactic Acid

84. Which of the following power is most hazardous for us?

(a) Thermal (b) Hydro (c) Nuclear (d) Geo-Thermal

85. The ultimate source of energy is
(a) Biomass (b) Wind (c) Solar (d) Hydro

86. The Turbine converts Potential/Kinetic Energy into
(a) Chemical energy (b) Mechanical energy
(c) Pressure energy (d) Electrical energy

87. Which country produces maximum Water Power?
(a) China (b) Brazil (c) Canada (d) Russia

88. In oceans how many high tides occur in about 24 Hours.
(a) 4 (b) 2 (c) 1 (d) 8

89. Which of the following energies is non-polluting
(a) Thermal (b) Nuclear (c) Hydro (d) None of these

90. The Green Revolution in India started in the year
(a) 2014 (b) 1950 (c) 1981 (d) 1965

91. Father of Indian Green Revolution is
(a) Swaminathan (b) Medha Patekar
(c) H.N. Bahuguna (d) Narendra Modi

92. Apiculture is related to
(a) Silk worm (b) Flower (c) Bees (d) Vegetable

93. Science of cultivating 'forest trees' is called
(a) Sericulture (b) Silviculture (c) Viticulture (d) Floriculture

94. According to Geological Survey of India, flood prone area in India is,
(a) 12.5% (b) 25% (c) 33% (d) 40%

95. The traditional agriculture causes
(a) Deforestation (b) Soil Erosion
(c) Depletion of Nutrient (d) All these

96. Which is the most important cause of air pollution?
(a) Industrial emission (b) Burning of Fossil Fuel
(c) Wild fires (d) Microbial decay

97. Which of the following is a Secondary air pollutant?
(a) SO_2 (b) CO_2 (c) SPM (d) O_3

98. Which of the following fuels causes minimum air pollution?
(a) Kerosene oil (b) Hydrogen (c) Coal (d) Diesel

99. Which of the following gases can deplete ozone layer in upper atmosphere?
(a) Ammonia (b) Carbon monoxide
(c) Methane (d) Sulphure dioxide

100. Air is a/an
(a) Element (b) Mixture (c) Compound (d) Solution

101. Which is a natural cause of air pollution?
(a) Volcanoes (b) Coal burning (c) Flood (d) Automobiles

102. Smog is
(a) Ozone + Smoke (b) Vehicular Exhaust
(c) Fog + Ozone (d) Smoke + Fog

103. According to a Greenpeace Report of 2019, India is highest emitter of
(a) Carbon monoxide (b) Ammonia
(c) Sulphur dioxide (d) Nitrogen oxide

104. The Air Quality Index (AQI) of 150 falls under category
(a) Very poor (b) Moderate (c) Poor (d) Good

105. Respiratory diseases are caused due to
(a) Air pollution (b) Water pollution (c) Soil pollution (d) All these

106. Which is the main Greenhouse Gas?
(a) Sulphur dioxide (b) Methane (c) Nitrous oxide (d) Carbon dioxide

107. Major source of carbon monoxide (CO) as air pollution is
(a) Industrial sector (b) Domestic sector (c) Transport sector (d) Thermal Power sector

108. Rain is called Acid Rain when its pH-value is
(a) 1 to 3 (b) Less than 5.6 (c) 5.6 to 7.0 (d) More than 7

109. Water Pollution causes the disease
(a) Conjunctivitis (b) Respiratory infection
(c) Diarrhoea (d) Bronchitis

110. Which of the following is not a water pollutant?
(a) Glacier (b) Silt (c) Arsenic (d) Chromium

111. According to WHO the Arsenic in drinking water should not exceed
(a) 6 μg/L (b) 5 μg/L (c) 1 μg/L (d) 10 μg/L

112. In wastewater treatment first step is removal of
(a) Micro-Organism (b) Organics (c) Solids (d) Colloids

113. The greatest volume of waste discharge to water body is from
(a) Spillage from oil pipelines (b) Sewage
(c) Nuclear waste (d) Spillage from oil tankers

114. The main reason for thermal pollution is
(a) Bio fuels (b) Organic forming
(c) Eco-friendly vehicles (d) Power plants

115. Radiation can cause
(a) Heart disease (b) Haemophillia (c) Leukemia (d) Bone Marrow disease

116. Which of the following element is most hazardous?
(a) Uranium (b) Carbon (c) Sulphur (d) Phosphorous

117. The radiation hazard to which general people are exposed is generally to level
(a) 5 to 10 rem (b) 0.1 to 5 rem (c) 10 to 15 rem (d) About 100 rem

118. Normally the cell phones have Specific Absorption Rate (SAR) is below
(a) 100.3 (b) 10.3 (c) 5.3 (d) 1.3

119. Municipal Solid Waste (MSW) in India has maximum
(a) Ash (b) Rubbish (c) Garbage (d) Dust

120. The burning of Municipal Solid Waste (MSW) is called
(a) Incineration (b) Composting (c) Shredding (d) Land filling

121. Which of the following is bio-degradable waste?
(a) Plastic (b) Kitchen waste (c) Synthetic fibre (d) Iron

122. Crushing and grinding of Municipal solid waste is called
(a) Shredding (b) Composting (c) Pulverization (d) Vermi-composting

123. Which of the following are the best worms used in composting?
(a) Maggots (b) Pink worms (c) Red wigglers (d) Griseofulvin

124. Which one of the following is a component of the hospital waste?
(a) Scrap (b) Toxic Metals (c) Mobiles (d) Pathogens

125. Which of the following is a natural way of composting?
(a) Vermi-composting (b) Land fill
(c) Incineration (d) Sanitary land fill

126. Ideal temperature for Vermi-composting is
(a) 20° to 25° C (b) 25° to 30° C (c) 30° to 35° C (d) 35° to 40° C

127. Medical waste is classified as
(a) Combustible (b) Hazardous (c) Compostable (d) Recyclable

128. A sewage Treatment Plant (STP) has
(a) UV-source (b) Ion Exchanger (c) Cyclone (d) Grit chamber

129. Which of the following is not a part of 3-Rs?
(a) Reduce (b) Reuse (c) Recover (d) Recycle

130. In India the fraction of food waste is
(a) Between 10-15% (b) Between 20-30%
(c) Between 25-35% (d) Greater than 40%

131. On Richter Scale an extremely severe earthquakes has a value of
(a) 8 to 8.9 (b) 4 to 4.9 (c) 6 to 6.9 (d) 7 to 7.9

132. Bhuj Earthquake of 2001 in which more than 20,000 people died occurred on
(a) Independence Day (b) Republic Day
(c) Gandhi Jayanti (d) Christmas Day

133. Which of the following Earthquake zone does not exist?
(a) 4 (b) 2 (c) 1 (d) 3

134. Which is not a class of Flood Zoning?
(a) Prohibitive zone (b) Restrictive zone (c) Warning zone (d) Lower zone

135. Which of the following is not a Natural disaster?
(a) Nuclear (radiation) (b) Tsunami
(c) Forest fire (d) Flood

136. A twisting powerful wind-storm is called
(a) Cyclone (b) Tornado (c) Tsunami (d) Dust storm

137. The maximum height of Tsunami wave measured till now
(a) 85 m (b) 75 m (c) 65 m (d) 55 m

138. The most recent cyclone (2022) hitting Andhra Pradesh is named
(a) Nargis (b) Jawad (c) Gulab (d) Asani

139. According to USGA which is the most unpredictable natural disaster
(a) Flood (b) Earthquake (c) Cyclone (d) Avalanch

140. What is the share of Natural resources of Developed countries?
(a) 80% (b) 70% (c) 60% (d) 50%

141. Which of the following tree/plant is not worshipped in India?
(a) Tulsi (b) Peepal (c) Banyan (d) Jamun

142. The sustainable Development Goals comprises of how many goals
(a) 10 (b) 15 (c) 17 (d) 11

143. Which of these is not a Goal for Sustainable Development Goal (SDG) to be achieved by 2030?
(a) Quality education (b) Space research (c) Gender equality (d) Zero hunger

144. Agenda 21 of Rio Summit (1992) is related to
(a) Sustainable development (b) Polluter-Pays principle
(c) Environmental education (d) Restoration of ozone layer

145. What is the theme of World Sustainable Development Summit 2021?
(a) Attaining 2030 Agenda (b) Partnership for Resilient Planet
(c) Redefining our common future (d) Toward 2030 Goals

146. The first airport powered by solar energy
(a) Banglore (b) Frankpurt (c) Singapore (d) Cochin

147. World Environmental day is observed on
(a) March 22 (b) June 5 (c) Sept 15 (d) April 22

148. Which Law was enacted after Bhopal Gas Tragedy?
(a) Environmental Protection Act (b) Pollution Control Act
(c) Disaster Management Act (d) Air Act

149. Environmental Protection Act (EPA) is also known as
(a) Environmental Legislation (b) Umbrella Legislation
(c) Eco Safety Law (d) Disaster Management Act

150. How many Ramsar sites have been declared so far in India?
(a) 95 (b) 85 (c) 75 (d) 65

TRUE OR FALSE QUESTIONS

1. The atmospheric layers are based on vertical profile of Air Temperature. (T or F)
2. Water vapour is tiny raindrops. (T or F)
3. Sky looks blue due to scattering of light by air molecules. (T or F)
4. Accumulation of salt in soil is known as salinization. (T or F)
5. When air temperature increases with height in the atmosphere is called Temperature Inversion. (T or F)
6. Wind direction is the direction to which wind is blowing? (T or F)
7. The Pyramid of energy in an ecosystem which is always inverted (T or F)
8. Bacteria is an abiotic component. (T or F)
9. Particulate Pollutants are small solid particles and liquid droplets. (T or F)
10. Atmospheric temperature is maintained by presence of Nitrogen. (T or F)
11. Harmful metals like antimony, arsenic and cadmium are found in Fly Ash. (T or F)
12. In India tropical evergreen forests occur in the State of Madhya Pradesh. (T or F)
13. Day and night are equal at Poles. (T or F)
14. Western Ghats are biodiversity hot spots in India. (T or F)
15. The unfavourable change in environment due to human activity is called Pollution. (T or F)
16. The complex network of inter-connected food chain is called Ecology Chain. (T or F)
17. Red data book contains data of threatened species. (T or F)
18. When two ecosystems overlap, the area is called Ecotone. (T or F)
19. Most abundant gas emitted from volcanoes is carbon-di-oxide. (T or F)
20. A river with high BOD is relatively clean. (T or F)
21. Animal life occurring in a particular region or time is called Fauna. (T or F)
22. Mulching helps in moisture conservation. (T or F)
23. The warmest time of the day is between 12 N to 2 PM. (T or F)
24. Ozone hole was first discovered in India. (T or F)
25. World's biggest Green House Gas emitter is USA. (T or F)
26. Electromagnetic radiation of wave length (λ) between 0.4 and 0.7 micrometers is called Visible Light. (T or F)
27. Violet light has shorter wavelength than Red Light. (T or F)

28. Most of the radiation emitted by a human body is in the form of Infra-red radiation and invisible. (T or F)
29. Carbon monoxide is a pollutant because it reacts with hemoglobin. (T or F)
30. Environmental education is interdisciplinary. (T or F)
31. Low level ozone is referred to as ionospheric ozone. (T or F)
32. Water used for paper mills should not contain magnesium. (T or F)
33. Blue revolution is associated with Iron & Steel. (T or F)
34. The climate of a given region is average temperature and precipitation conditions for every day of the year. (T or F)
35. The Red Data Book which lists endangered species is maintained by UNO. (T or F)
36. The Unit of Noise (Sound) is Decibel. (T or F)
37. World wetland day is on Feb. 2. (T or F)
38. The city of Ramsar is in Turkey. (T or F)
39. Asiatic lion is found only in Sundarbans. (T or F)
40. Himanchal Pradesh is the first Indian state to propose ban on all types of polythene packing. (T or F)
41. Lead in water can damage Kidney. (T or F)
42. Salim All Centre for Ornithology and Natural History is at Pune. (T or F)
43. Oil spills pollute soil and water. (T or F)
44. Microscopic floating organisms are called Plankton. (T or F)
45. Molten rocks deep inside earth are called lava. (T or F)
46. Lichen are the best indicator of SO_2 pollution. (T or F)
47. Photochemical smog is also called London smog. (T or F)
48. Maximum daily water demand is 1.8 times the average per capita daily demand. (T or F)
49. Hard water has hardness of the order of 100 ppm. (T or F)
50. Total Dissolved Solids (TDS) test indicates the intensity of water pollution. (T or F)
51. MPN Index is the measure of BOD. (T or F)
52. Detention time for water sedimentation tanks should be 2 to 4 Hours. (T or F)
53. Slow sand filters should be cleaned from one to three months. (T or F)
54. Aeration of water is done to remove odour. (T or F)
55. For defluoridation of water lime is used. (T or F)
56. Pathogens are removed by chlorination. (T or F)
57. A sewer is commonly designed to attain self cleansing velocity at average hourly rate of flow. (T or F)
58. The pH of fresh sewage is usually more than 7. (T or F)
59. Minimum DO in rivers to save fish should be 3 ppm. (T or F)

60. Most of bacteria in sewage are Pathogenic. (T or F)
61. The detention period adopted for grit chambers is around one minute. (T or F)
62. Dissolved organic solids in wastewater treatment are removed by reverse osmosis. (T or F)
63. The gas coming out of a sludge digestion tank is only CH_4. (T or F)
64. Activated sludge is the sludge in the secondary tank after aeration. (T or F)
65. Sludge digestion is a process of sewage screening. (T or F)
66. Blue green algae can fix atmospheric nitrogen. (T or F)
67. The lower limit of bacteria removal through activated sludge process is 70%. (T or F)
68. Zeolite is Sodium silicate. (T or F)
69. Zeolite process is also known as cation exchange process. (T or F)
70. A 30 dB increase in noise level represents 100 fold increase in sound intensity. (T or F)
71. The human hearing is sensitive to sounds from frequencies 20Hz to 20,000 Hz. (T or F)
72. Light pollution can cause serious road accidents. (T or F)
73. Greenhouse gases absorb heat. (T or F)
74. Biomass is non-renewable source of energy. (T or F)
75. Ozone is one of the primary pollutants. (T or F)
76. Tomato is extremely sensitive to SO_2. (T or F)
77. The waste found in oceans is mostly plastic. (T or F)
78. Salty water is mostly found in Lakes. (T or F)
79. Tajmahal is getting yellowish due to sulphur dioxide (SO_2) gas. (T or F)
80. Plastic bottles cannot be recycled. (T or F)
81. Incineration of solid waste helps in reducing air pollution. (T or F)
82. Deforestation is caused more by commercial farming. (T or F)
83. Carbon monoxide is **not** a greenhouse gas. (T or F)
84. Incomplete burning of fossil fuel emits Carbon dioxide. (T or F)
85. Silent valley of Kerala has Deciduous Forests. (T or F)
86. Corbett National Park was established to protect Tigers. (T or F)
87. Bharatpur bird Sanctuary is in Bihar. (T or F)
88. Andaman & Nicobar Islands have maximum Number of National Parks. (T or F)
89. Gases of Volcanic Eruption are natural air pollutants. (T or F)
90. Coir is used for sound-proofing of auditoriums etc. (T or F)
91. CFC are responsible for Ozone Layer depletion. (T or F)
92. Hydrocarbons are Mutagenic pollutants. (T or F)
93. According to CPCB, particles which can cause great harm to human health are of size 5 microns. (T or F)
94. Particulate matter can be removes by Radioactive Converters. (T or F)

95. Grass, Goat and Human constitute a human chain. (T or F)

96. Tobacco addiction is due to cocaine. (T or F)

97. India should have minimum of 50% forest area. (T or F)

98. The rain is generally considered as Acid Rain when pH falls below 5. (T or F)

99. Zero hardness of water can only be achieved by ion-exchange method. (T or F)

100. The minimum Dissolved Oxygen (DO) which is necessary to be present in water for survival of aquatic life is 10 ppm. (T or F)

101. Spherical shaped bacteria is called cocci. (T or F)

102. The effluent from septic tank is discharged into drainage. (T or F)

FILL IN THE BLANKS

1. Surroundings of any object is called ....
2. Interaction between different forms of life are known as .... Factors.
3. The development that meets the needs of present without compromising the needs of future generation is called .... developments.
4. The Earth Summit is 1992 was held at ....
5. Which Ministry in India looks after environmental issues ....?
6. Besides Atmosphere and Hydrosphere which is the 3rd component of Environment ....
7. Lithosphere, Hydrosphere and Atmosphere together are known as ....
8. Which element is in maximum quantity in Earth's crust ....?
9. The moving masses of ice are called ....
10. Ozone Layer is found in which atmospheric layer ....
11. Which type of rocks are formed from magma .... ?
12. The highest mountain peak in India is named ....
13. The science of all relations of organisms to all their environment is called ....
14. The place where an organization lives is called its ....
15. An ecological community of organisms maintained under a climate zone is called ....
16. Birth-Death Ratio is called ....
17. The specie which derives benefit at the expense of other (host) is called ....
18. Among a Frog and insect, which is predator? ....
19. Name one Carnivore plant ....
20. Large cities are known as ....
21. Symbiosis means ....
22. In India minimum size of population for an Urban settlement is ....

23. Some people move from place to place in search of their living without having permanent residence are called ....
24. 'Natural Selection' and 'Survival of fittest' theory was given by ....
25. Science of managing forests on sustainable basis is called ....
26. Out of about 3% freshwater on earth the percentage as groundwater is about ....
27. Which sector consumes maximum quantity of water ....
28. Sardar Sarovar Project is on which river ....?
29. The Karnataka-Tamilnadu water dispute is on which river waters ....
30. What percentage of India's land is under forest cover ....?
31. Red soil is due to presence of which substance ....?
32. Landslides are mainly triggered by ....
33. The global rate of soil erosion annually is 2.4 t/ha. In India it is about ....
34. According to the Ministry of Power, What percentage of Electricity is produced through Fossil Fuels ....?
35. According to UN Convention on Law of the Sea, the territorial limit from shoreline into sea is ....
36. Marshes and Swamps are known as .... wetlands.
37. Ozone depletion can cause the disease of ....
38. The decolouring of Taj Mahal is due to ....
39. The part of plant through which water is evaporated is called ....
40. Humidity is measured by ....
41. The most talked about soil pollutant is ....
42. Paper is made up of ....
43. The chemical responsible for destroying chlorophyll in plant leaves is ....
44. Hypertension, Hearing loss, stress and headache can be due to high ....
45. Which country has maximum number of tube wells ....?
46. The first National Parks 'Jim Corbett' is in the district of Uttrakhand ....
47. First trophic level in the food chain are ....
48. The energy stored in the Earth is called .... Energy.
49. Fine organic or inorganic particles suspended in air are called ....
50. Ramsar convention refers to the conservation of ....
51. Pollutants caused by human activity are called ....
52. Association between members of two species is called ....
53. Ecosystem of flowing water is called .... Ecosystems.
54. Major cause of diminishing wildlife number is loss of ....
55. Plant life occurring in a particular region or time is called ....

56. The combustion of waste in absence of oxygen is called ....
57. Ornithology is the study of ....
58. Zoological Survey of India is in city of ....
59. Shallow ponds with DO at all depths are called .... ponds.
60. Excessive discharge of nutrients into a water body may cause ....
61. Composting and lagooning are the methods of .... disposal.
62. Day time noise prescribed for residential areas in India is ....
63. Bio-degradable solid waste is decomposed by earthworms is called .... composting.
64. Most efficient air pollution control device is .... precipitator.
65. How may Bio-diversity hot-spots exits in the World? ....
66. The instrument which can measure temperature, pressure and humidity at numerous altitudes is called ....
67. Brackish water ecosystems are found in ....
68. The upper portion of an Egg-shaped sever line is known as ....
69. The reduction of organic acids produce ....
70. Chipko movement was to save ....
71. The theoretical oxygen demand of 300 mg/L glucose solution will be ....
72. Sloughing occurs in .... filter.
73. Detention period for a septic tank in India is about ....
74. The trap used for a water closet is called ....
75. The most important environmental issue threatning mankind is ....
76. World environment day is celebrated on ....
77. Reducing the degree or intensity of, or eliminating pollution is called ....
78. Ozone layer absorbs .... radiation.
79. The amount of energy transfer from one trophic level to another in food chain is about .... %.
80. The primary consumers are ....
81. First National Park of India is ....
82. Indian state having the largest forest cover ....
83. Which state has highest percentage of forest-cover ....
84. 'Valley of Flowers' National Park is in the state of ....
85. The agents which bring undesirable changes in environment are called ....
86. Fluoride pollution mainly affect the ....
87. The major source of noise pollution is .... system.
88. Chlorofluorocarbon (CFC) are responsible for .... Inversion.
89. Algal bloom is caused due to presence of large amount of .... in water bodies.
90. Acid rain mainly causes .... pollution.

91. Bhopal Gas tragedy took place because Methyl isocyanate reacted with ....
92. The natural place of an organism or community is known as ....
93. The plant growing under direct sunlight are known as ....
94. The upper part of an aquatic system is called ....
95. Which place in India is known as Biodiversity hotspot ....?
96. Full form of EIA? ....
97. Accumulation of salts in soil is called ....
98. The cause of Melanorina (skin cancer) is depletion of ....
99. The part of the plant which evaporated water is ....
100. Winter sleep of animals is called ....
101. Coagulation is required when particles are of .... size.
102. The total domestic water demand per capita per day may be taken as ....
103. To remove colour, odour and taste the process is ....
104. The Yield of Rapid Sand Filter is .... times more than the slow sand filter.
105. Hypo-Chlorine of lime is known as ....

ANSWERS

Multiple Choice Questions

1. b	2. c	3. d	4. a	5. a	6. b	7. d	8. c	9. a
10. b	11. a	12. d	13. c	14. b	15. b	16. c	17. a	18. a
19. d	20. b	21. c	22. a	23. d	24. d	25. b	26. c	27. d
28. a	29. d	30. b	31. c	32. b	33. d	34. a	35. c	36. d
37. a	38. b	39. b	40. c	41. b	42. d	43. a	44. c	45. d
46. a	47. b	48. c	49. d	50. d	51. a	52. b	53. c	54. c
55. b	56. a	57. d	58. d	59. a	60. b	61. c	62. d	63. b
64. c	65. a	66. c	67. c	68. d	69. d	70. a	71. b	72. d
73. a	74. c	75. b	76. c	77. d	78. a	79. b	80. c	81. b
82. a	83. d	84. c	85. c	86. b	87. a	88. b	89. c	90. d
91. a	92. c	93. b	94. a	95. d	96. b	97. d	98. b	99. c
100. b	101. a	102. d	103. c	104. b	105. a	106. d	107. c	108. b
109. c	110. a	111. d	112. c	113. b	114. d	115. c	116. a	117. b
118. d	119. c	120. a	121. b	122. c	123. c	124. d	125. a	126. c
127. b	128. d	129. c	130. d	131. a	132. b	133. c	134. d	135. a
136. b	137. c	138. d	139. b	140. a	141. d	142. c	143. b	144. a
145. c	146. d	147. b	148. a	149. b	150. c			

True & False

1. T	2. F Gas	3. T	4. T	5. T
6. F Direction From	7. F Upright	8. F Biotic	9. T	
10. F CO_2	11. T	12. F Assam (Asom)	13. F Equator	14. T
15. T	16. F Food web	17. T	18. T	
19. F Water vapour	20. F Very Polluted	21. T	22. T	
23. F 3 PM to 5 PM	24. F Antarctica	25. F China	26. T	
27. F $\lambda_v > \lambda_r$	28. T λ = 12 micron	29. T	30. T	
31. F Tropospheric	32. F Iron	33. F Fishing	34. T	
35. F IUCN	36. T	37. T	38. F Iran	
39. F Gir Forest	40. T	41. T	42. F Coimbtore	43. T
44. T	45. F Magma	46. T	47. F Los Angles Smog	
48. T	49. F 200 ppm	50. TRenewable	51. F Coliform Bacteria	
52. F 4 to 8 Hours	53. T	54. T	55. F Alum	56. T
57. F At minimum rate of Flow		58. T	59. F 4 ppm	
60. F Saprophytic	61. T	62. T	63. F $CH_4 + CO_2$ Ratio 70 : 30	
64. T	65. F Treatment	66. T	67. F 90%	68. T

69. T **70.** F 1000 fold **71.** T **72.** T **73.** T
74. F **75.** F Secondary **76.** T **77.** T
78. F Oceans **79.** T **80.** F **81.** F **82.** T
83. T **84.** F **85.** F Tropical Rain Forests **86.** T
87. F Rajasthan **88.** T **89.** T **90.** F Styrofoam **91.** T
92. T **93.** F 2.5 Micron **94.** F Electrostatic Converter **95.** T
96. F Nicotine **97.** F 33% **98.** T **99.** T
100. F4 ppm **101.** T **102.** F Soak Pits

Fill in the Blanks

1. Environment **2.** Biotic **3.** Sustainable **4.** Rio-de-Janero
5. Ministry of Forest and Climate change **6.** Lithosphere **7.** Biosphere
8. Oxygen **9.** Glaciers **10.** Stratosphere **11.** Igneous
12. Kanchen-Jangha **13.** Ecology **14.** Habitat **15.** Biome
16. Vital Index **17.** Parasite **18.** Frog **19.** Pitcher Plant
20. Metropolis **21.** Living Together **22.** 5000 **23.** Nomads
24. Charles Darwin **25.** Silviculture **26.** 0.75% **27.** Agriculture
28. Narmada **29.** Cauvery **30.** 21.71% **31.** Iron oxide
32. Heavy Rainfall **33.** 16 t/ha/yr **34.** 58.5% **35.** 22.2 km
36. Freshwater **37.** Cancer **38.** Acid Rain **39.** Stomata
40. Hygrometer **41.** Plastic **42.** Wood (Grass/Bamboo)
43. Sulphar dioxide (SO_2) **44.** Noise **45.** India **46.** Nanital
47. Green plants **48.** Geothermal **49.** Aerosol **50.** Wetlands
51. Anthropogenic **52.** Symbiosis **53.** Lotic **54.** Habitat
55. Flora **56.** Pyrolysis **57.** Birds **58.** Kolkata
59. Aerobic **60.** Eutrophication **61.** Sewage **62.** 55 dB
63. Vermi **64.** Electrostatic Precipitator **65.** Thirty-six (36)
66. Radiosonde **67.** Deltas **68.** Arch **69.** Methane
70. Trees **71.** 320 mg/L **72.** Trickling **73.** 24 Hours
74. P-Trap **75.** Global warming **76.** 5th June **77.** Abatement
78. Ultraviolet (UV) **79.** 10% **80.** Herbivores **81.** Corbett National Park
82. Madya Pradesh (MP) 77.5k km^2 **83.** Mizoram 90% **84.** Uttrakhand
85. Pollutants **86.** Teeth **87.** Transport **88.** Thermal
89. Nutrients **90.** Soil **91.** Water **92.** Habitat **93.** Heliophates
94. Plankton **95.** Western Ghats **96.** Environmental Impact Assessment
97. Salinization **98.** Ozone **99.** Stomata **100.** Hibernation
101. Fine **102.** 135 Litres **103.** Aeration **104.** 30 times
105. Bleaching Powder

SECTION 8

Fluid Mechanics and Hydraulic Machines

Definitions – 31

Multiple Choice Questions – 103

True & False – 78

Fill in the Blanks – 75

DEFINITIONS

Absolute Pressure: The pressure intensity measured from the state of vacuum or zero pressure is called absolute pressure.

Acceleration: In fluid acceleration can either be local or convective. Local acceleration or temporal acceleration is defined as the rate of change of velocity with respect to time at a given location in a flow field. Convective acceleration is defined as the rate of change of velocity due to change of position of fluid particles in a fluid flow.

Adhesion: It is the tendency of dissimilar particles (molecules) to cling to one another.

Adiabatic Process: It is the process in which there is o exchange of heat from the system to its surroundings.

Atmospheric Pressure: It is the force per unit area exerted by an atmospheric column (i.e. the entire body of air above specific area).

Standard sea-level pressure equals each of the following, 760 mm of mercury, $\frac{1013.25 \times 10^3 \text{ dynes}}{\text{cm}^2}$, 1013.25 milli bar, 101.325 kilo pascal, or simply one standard atmosphere.

Bernoulli's Equation: It is based on the principle of conservation of energy. For a steady flow of inviscid, incompressible fluid along a streamline or stream tube it is expressed as $\frac{p}{\gamma} + \frac{V^2}{2g} + z = \text{constant}$ (Each term has units of $N - \frac{m}{N}$ or m).

Boundary Layer: The extensive fluid medium around bodies moving in fluids or fluids moving over a stationary solid boundary can be divided into two regions-a thin layer adjoing the boundary called boundary layer, where viscous shear takes place, and the other, the region outside this layer where the flow behaviour is quite like that of an ideal fluid.

Bouyancy (or upthrust): It is net upward force exerted by a fluid that opposes the weight of a fully or partially submerged body. According to Archimedes "an object, wholly or partially submerged in a fluid is buoyed up by a force equal to the weight of the fluid displaced by the object".

Buckingham's π-Theorem: It states that if in an equation of a physical phenomenon there are n-variables involving m-fundamental dimensions then the equation can be expressed in terms of (n – m) non-dimensional parameters called π-terms. (Fundamental Dimensions are Mass, Length, Time; MLT)

Caplillarity (or Capillary Rise): It is a phenomenon in which liquid spontaneously rises or falls in a narrow space such as a thin tube or in the voids of a porous material. Surface Tension is an important factor in the phenomenon of capillarity.

Circulation and Vorticity: These are the two primary measures of rotation in a fluid. Circulation denotes the integral of the fluid velocity around a closed curve. It is a scalar quantity. Vorticity is a vector quantity which gives microscopic measure of rotation at any point in the fluid.

Compressibility of Fluids: It is the measure of the relative volume change in response to a pressure change. For a given mass of fluid and increase in pressure ($\Delta p > 0$), will cause a decrease in volume ($\Delta V < 0$).

Continuity Equation: It is mathematical expression of the Law of Conservation of mass. For incompressible flow volume rate of flow (discharge) remains constant from section to section. Change in cross-section only affects the velocity. Mathematically, $Q = A_1V_1 - A_2V_2$.

Darcy-Weisbach Friction Factor: This is one of the most general friction head loss equations for a pipe segment. It is expressed as, $hf = f\frac{L}{D}\frac{V^2}{2g}$, where f is the friction factor which is a function of Reynolds Number and equivalent sand grain roughness $\frac{\in}{D}$.

Density of Fluids: It is defined as the ratio of the mass of the fluid to its volume. The density of gases depends on pressure and temperature, while the density of liquid remains almost constant. The density of water is $1000 kg/m^3$ and that of air is $1.225 kg/m^3$.

Dimensions and Units: Physical quantities have dimensions and units of measurement. The three fundamental dimensions generally used are F L T or M LT (Force, Mass, Length and Time). The units of measurement in SI system are– Force-Newton (N), Mass-Kilogram (Kg), Length (m) and Time (s).

Dimensions for Force: $M L T^{-2}$, Density ML^{-3}, Dynamic Viscosity $ML^{-1} T^{-1}$, Kinematic Viscosity $L^2 T^{-1}$.

Drag and Lift: A body wholly immersed in **real fluid** may be subjected to two kinds of forces due to relative motion between the body and the fluid. The component of force in the direction of flow is called Drag Force (F_D) and component right angle to the direction of flow is called Lift Force (F_L).

Dynamic Viscosity (μ): Viscosity is that property of a fluid by which it offers resistance to the movement of one layer over an adjacent layer of fluid. The layers experience a shear stress. Newton gave a formula for shear stress (τ) as, $\tau = \frac{\mu\, du}{dy}$, where $\frac{du}{dy}$ is velocity gradient and μ coefficient of Dynamic Viscosity. The units of μ are $\frac{kg}{m-s}$ and $\frac{g}{cm-s}$ (Poise).

Equipotential Lines: A function ϕ has been defined such that its derivative in any direction (with a negative –ve sign) gives the velocity in that direction. Equipotential lines are those lines in flow field where ϕ is constant.

Equivalent Sand Roughness: The sand-grain roughness height is the roughness height which gives the same actual resistance coefficient as that caused by the real non-uniform wall roughness. The height of this Equivalent Sand Roughness is often specified in terms of relative roughness (e.g. equivalent protrusion height-to-pipe diameter).

Euler's Equation of Motion: Euler considered only gravity and pressure forces for analysis of fluid motion neglecting viscous and other forces.

($f_{gx} + f_{px} = \rho_{ax}$), Integration of Euler's equation gives Bernoulli's equation.

Flow Net: It is composed of a family of equipotential lines and corresponding family of stream lines (flow lines). Flow lines indicate the path along which water flows. Equipotential lines connect points with equal total head (potential). Both these lines intersect each other at right angles.

Forced and Free Vortex : Free vortex is a flow in horizontal plane. In this flow the product of radius (r) and tangential velocity (V) is constant (C). (C = Vr). In a forced vortex the velocity varies directly with the radius $\left(\frac{V}{r}=C\right)$. Examples,

Free Vortex: Flow in a wish basin.

Forced Vortex: Rotating tank filled with liquid.

Hagen-Poiseuille Equation: The steady state laminar flow of an incompressible fluid of viscosity (μ), through a tube of circular cross section, diameter (D) is described by Hagen – Posiuille equation which gives the loss of head due to viscous friction. It is written as, $h_f = \frac{32\,\mu\,VL}{\gamma\,D^2}$. For a particular pipe of length L, $h_f \propto V$.

Hydraulic Grade Line (HGL): Referring to Bernoulli's equation, the sum of elevation head and pressure head is called the piezometric head $\left(Z+\frac{p}{\gamma}\right)$ or static head. The line joining the tops of piezometric columns of liquid is called Hydraulic Grade Line.

Ideal Fluids and Real Fluids: Real fluids in nature have viscosity (μ), compressibility (κ) and Surface Tension (σ). Ideal fluids are inviscid (non-viscous), incompressible and without surface tension. These are imaginary and do not exist in nature. Ideal fluid concept is used to simplify mathematical analysis of fluid flow problems. Water and air approximate to ideal fluid.

Impulse Momentum Equation: It is based on the Principle of conservation of momentum. It states that net force acting on a fluid mass is equal to rate of change of momentum with respect to time. Mathematically, $F = m \times a = m\,.\,\frac{dv}{dt}$.

Irrotational Flow: A flow is said to be irrotational if the fluid particle while moving in the direction of flow do not rotate about their mass centre. It may however be ralised that such a flow can occur in ideal fluids only. If the viscosity is low (like in water), flow can be assumed to be irrotational.

Kinetic Energy Correction Factor (α): It is the ratio of kinetic energy of flow per second on the actual velocity across a section to kinetic energy of flow per second based on average or mean velocity. Mathematically $\propto = \frac{1}{AV^3}\int_A V^3 dA$.

For uniform velocity distribution $\propto = 1$.

For non-uniform velocity distribution $\propto > 1$.

For turbulent flow $\propto \approx 1$. For laminar flow $\propto = 2$.

Kinematic Similarity: In dimensional analysis kinematic similarity the streamline pattern and velocity directions or triangles of velocity component, at homologous points are similar and bear the same ratio between model and prototype. Like Velocity Ratio $V_r = \frac{V_m}{P_p}$ etc.

Laminar Flow: It is the flow in which fluid particles follow smooth paths in layers, with each layer moving smoothly past the adjacent layers with little or no mixing. There are no cross-currents perpendicular to the direction of flow. Laminar flow is characterized by low Reynold's Number, which indicate low inertia and high viscosity effects.

MULTIPLE CHOICE QUESTIONS

1. The unit weight of Pure Water in N/m^3 is:
   (a) 1000 (b) 981 (c) 9810 (d) 9.81
2. If Froude Number $F_1 = 6$, the ratio between conjugate depths of hydraulic Jump $\frac{D_2}{D_1}$ will be,
   (a) 10 (b) 9 (c) 7 (d) 8
3. In a Hydraulic Jump the head loss is given by formula,
   (a) $\frac{(y_2 - y_1)}{4\, y_1 y_2}$ (b) $\frac{(y_2 - y_1)^2}{4\, y_1 y_2}$ (c) $\frac{(y_2 - y_1)^3}{4\, y_1 y_2}$ (d) $\frac{(y_2 - y_1)^3}{y_1 y_2}$
4. Maximum Vacuum Created at the Summit of a siphon is
   (a) 2.7 m of water (b) 5.5 m of water (c) 7.4 m of water (d) 13.6 m of water
5. The weight of one litre of oil of density 0.8 will be
   (a) 78.4 N (b) 784 N (c) 7840 N (d) 7.84 N
6. Each term of Bernoulli's Equation has the unit of
   (a) $\frac{N}{m}$ (b) $\frac{m}{N}$ (c) $\frac{N_m}{N}$ (d) $\frac{N_m}{m}$
7. Which of the following is not applied to Bernoulli's theorem
   (a) Fluid is homogenous (b) Fluid is viscous
   (c) Flow is along stream line (d) Steady flow
8. Kinematic viscosity (ν) has the unit of
   (a) $\frac{m^2}{S}$ (b) $\frac{N.S}{m^2}$ (c) $\frac{N.S}{m}$ (d) $\frac{Kg.S}{m^2}$
9. The ratio of maximum velocity to mean velocity of viscous flow through a circular pipe is
   (a) 0.5 (b) 2 (c) 1.25 (d) 1.5
10. Reynold's Number is the ratio between Inertia Force to
    (a) Pressure Force (b) Surface Tension Force
    (c) Viscous Force (d) Elastic Force
11. Inside pressure in a hollow soap bubble of diameter d is
    (a) $\frac{2\sigma}{d}$ (b) $\frac{4\sigma}{d}$ (c) $\frac{6\sigma}{d}$ (d) $\frac{8\sigma}{d}$
12. Piezometer is designed to measure
    (a) Velocity head (b) Total pressure (c) Negative pressure (d) Static pressure
13. In Newton's Law of viscosity, $\tau = \mu \frac{du}{dy}$, μ has the units of
    (a) $Kgm^{-1} S^{-1}$ (b) $Kgm\ S^{-2}$ (c) $Kg\ S^{-1}$ (d) Kg/Sm^{-1}

14. In a fully submerged body the 'weight loss' is equal to
(a) Weight of the body
(b) Weight of the fluid displaced by the body
(c) Weight of body + weight of fluid displaced
(d) Weight of body – Weight of fluid displaced

15. Equivalent height of water column with respect to 15 cm Mercury column will be, about
(a) 1.5 m (b) 2.72 m (c) 2 m (d) 1.36 m

16. Poise is the unit of
(a) Stress (b) Viscosity (c) Surface tension (d) Pressure

17. For laminar flow in pipes the Reynold's Number should be
(a) < 1000 (b) < 1500 (c) < 2000 (d) < 2500

18. The horizontal component of buoyant force is
(a) Zero (b) Negligible
(c) Equal to weight of body (d) Equal to buoyant force

19. In a Free vortex motion, the radial component of velocity everywhere is
(a) High (b) Zero (c) Low (d) Non-zero and finite

20. Any fluid which Obey's the Newton's Law of viscosity is called
(a) Real Fluid (b) Ideal Fluid (c) Newtonian Fluid (d) Non-Newtonian Fluid

21. Hygrometers are used to measure
(a) Humidity (b) Soil moisture
(c) Infiltration capacity (d) Vapour pressure

22. The Law of conservation of mass is represented by
(a) Bernoulli's Equation (b) Momentum Equation
(c) Continuity Equation (d) Euler's Equation

23. Stagnation point is fluid mechanics is, where velocity is
(a) Infinite (b) Unity (c) Zero (d) Constant

24. Probable Laminar Flow may occur if
(a) Viscosity is low (b) Viscosity is high (c) Density is high (d) Temperature is high

25. Absolute pressure of a fluid cannot be
(a) –1 bar (b) 0 (c) 9.8 bar (d) 101 bar

26. Fluid offers no resistance to change in
(a) Pressure (b) Flow (c) Volume (d) Shape

27. An ideal fluid is
(a) Incompressible (b) Non-Viscous
(c) Non-Viscous but compressible (d) Non-Viscous and incompressible

28. If no resistance is offered to displacement, such substance is
(a) Fluid (b) Gas (c) Ideal Fluid (d) Plastic

29. Density of water is maximum at temperature

(a) 0° C (b) 4° C (c) 100° C (d) Zero Kelvin

30. Property of liquid by which its own molecules are attracted is called

(a) Cohesion (b) Adhesion (c) Viscosity (d) Surface tension

31. When the flow characteristics at a given instant remain same at every location, the flow is said to be

(a) Laminar (b) Uniform (c) Steady (d) Quasi-Static

32. The tendency of liquid surface to curve is due to

(a) Cohesion (b) Adhesion (c) Elasticity (d) Surface tension

33. The stress-strain relation in Newtonian fluid is

(a) Parabolic (b) Hyperbolic (c) Linear (d) Unpredictable

34. Which of the following functions satisfy the continuity equation

(a) $\phi = A(x^2 - y^2)$ (b) $\phi = \log xy$ (c) $\phi = x + 3y$ (d) $\phi = x^2 + y^2$

35. A flow through an expanding tube at a constant rate is called

(a) Steady-uniform (b) Steady-non uniform

(c) Unsteady-uniform (d) Unsteady-non uniform

36. A point in compressible flow where the velocity of fluid is zero, is called

(a) Critical point (b) Vena-contracta (c) Stagnation point (d) None of these

37. The Momentum Correction factor for the following velocity profile is

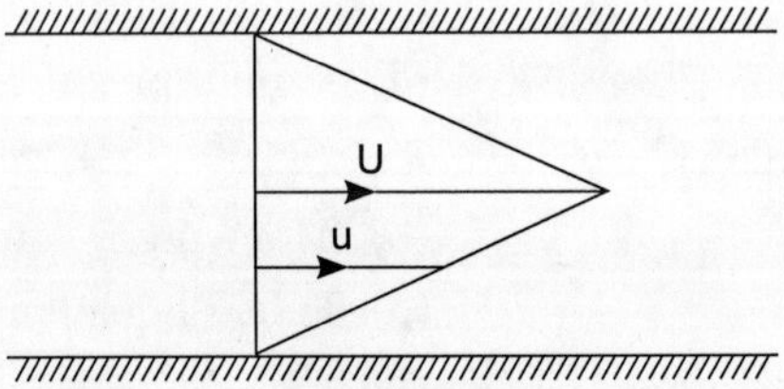

U

u

(a) 2.0 (b) 1.50 (c) 1.33 (d) 1.00

38. The Toricelli gives velocity of jet as

(a) $\sqrt{gh}$ (b) $\sqrt{2gh}$ (c) $\frac{1}{\sqrt{2gh}}$ (d) $\frac{\sqrt{2gh}}{3}$

39. The flow of water in a wash hand basin when it is being emptied through a central opening is an example of

(a) Free vortex (b) Forced vortex (c) Rotational vortex (d) Rankine vortex

40. Apart from Inertia Force which of the following forces is most important in motion of submarines underwater

(a) Viscous (b) Gravity (c) Compressive (d) Surface tension

41. The relationship $\frac{dp}{dx} = \frac{d\tau}{dy}$ is valid for the flow

(a) Irrotational (b) Non uniform (c) Uniform (d) Unsteady

42. The basic principles of fluid flow are

(a) Conservation of Mass (b) Conservation of Energy

(c) Conservation of Momentum (d) All these

43. In a model similarity, if gravitational and inertial forces are the only forces, then the discharge ratio in terms of Length dimension ratio L_r is

(a) $L_r^{3/2}$ (b) $L_r^{1/2}$ (c) $L_r^{5/2}$ (d) $L_r^{1/3}$

44. Reynolds' number is the ratio of inertial to

(a) Elastic Force (b) Viscous Force

(c) Surface Tension Force (d) Gravity Force

45. A stream lined body is defined as the body about which

(a) Flow is laminar (b) Flow is along streamline

(c) The flow separation is suppressed (d) The drag is zero

46. Euler's equation of motion

(a) Is applicable to any fluid

(b) Considers only gravity and pressure forces acting on an inviscid fluid

(c) Considers gravity and viscous forces acting on a real fluid

(d) Considers only pressure force on a real fluid

47. A model of a reservoir is emptied in 10 minutes. If the model scale is 1 : 25, the time taken by prototype to empty itself, would be

(a) 4h 10 m (b) 50 m (c) 24 h 30 m (d) 10 m

48. A stream function is defined only

(a) For incompressible flow (b) For irrotational flow

(c) For continuous flow (d) For steady flow

49. Which of the following velocity potential (ϕ) satisfies continuity equation?

(a) x^2y (b) $x^2 - y^2$ (c) $\cos x$ (d) $x^2 + y^2$

50. A two dimensional flow field is given by stream function, $\psi = x^2 - y^2$, the magnitude of absolute velocity at a point (1, 1) is

(a) 4 (b) 8 (c) 2 (d) $2\sqrt{2}$

51. In an iceberg 15% its volume projects above the sea surface. Specific weight of sea water is $10.5\,\frac{KN}{m^3}$. What is the specific weight of ice berg?

(a) $12.52\,\frac{KN}{m^3}$ (b) $9.81\,\frac{KN}{m^3}$ (c) $8.93\,\frac{KN}{m^3}$ (d) $7.83\,\frac{KN}{m^3}$

52. The Circulation (π) around a circle of radius 2 units for velocity field $u = 2x + 3y$ and $v = -2y$ is

(a) -6π units (b) -12π units (c) -18π units (d) -24π units

53. The reading of a differential manometer of a venturimeter placed at 45° to horizontal is 11 cm. Now the venturimeter is turned to horizontal position, the reading will be

(a) Zero (b) $\frac{11}{r^2}$ (c) 11 cm (d) 1112 cm

54. The Kinetic Energy Correction Factor for a fully developed laminar flow through a circular pipe is
(a) 2 (b) 1 (c) 1.33 (d) 1.5

55. In a pipe flow the head loss due to sudden expansion is expressed by the formula
(a) $\frac{V_1^2 - V_2^2}{2g}$ (b) $\left(\frac{V_1 - V_2}{2g}\right)^2$ (c) $\frac{(V_1 - V_2)^2}{g}$ (d) $\frac{(V_1 - V_2)^2}{2g}$

56. A fire protection system supplies water from a tower with bent pipe as shown. The pipe friction factor f is 0.03. Ignoring minor losses, the maximum discharge Q will be
(a) $31.7 \frac{lit}{s}$ (b) $24.0 \frac{lit}{s}$ (c) $15.9 \frac{lit}{s}$ (d) $12.0 \frac{lit}{s}$

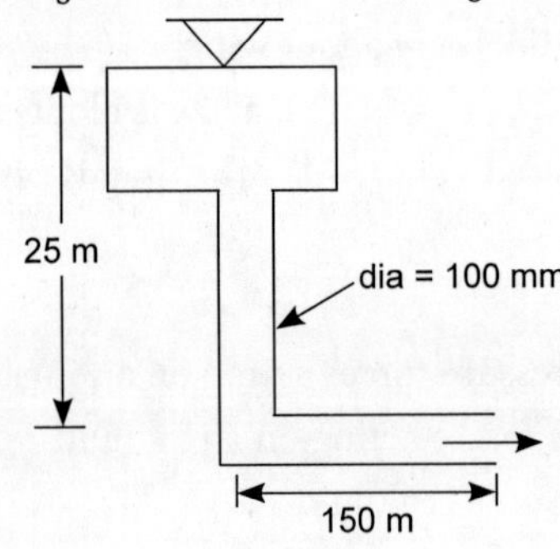

57. For a steady incompressible flow through a closed conduct of Uniform Cross-section, the direction of flow will always be
(a) From higher to lower elevation (b) From higher to lower pressure
(c) From higher to lower velocity (d) From higher to lower piezemetric head

Common Data for Q 58 and 59

A pipeline (diameter 0.3 m, length 3 km) carries water from points P to R. The pezometric heads at P and Q are to be maintained at 100 m and 80 m respectively. To increase the discharge, a second pipe is added in parallel from Q to R of length 2 km. Assume the friction factor f = 0.04

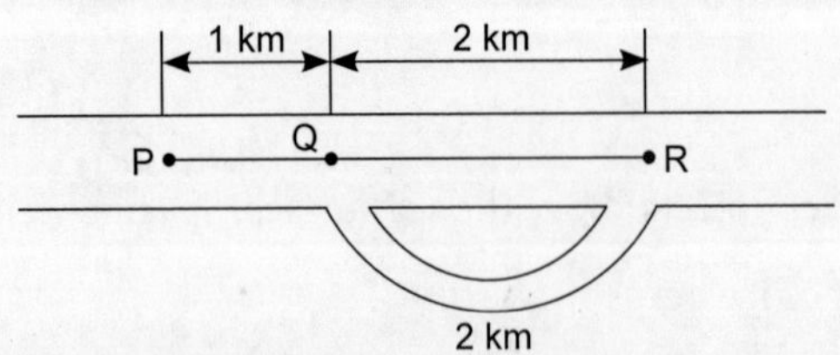

58. What will be increase in discharge if the additional pipe has same diameter (0.3 m)
(a) 0% (b) 33% (c) 41% (d) 67%

59. If there is no restriction on the diameter of additional pipe, what would be maximum discharge theoretically possible from this arrangement?
(a) 0% (b) 50% (c) 67% (d) 73%

60. The rise of liquid of specific weight (γ) and surface tension (σ) in a capillary tube of radius (r) is given by

(a) $\frac{\sigma}{2r\gamma}$ (b) $\frac{2\sigma}{r}$ (c) $\frac{2\sigma}{r\gamma}$ (d) $\frac{\gamma\sigma}{2r}$

61. Cavitation is caused by

(a) High pressure (b) Low pressure (c) High velocity (d) High temperature

62. On an immersed body in a flowing fluid, the lift force is

(a) Due to buoyant force

(b) Always in opposite direction to gravity

(c) Due to wake formation

(d) The dynamic fluid force component normal to approach velocity

Linked Answer Question 63 and 64

An automobile with projected area 2.6 m^2 is running on a road with speed of 120 km per hour. The mass density and kinematic viscosity of air are $\frac{1.2\ kg}{m^3}$ and $1.5 \times 10^{-5}\ \frac{m^2}{s}$. The drag coefficient is 0.30.

63. The drag force on automobile is

(a) 620 N (b) 600 N (c) 580 N (d) 520 N

64. The metric horse power required to overcome the drag force is

(a) 33.23 (b) 31.23 (c) 23.23 (d) 20.23

65. The thickness of the Boundary Layer on a flat plate at a point A is 2 cm and at a point B 1.0 m downstream of A is 3 cm. What is the distance of A from the leading edge of the plate?

(a) 0.5 m (b) 0.8 m (c) 1.0 m (d) 1.25 m

66. The Prandtl mixing length for turbulent flow through pipes is

(a) Dependent on shear stress at the wall (b) A universal constant

(c) Zero at the pipe wall (d) Independent of radial distance from pipe axis

67. The number of π- terms needed to express the function; f (A, V, t, μ, L) = 0 are

(a) 5 (b) 4 (c) 3 (d) 2

68. The repeating variable in dimensional analysis should

(a) Include the dependent variable (b) Have among themselves all basic dimensions

(c) Be derivable from one another (d) Exclude one of the basic dimensions

69. Both Reynolds' Number and Froude Numbers assume significance in one of the following examples

(a) Motion of Submarine at large depth (b) Motion of Ship in deep seas

(c) Cruising of a missile in air (d) Flow over spillways

70. In $\frac{1}{50}$ model of a spillway the discharge was 0.3 $\frac{m^3}{s}$. The corresponding discharge in the prototype will be $\left(\frac{m^3}{s}\right)$.

(a) 2.0 (b) 15.0 (c) 106.0 (d) 5303

71. The relationship between the length scale ratio (L_r) and the velocity scale ratio (V_r) in hydraulic models, in which Froude dynamic similarity is maintained is

(a) $V_r = L_r$ (b) $Lr = \sqrt{V_r}$ (c) $V_r = L_r^{1.5}$ (d) $V_r = \sqrt{L_r}$

72. The drag force F_D, on a sphere kept in a uniform flow field depends upon the diameter of sphere D, flow velocity V, fluid density ρ and dynamic viscosity μ which of the following options represents the non-dimensional parameters which could be used to analyze the problem?

(a) $\frac{F_D}{V_D}, \frac{\mu}{\rho VD}$ (b) $\frac{F_D}{\rho VD^2}, \frac{\rho VD}{\mu}$ (c) $\frac{F_D}{\rho V^2D^2}, \frac{\rho VD}{\mu}$ (d) $\frac{F_D}{\rho V^3D^3}, \frac{\mu}{\rho VD}$

73. In networks of pipes

(a) The algebraic sum of discharges around each circuit is zero
(b) The algebraic sum of discharges around each circuit is not zero
(c) The elevation of hydraulic grade line is assumed for each junction point
(d) Elementary circuit are replaced by equivalent pipes

74. Three reservoirs A, B and C are interconnected by pipes as shown in the figure. Water surface elevations in the reservoirs and Piezometric head at junction J are indicated in the figure.

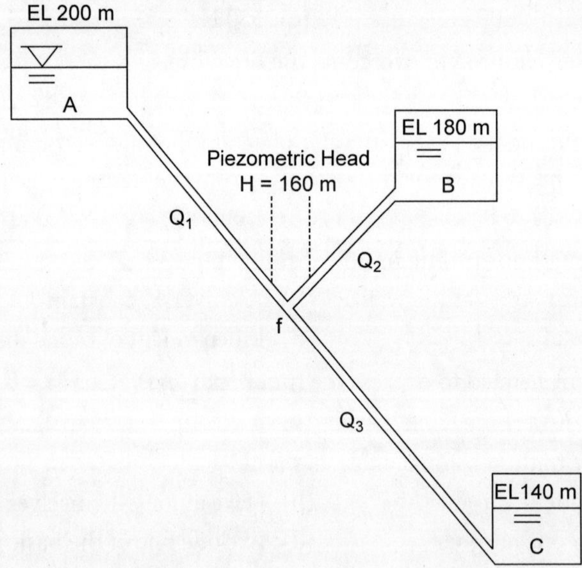

Discharges Q_1, Q_2, Q_3 are related as

(a) $Q_1 + Q_2 = Q_3$ (b) $Q_1 = Q_2 + Q_3$ (c) $Q_2 = Q_1 + Q_3$ (d) $Q_1 + Q_2 + Q_3 = 0$

75. A jet of water 10 cm diameter impinges normally on a fixed plate with a velocity $\frac{25m}{s}$. The Force exerted on plate will be (approximately)

(a) 7000 N (b) 6000 N (c) 5000 N (d) 4000 N

76. A jet of water 100 mm diameter velocity $\frac{20m}{s}$ strikes a stationary plate. The normal force on the plate

(i) When plate is normal to jet F_n

(ii) When plate is inclined at 45° F_i. The ratio $\frac{F_n}{F_i}$ will be

(a) 1.4 (b) 0.7 (c) 1.0 (d) 2.0

77. A jet of water 10 cm diameter moving with a velocity $\frac{12m}{s}$ impinges on a series of Vanes moving with a velocity of $\frac{8m}{s}$.

The efficiency of the system is

(a) 55.5% (b) 44.4% (c) 33.3% (d) 22.2%

78. A jet of water 40 mm in diameter enters a fixed curved Vane at an angle 20° to the horizontal and velocity $\frac{50m}{s}$, the ratio of Normal to tangential forces will be, if it leaves the vane at 15° to the horizontal.

(a) 43 (b) 33 (c) 23 (d) 19

79. The Force F is given by

(a) $F = \frac{waV}{g}$ (b) $F = \frac{wa^2V}{g}$ (c) $F = \frac{aV}{wg}$ (d) $F = \frac{waV^2}{g}$

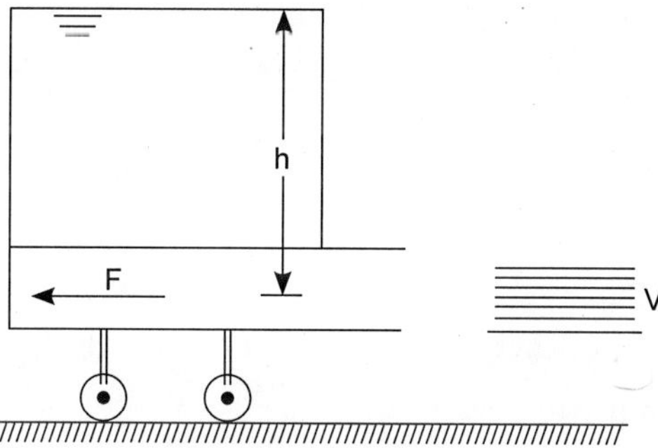

where, w = specific weight of liquid.

a = area of jet.

V = velocity of jet.

80. A Pelton wheel develops 5000 HP under a head of 100m and with overall efficiency 85%. Assume coefficient of velocity for the nozzle as 0.98, the discharge will be

(a) 3.90 cumec (b) 4.50 cumec (c) 4.10 cumec (d) 3.57 cumec

81. Pelton wheel is a

(a) Tangential flow impulse turbine (b) Inward flow impulse turbine

(c) Outward flow impulse turbine (d) Inward flow reaction turbine

82. The hydraulic efficiency of an impulse turbine is maximum when velocity of wheel is .... of the jet velocity

(a) $\frac{1}{4}$ (b) $\frac{1}{2}$ (c) $\frac{3}{2}$ (d) Double

83. In a Kaplan turbine runner the number of blades are generally between

(a) 2 to 4 (b) 4 to 8 (c) 8 to 16 (d) > 16

84. The runaway speed of a hydraulic turbine is the speed

(a) At full load
(b) Corresponding to maximum overload allowed
(c) At which there is no damage to runner
(d) At which turbine runs freely without load

85. Braking jet in Impulse turbine is used to

(a) Break the jet of water
(b) To bring the runner at rest quickly
(c) To reverse the direction of wheel
(d) None of these

86. Which of the following pump is good for small discharge and high head

(a) Centrifugal (b) Axial flow (c) Reciprocating (d) Mixed flow

87. The jet ratio is defined as the ratio of

(a) Diameter of jet and diameter of Pelton wheel
(b) Velocity of jet to velocity of Pelton wheel
(c) Diameter of Pelton wheel to that of jet
(d) Velocity of Pelton wheel to that of jet

88. The discharge of a double acting reciprocating pump is

(a) LAN (b) 2 LAN (c) $\frac{\text{LAN}}{60}$ (d) $\frac{2\text{ LAN}}{60}$

where L = Length of stroke, A = Area of cross section of Piston, N = Speed of crank in RPM

89. Power required to drive a centrifugal pump is directly proportional to

(a) Diameter of Impeller
(b) Square of diameter of impeller
(c) Cube of the diameter of impeller
(d) Fourth power of diameter of impeller

90. Multi-stage centrifugal pumps are used to

(a) Give high discharge
(b) Produce high heads
(c) Pump viscous fluids
(d) All these

91. When the ratio of Forces in model and prototype at corresponding points are same it is said to be

(a) Dynamic similarity
(b) Kinematic similarity
(c) Geometric similarity
(d) None of these

92. Discharge of a centrifugal pump is

(a) Directly proportional to N
(b) Inversely proportional to N
(c) Directly proportional to N^2
(d) Inversely proportional to N^2

93. The specific speed of a hydraulic turbine depends upon

(a) Speed and Power developed
(b) Discharge and Power developed
(c) Speed and Head of water
(d) Speed, Power and Head of water

94. Which of the following turbines is preferred for a specific speed of 60 to 300 RPM?

(a) Pelton Wheel (b) Francis Turbine (c) Kaplan Turbine (d) None of these

95. Which of the following pumps can successfully lift water from deep wells?

(a) Centrifugal pump (b) Reciprocating pump

(c) Jet pump (d) Air lift pump

96. A hydraulic ram is a device used to

(a) Store energy of water

(b) Increase water pressure

(c) Lift water from deep wells

(d) To lift small quantity of water to greater heights when large quantity of water is available at lower heights

97. The speed of an imaginary turbine, identical with the given turbine, which develops unit power under a unit head is known as

(a) Normal speed (b) Unit speed (c) Specific speed (d) Model speed

98. The specific speed of a pump is an index of its

(a) Power consumption (b) Size

(c) Type and shape (d) Efficiency

99. Which of the following is true for an ideal fluid, it?

(a) Is incompressible and viscous (b) Is incompressible and inviscid

(c) It has surface tension (d) Is compressible and inviscid

100. Which one of the following pair is correctly matched?

(a) Prandtl, Flow through channels

(b) Continuity equation, Law of Energy conservation

(c) Mixing length, Laminar flow

(d) Karman's vortex street, Flow past a cylinder

101. The unit power developed by a turbine under a head of water (H) and power developed by turbine

(a) $\frac{P}{\sqrt{H}}$ (b) $\frac{P}{H}$ (c) $\frac{P}{H^{3/2}}$ (d) $\frac{P}{H^2}$

102. The draft tube efficiency is defined as the Ratio of

(a) Kinetic energy recovered to the total energy supplied

(b) Potential energy recovered to the total energy supplied

(c) K.E at the entrance to the K.E at exit

(d) None of these

103. Which of these is Newtonian fluid

(a) Blood (b) Honey (c) Water (d) Ketchup

TRUE OR FALSE QUESTIONS

1. If y_1 and y_2 are conjugate depths in a hydraulic jump the head loss is $\frac{(y_2 - y_1)^3}{4 y_1 y_2}$. (T or F)
2. A Currentmeter directly measures discharge. (T or F)
3. Pressure of 1 bar is equal to 100 $\frac{N}{m^2}$. (T or F)
4. By anemometer wind speed can be measured. (T or F)
5. Viscosity helps fluid to flow faster. (T or F)
6. With rise in temperature viscosity of gases increase. (T or F)
7. Milk, blood, paper pulp are the examples of Pseudoplastics. (T or F)
8. Even if a single parameter changes with respect to time, the flow becomes unsteady. (T or F)
9. Continuity equation is based on conservation of Energy Law. (T or F)
10. Vorticity is inversely proportional to circulation. (T or F)
11. The negative derivative of ϕ gives velocity, ϕ is called Potential function. (T or F)
12. The other name of velocity potential function is stream function. (T or F)
13. Difference between Total Energy Line (TEL) and Hydraulic Gradient Line (HGL) gives velocity Head. (T or F)
14. In a siphon flow, the pressure should not fall below the vapour pressure. (T or F)
15. In a pipe, for Laminar Flow, Reynolds Number > 2000. (T or F)
16. There is no flow across a streamline. (T or F)
17. 1 Pa – s = 10 poise. (T or F)
18. In a capillary tube liquid rises due to Elastic Property. (T or F)
19. A whirlpool in a river is an example of Forced Vortex. (T or F)
20. Bernoulli's Equation is for Ideal Fluids. (T or F)
21. Each term of Bernoulli's Equation $\left(\frac{P}{\rho g} + \frac{V^2}{2g} + Z = C\right)$ has the Units of $\frac{N - m}{N}$. (T or F)
22. In unsteady flow $\frac{\partial v}{\partial t} = 0$. (T or F)
23. Liquids are more compressible than solids and gases. (T or F)
24. Hagen-Poseuille equation is for estimating head loss in a pipe flow. (T or F)
25. Frictional loss in pipe flow $h_f = \frac{f\,l\,V^2}{2g\,d}$ was given by Darcy-Weisbach. (T or F)
26. Relative roughness in a pipe of diameter D and height of roughness ϵ is given by $\frac{D}{\epsilon}$. (T or F)
27. Continuity equation for compressible fluids is $\rho_1 A_1 V_1 = \rho_2 A_2 V_2$. (T or F)

28. In a frictionless adiabatic or Isentropic flow p v^k = constant where $k = \frac{C_V}{C_p}$. (T or F)

29. Mach Number is the ratio of Inertial force to elastic force. (T or F)

30. In hypersonic flow Mach number is $1 < M < 3$. (T or F)

31. Compressibility of fluids is expressed quantitatively by means of Bulk Modulus of Elasticity (K). (T or F)

32. Bulk Modulus of elasticity is defined as $\left(-\frac{\Delta p}{\Delta V/V}\right)$. (T or F)

33. The bulk modulus of elasticity for water is about (at NTP) $K_{water} = 2 \times 10^6 \frac{KN}{m^2}$. (T or F)

34. The Bulk Modulus of elasticity is more for gases than water. (T or F)

35. In Irrotational flow $V \propto r$ or $\frac{V}{r}$ = constant. (T or F)

36. In free vortex $V \propto r$. (T or F)

37. The rate of flow of water in a 100 mm dia pipe with an average velocity of $\frac{4\ m^3}{s}$ is 100 π. (T or F)

38. The pattern of streamlines and equipotential lines is called a flow net. (T or F)

39. In a sink the flow is radially outwards. (T or F)

40. The lift force per unit length on a body depends upon the density of fluid. (T or F)

41. The hydraulic mean depth for a circular pipe of diameter d is $\frac{d}{4}$. (T or F)

42. When two pipes are in parallel, the total head loss is equal to the sum of the head loss in each pipe. (T or F)

43. A siphon is used to connect two reservoirs at different levels intervening by a high ridge. (T or F)

44. If velocity potential exists in a flow, the flow is rotational. (T or F)

45. The value of coefficient of velocity C_V for a sharp edged Orifice increases with head of water. (T or F)

46. The loss of head at the entrance in a pipe is $\frac{V^2}{2g}$. (T or F)

47. All gases are considered as compressible and liquids incompressible. (T or F)

48. The hydraulic gradient line is always parallel to centre line of the pipe. (T or F)

49. Friction factor for Laminar flow is directly proportional to Reynold's number (R_e). (T or F)

50. Due to aging of pipeline its carrying capacity decreases due to increase in Darcy-weisbach friction factor. (T or F)

51. A flow through an expanding tube at increasing rate is called unsteady non-uniform flow. (T or F)

52. The flow net is an analytical representation of irrotational flow in two dimensions. (T or F)

53. For a flow to be irrotational the vorticity is zero. (T or F)

54. If a Pitot tube is placed with its nose upstream or downstream, the reading will be same. (T or F)

55. A nozzle is generally made of divergent shape. (T or F)

56. The D-Alembert's principle is used for changing the dynamic equilibrium of a fluid mass into a static equilibrium. (T or F)
57. A jet of water enters and leaves a fixed curved vane tangentially. The force of jet along perpendicular to vane is zero. (T or F)
58. The Principle of Jet Propulsion is used in driving the ships and aeroplanes. (T or F)
59. Work done by a turbine depends upon the weight of water flowing per second. (T or F)
60. The gross or total head of turbine is the sum of the water levels at head race and tail race. (T or F)
61. The width of the bucket for the Pelton Wheel is generally 3 times the diameter of jet. (T or F)
62. The Jet ratio is the diameter of Pelton Wheel to diameter of Jet. (T or F)
63. Product of Mechanical efficiency and hydraulic efficiency is known as Overall efficiency. (T or F)
64. Among Pelton, Kaplan, Francis and Propeller turbine Kaplan turbine has least efficiency. (T or F)
65. The height of rise or fall of a liquid in a capillary tube is given by formula $\frac{h = 4\,\sigma \cos \alpha}{wd}$. (T or F)
66. In a Pelton Wheel speed factor is the ratio of Bucket Speed (u) and theoretical velocity $\sqrt{2\,gH}$. (T or F)
67. To generate 10,000 HP under a head of 81 m while working at a speed of 500 RPM, the most suitable type of turbine is Kaplan turbine. (T or F)
68. A centrifugal pump gives maximum efficiency when its impeller blades are bent forward. (T or F)
69. Draft tubes are used only in reaction turbines. (T or F)
70. The action of a centrifugal pump is that of a reversed reaction turbine. (T or F)
71. Multi-stage pumps are used to give high discharge. (T or F)
72. A hydraulic accumulator is a device used to store pressure energy which is supplied to a machine later on. (T or F)
73. The inlet energy in a hydraulic turbine is always greater than the outlet energy. (T or F)
74. The sprinkler shown here discharges 30 liters of water per minute through each nozzle of area 0.0001 m^2, neglecting friction the speed of rotation will be 92 RPM. (T or F)

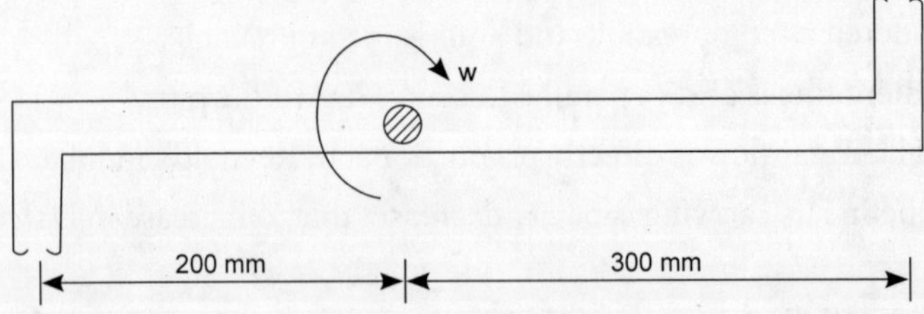

75. Between the conical, hydrocone and elbow type of draft tubes the efficiency of conical type is maximum. (T or F)
76. For handling oils and other viscous liquids Gear Pumps are most suitable. (T or F)
77. A high head turbine has high specific speed and speed of rotation. (T or F)
78. Priming is necessary in centrifugal pumps but not in reciprocating or gear pumps. (T or F)

FILL IN THE BLANKS

1. Water is said to be .... solvent.
2. Specific density of water is ....
3. 1000 kg of water will weigh .... N.
4. Unit of Surface Tension is ....
5. Specific volume of a liquid is reciprocal of ....
6. One litre of water occupies .... ml of water.
7. The pressure exerted by vapour is called ....
8. Rate of change of temperature in atmosphere with height is called ....
9. Wind speed is measured by ....
10. The upper size of a water drop before breaking is .... mm.
11. Wind velocity at height (h) is given by formula, ...
12. Units of Dynamic Viscosity are ....
13. Units of Kinematic Viscosity are ....
14. Relation between Kinematic Viscosity (ν) and Dynamic Viscosity (μ) is, ...
15. There are two methods of describing fluid motion, one, is Lagrangian method, other is ....
16. Bernoulli's Equation is on the principle of ....
17. In a static fluid the pressure at depth *h* is ....
18. The specific weight of a liquid is 8.2 $\frac{KN}{m^3}$, its mass density is ....
19. The velocity profile in Laminar pipe flow has a shape of ....
20. An ideal fluid has no ....
21. The real fluid is of .... in nature.
22. Pascal (Pa) is the unit of ....
23. Gauge pressure is above ....
24. The pressure in a fluid at rest is equal to P_{atm} + ...
25. In Bernoulli's Equation $p + \frac{1}{2}\rho V^2 + \rho gh$ = constant, the middle term has the units of ....
26. A 3 kg block of silver (density $=10.5 \times 10^3 \frac{kg}{m^3}$) is immersed in water, the Buoyant Force will be ....
27. The property of fluid by virtue of which it offers resistant to its motion is ....
28. The fluid which follows the equation $\left(\tau = \mu \frac{du}{dy}\right)$ is known as ....
29. The pressure acting on a water droplet is ....

30. The pressure acting on a soap bubble is ....
31. Pseudoplastic, Thixotropic Fluid etc. are .... fluids.
32. Time independent flow is called .... flow.
33. If p represents all parameters of fluid flow and $\frac{\partial p}{\partial t} = 0$, the flow is classified as ....
34. $\frac{V^2}{2g}$ is called ....
35. Cavitation may occur if pressure falls below ....
36. The critical Reynold's Number for flow over a flat plate is approximately ....
37. In a pipe flow the Laminar Flow starts changing into Turbulent flow at Reynolds Number ....
38. Part of the Bernoulli's equation $(z + \frac{p}{\gamma})$ is know as .... head.
39. The streamlines in a free vortex are .... circles.
40. The type of flow represented by, $u = (3x - 3y)\,t$ and $v = (x - 2y)\,t$ is .... and ....
41. The pressure at the summit of a syphon is .... than atmospheric pressure.
42. The power transmitted through a pipeline is maximum when the head lost due to friction in the pipe is equal to, ... of the total supply head.
43. Flow net is the combination of flow lines and .... lines.
44. $F = \frac{d(mV)}{dt}$ is the .... equation.
45. When water rises in a capillary tube the upper surface is ....
46. The centre of pressure acts .... the centre of gravity of immersed surface.
47. A flow in which the quantity of flowing liquid per unit time is constant is called .... flow.
48. The relation $\frac{\partial^2 \phi}{\partial x^2} + \frac{\partial^2 \phi}{\partial y^2} + \frac{\partial^2 \phi}{\partial z^2} = 0$ is referred to as .... equation.
49. The discharge over a rectangular notch is proportional to H^n, where n is equal to ....
50. Poise is the CGS system unit of ....
51. When Mach Number is more than 6, the flow is known as ....
52. A point in a compressible flow where as velocity of flow is zero is called .... point.
53. The reciprocal of Euler's Number is called .... number.
54. A Pelton wheel is a .... flow impulse turbine.
55. A double overhung Pelton wheel has .... runners.
56. The ratio of actual work available at the turbine to the energy imparted to the wheel is known as .... efficiency.
57. The speed of a turbine runner is directly proportional to H^x where x is equal to ....
58. A turbine runs at 240 RPM under a head of 9 m. If operating head is 16 m, the speed of the turbine will be ....

59. In a centrifugal pump, the regulating valve is provided on .... pipe.
60. A reciprocating pump is also known as a .... pump.
61. The specific speed of a turbine in terms of N, P and H is given by ....
62. In a mixed flow reaction turbine, the flow of water is partly radial and partly ....
63. Power required to drive a centrifugal pump is .... proportional to cube of the speed (N).
64. The flow where no heat (q) enters or leaves is called .... flow.
65. With rise in temperature viscosity of liquids ....
66. In order to avoid tendency of separation at throat in a Venturimeter, the ratio of the diameter at throat to the diameter of main pipe should be ....
67. Coefficient of velocity C_V is defined as the ratio of actual velocity at .... to the theoretical velocity.
68. Coefficient of discharge is the product of coefficient of contraction and coefficient of ....
69. an axial flow pump is suitable for low head, high discharge and has .... specific speed.
70. If K is Bulk Modulus and ρ is density of fluid then celerity (velocity) of pressure wave in the fluid is given by ....
71. The maximum thickness of the boundary layer in a pipe of radius R is ....
72. If δ_1 is the thickness of Laminar Boundary Layer at a distance x_1 from the leading edge of the flat plate, the thickness δ_2 at $x_2 = 2x_1$ will be ....
73. The existence of Boundary layer is due to fluid property of .... of fluid.
74. A hydraulic press has a ram of 15 cm diameter and a plunger of 15 mm diameter. It is required to lift a weight of 1.0 tonne. The force required on the plunger is .... kg.
75. The unit discharge (Q) and unit speed (N) curves for different turbines are given in figure, the curve which represents to Francis Turbine ....

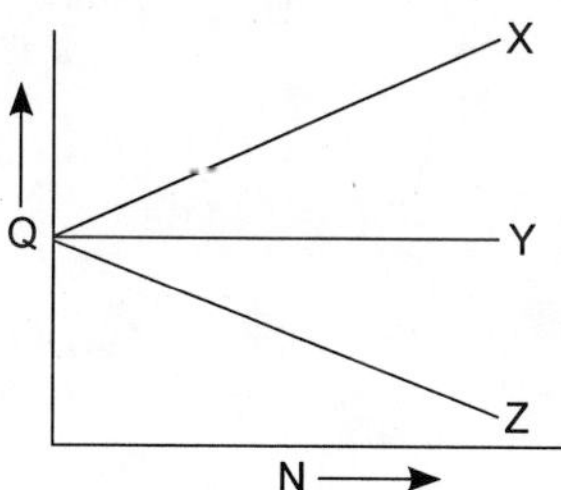

ANSWERS

Multiple Choice Questions

1. c	**2.** d	**3.** c	**4.** c	**5.** d	**6.** c	**7.** b	**8.** a	**9.** b
10. c	**11.** d	**12.** d	**13.** a	**14.** b	**15.** c	**16.** b	**17.** c	**18.** a
19. b	**20.** c	**21.** a	**22.** c	**23.** c	**24.** b	**25.** a	**26.** d	**27.** d
28. c	**29.** b	**30.** a	**31.** b	**32.** d	**33.** c	**34.** a	**35.** b	**36.** c
37. c	**38.** b	**39.** a	**40.** a	**41.** c	**42.** d	**43.** c	**44.** b	**45.** c
46. b	**47.** b	**48.** a	**49.** b	**50.** d	**51.** c	**52.** b	**53.** c	**54.** a
55. d	**56.** b	**57.** d	**58.** c	**59.** d	**60.** c	**61.** b	**62.** d	**63.** d
64. c	**65.** b	**66.** c	**67.** d	**68.** b	**69.** b	**70.** d	**71.** d	**72.** c
73. b	**74.** a	**75.** ??	**76.** a	**77.** b	**78.** c	**79.** d	**80.** c	**81.** a
82. b	**83.** b	**84.** d	**85.** b	**86.** c	**87.** c	**88.** d	**89.** d	**90.** b
91. a	**92.** a	**93.** d	**94.** b	**95.** d	**96.** d	**97.** c	**98.** c	**99.** b
100. d	**101.** c	**102.** a	**103.** c					

True & False

1. T	**2.** F Velocity	**3.** F 1,00,000 Pa or N/m^2	
4. T	**5.** F	**6.** T	**7.** T
8. T	**9.** T	**10.** F Directly	**11.** T
12. F Orthogonal	**13.** T	**14.** T	**15.** F Less than 2000
16. T	**17.** T	**18.** F Adhesion	**19.** F Free vortex
20. T	**21.** T	**22.** F $\neq 0$	**23.** F
24. T	**25.** T	**26.** F $\frac{\in}{D}$	**27.** T
28. F $\frac{C_p}{C_v}$	**29.** T	**30.** F $M > 3$	**31.** T
32. T	**33.** T	**34.** F $Kw > ka$	**35.** T
36. F $V \propto \frac{1}{r}$	**37.** F $\frac{\pi}{100}$	**38.** T	**39.** F Inward
40. T	**41.** T	**42.** F Equal to each pipe	
43. T	**44.** F Irrotational	**45.** T	**46.** F $\frac{0.5\,v^2}{2g}$

47. T **48.** F **49.** F Inversely $\left(\frac{64}{R_e}\right)$ **50.** T

51. T **52.** F Graphical Representation **53.** T

54. F **55.** F Convergent **56.** T **57.** F

58. T **59.** T **60.** F Difference **61.** F 5 times

62. T **63.** T **64.** F Pelton Turbine **65.** T

66. T **67.** F Francis **68.** F Bent Backwards

69. T **70.** T **71.** F High head **72.** T

73. T **74.** F 184 RPM **75.** T **76.** T

77. F **78.** T

Fill in the Blanks

1. Universal **2.** One (1) **3.** 9810 N **4.** $\frac{N}{m}$

5. Mass Density **6.** 1000 ml **7.** Partial or Vapour Pressure

8. Lapse rate **9.** Anemometer **10.** 6.25 mm **11.** $V_h = Ch^{1/7}$

12. $\frac{Kg}{m-s}$ Pa – s Poise **13.** $\frac{m^2}{s}$ Stokes **14.** $\nu = \frac{\mu}{\rho}$ **15.** Eulerian

16. Energy conservation **17.** ρgh **18.** 836 $\frac{kg}{m^3}$

19. Parabola **20.** Viscosity **21.** Viscous **22.** Pressure

23. Atmospheric **24.** Patm + ρgh **25.** $\frac{N}{m^2}$ **26.** 2.8 N

27. Viscosity **28.** Newtonian` **29.** $\frac{4\sigma}{d}$ **30.** $\frac{8\sigma}{d}$

31. Non-Newtonian **32.** Steady **33.** Steady **34.** Velocity Head

35. Vapour Pressure **36.** 5,00,000 **37.** 2000 **38.** Piezometer

39. Concentric **40.** Steady Compressible **41.** Less

42. $\frac{1}{3}$ **43.** Equi-potential **44.** Impulse Momentum

45. Concave **46.** Below **47.** Steady **48.** Laplace's

49. $\frac{3}{2}$ **50.** Dynamic Viscosity **51.** Hypersonic **52.** Stagnation

53. Newton **54.** Tangential **55.** Two **56.** Mechanical

57. $\frac{1}{2}$	**58.** 320 RPM	**59.** Delivery Pipe	**60.** Displacement
61. $\frac{N\sqrt{P}}{H^5/4}$	**62.** Axial	**63.** Directly	**64.** Adiabatic
65. Decreases	**66.** $\frac{1}{3}$ to $\frac{1}{2}$	**67.** Vena-contracta	**68.** Velocity
69. Higher	**70.** $\sqrt{\frac{k}{\rho}}$	**71.** R	**72.** $\sqrt{2}\,\delta_1$
73. Viscosity	**74.** 10 kg	**75.** Z	

Section 9

Groundwater

Definitions – 38

Multiple Choice Questions – 21

True & False – 16

Fill in the Blanks – 15

DEFINITIONS

Aquifer: It is a saturated formation of earth material which not only stores water but yields it in sufficient quantity. Unconsolidated deposits of sand and gravel form good aquifers.

Aquitard: It is a saturated geological formation which is poorly permeable and hence does not yield water freely to wells. Sandy clay is an example of aquitard.

Aquiclude: It is a geological formation which is essentially impermeable to the flow of water. The yield is almost nil. Clay is an example of aquiclude.

Aquifuge: It is a geological formation which is neither porous nor permeable. It does not transmit any water.

Artesian Aquifer: It is also called confined aquifer which is confined between two impervious beds such as aquicludes or aquifuges. The water in it is under pressure.

Artificial Recharge: The process of enhancing the amount of water recharging the aquifer in a given location is known as artificial recharge.

Coefficient of Permeability (Hydraulic Conductivity): It is the rate of flow of water in a porous medium through a unit cross-sectional area, under a unit hydraulic gradient and standard temperature, usually 20°C.

Confined Aquifer: It is the same as artesian aquifer.

Depression Head: It is the difference in the water table elevation and level inside the well.

Drawdown: The drop in the water table elevation at any point from its previous static level is called drawdown.

Dug Well: Also called 'Open well'. These wells tap water in unconfined aquifers under atmospheric pressure condition unlike artesian wells.

Dupit's Parabola: When underground water flows from one ditch to other in unconfined aquifer the surface shape is known as Dupit's parabola.

Effluent Streams: When the bed of the stream is below the groundwater table, the groundwater contributes to the flow in stream. Such stream is called effluent stream or gaining stream.

Influent Streams: When groundwater table fall below the bed of the stream, the stream water percolates into the groundwater storage, such a stream is called influent or losing stream.

Leaky Aquifer: A confined aquifer is called leaky aquifer if its both confining strata are aquitards.

Maximum Yield: The discharge from a well corresponding to critical head is called maximum or critical yield.

Perched Water Table: A lens or localised patch of impervious stratum can occur inside an unconfined aquifer in such a way that it retains a water table above the general water. Such a water table is known as perched water table.

Percolation: It is the process by which water moves downward through the soil under gravitational forces.

Perennial Rivers: These are those rivers which receive water from groundwater and snowmelt. They have water throughout the year.

Phreatic Surface: In unconfined aquifers the free surface of the water table is known as phreatic surface. It has atmospheric pressure all over.

Piezometric Surface: It is an imaginary surface to which the water level would rise if a piezometer was inserted in the aquifer. Thus, it indicates the pressure of the water in the aquifer. Hence, a piezometric surface is water table equivalent of the confined aquifer.

Porosity: The amount of pore space per unit volume of the aquifer material is called

$$\text{Porosity (n)} = \frac{\text{Volume of voids}}{\text{Volume of Porous media}} \left[\frac{V_V}{V_O}\right]$$

Radius of Influence: When pumping takes place in a well, lowering of water level takes place. Due to radial flow in the well, aquifer water table takes a conical shape called **cone of depression**. The areal extent of the cone of depression is called **area of influence** and its radial extent Radius of Influence.

Recharge: When water seeps into the ground to replenish underground aquifers. It may be natural or artificial recharge of groundwater.

Recuperation or Recovery: When pumping of well stops, there is a gradual accumulation of storage till the original (static) level is reached. The stage is called recuperation or recovery. It is an unsteady flow phenomenon.

Reynold's Number: Darcy's Law is valid for Laminar flows only. For practical use Reynold's Number for underground flow should be less than 1.

Safe Yield: The maximum rate at which the withdrawal of groundwater in a basin can be carried without producing any undesirable effects is termed as safe yield (S_y).

Undesirable effects : manifest as land subsidence due to reduced pore water pressure and reduced discharge.

Seepage: It is the flow of water through the soil or ground. The flow occurs through the pores or interstices at a very slow velocity.

Specific Capacity: The discharge per unit drawdown at the well is known as specific capacity of a well. It is a measure of well's performance.

Specific Retention: The fraction of water held back in the aquifer is known as specific retention (S_r). (Porosity = $S_y + S_r$)

Specific Storage (S_s): It is defined as the volume of water that a unit volume of aquifer releases from the storage under unit decline in hydraulic head.

Springs: When water table intersects the land surface the groundwater comes out to the surface in the form of a spring.

Steady Flow: When the drawdown surface attains a constant position with respect to time, the well is known to be operating under steady flow condition.

Storage Coefficient 'S' (Storativity): Is the volume of water released by a column of a confined aquifer of unit cross-sectional area under a unit decrease in piezometric head. (Typical value of S in confined aquifers lie in the range 5×10^{-5} to 5×10^{-3}).

Transmissibility 'T' (Transmissivity): It is the product of the hydraulic conductivity (k) and the thickness of aquifer (b) and expressed as $\frac{m^2}{day}$ (T = kb).

Unconfined Aquifer (Water Table Aquifer): It is the one in which a free water surface i.e., a water table exists. Recharge takes place by infiltration of precipitation from the ground surface.

Water Divide: It is a hydraulic pressure boundary between two groundwater bodies from which flow occurs in different direction. This is the point of highest elevation on water table between streams.

Well Losses: Head loss due to flow through porous media, head loss due to turbulent flow near the screen and head loss due to flow through screens and casing together are called well losses.

MULTIPLE CHOICE QUESTIONS

1. Which of the following soils has maximum porosity?
   (a) Uniform Sand (b) Soft Clay (c) Sand & Gravel (d) Soft bentonite
2. If diameter of a well is doubled the Yield will increased by about
   (a) 100% (b) 200% (c) 50% (d) 10%
3. In a well in an unconfined aquifer the Yield is proportional to,
   (a) $H - h$ (b) $H^2 - h^2$ (c) $H + h$ (d) $H^2 + h^2$
4. In an aquifer, Porosity is 30% and specific retention 10%. The expected specific Yield will be
   (a) 40% (b) 20% (c) 30% (d) 10%
5. Darcy's Law is valid for Reynold's Number,
   (a) > 1 (b) $= 1$ (c) < 1 (d) ≤ 1
6. A soil strata may have a
   (a) Soil zone (b) Intermediate zone
   (c) Capillary zone (d) All the above
7. The freshwater on earth is 2.5% of the total, out of this percentage of groundwater is about
   (a) 20% (b) 12% (c) 1% (d) 30%
8. The discharge per unit drawdown of a well, is called
   (a) Specific storage (b) Specific Yield (c) Specific capacity (d) Specific retention
9. Main objective of groundwater exploration is to locate
   (a) Wilting zone (b) Capillary zone (c) Vadose zone (d) Phreatic zone
10. Freshwater running underground are called
    (a) Streams (b) Creeks (c) Aquifer (d) Artesian
11. The coefficient of Transmissibility (T) of an aquifer is given by
    (a) $T = \frac{k}{b}$ (b) $T = \frac{b}{k}$ (c) $T = bk^2$ (d) $T = bk$

12. The coefficient of permeability has the unit of
    (a) $\frac{m^2}{S}$ (b) $\frac{m}{S}$ (c) $\frac{m}{S^2}$ (d) m S
13. According to Dupit's Theory, Darcy's Law is not valid near
    (a) Top surface of well (b) Well face
    (c) Sides of well (d) Everywhere
14. Which of the following geological formation give negligible water?
    (a) Aquifuse (b) Aquifer (c) Aquitard (d) Aquiclude
15. The geological formation which contains good amount of water but gives no Yield
    (a) Aquifer (b) Aquifuse (c) Aquitard (d) Aquiclude
16. The relation between porosity (n), specific retention (Sr) and specific yield (Sy) is
    (a) n = Sr + Sy (b) n = Sr × Sy (c) n = Sr – Sy (d) n = Sy – Sr
17. Dupit's Theory was later modified by
    (a) Darcy (b) Thiem (c) Khosla (d) Buckingham
18. Which of the following is depicted by Dupit's Theory
    (a) Flow velocity (b) Yield from aquifer
    (c) Radial Flow (d) Coefficient of Transmissivitity
19. When two wells situated near to each other each discharging Q_1. If there was only one well discharging Q under the same drawdown, then
    (a) $Q > Q_1$ (b) $Q < Q_1$ (c) $Q = Q_1$ (d) $2Q = Q_1$
20. The specific Yield of a particular sand strata is 30 % and porosity is 50 %, then the specific retention of sand will be
    (a) 80 % (b) 55 % (c) 40 % (d) 20 %
21. Strainer-type tube-wells are unsuitable for
    (a) Coarse gravel (b) Fine grained strata
    (c) Clean gravel (d) All these

TRUE OR FALSE QUESTIONS

1. The fresh water as groundwater is much more than in rivers and lakes of the world. (T or F)
2. Water shortage problem is confined to Rural areas only. (T or F)
3. Well water exists in confined aquifer only. (T or F)
4. Area covered with vegetation has less infiltration compared to bare land. (T or F)
5. Deeper the Groundwater table more will be infiltration. (T or F)
6. An influent stream gains water from groundwater. (T or F)
7. Uniform sand has maximum porosity compared to other soil formations. (T or F)

8. Porosity is the sum of specific yield and specific retention. (T or F)
9. Groundwater table is at highest level near the end of rainy season. (T or F)
10. Flowing well cannot occur in confined aquifer. (T or F)
11. Perched aquifer is always an unconfined aquifer. (T or F)
12. The radius of influence has a fixed value of 500 m. (T or F)
13. Transmissibility has the unit of $m^{3/5}$. (T or F)
14. If the radius of a tube well is doubled the discharge will increase four times. (T or F)
104. The ratio of discharge of tube wells of radii 25 cm and 50 cm other parameters remaining the same, will be 1.1 (Taking R = 300 m). (T or F)
15. Groundwater system of irrigation is more popular in India compared to canal irrigation. (T or F)
16. For study of unsteady flow, it is not necessary to know the storage coefficient in Groundwater flow. (T or F)

FILL IN THE BLANKS

1. Geological formation with no water Yield are called ....
2. Porosity is equal to specific retention plus ....
3. In confined flow equation for well discharge Q, R, is called ....
4. In confined aquifer flow towards a well is proportional to (H – h), which is called ....
5. In Sichartt's equation for Radius of Influence, $R = 3000\ S\ \sqrt{k}\ k$, S is, ...
6. In flow towards well $k \times B$ is known as ....
7. In tube wells the strainer mesh size is kept between ....
8. Removing fine material from aquifer surrounding the strainer pipe in a tube well is the process called ....
9. Constant Level Pumping Test and Recuperation Test are done to find ....
10. Out cropping of water table results in a ....
11. When artesian aquifer outcrops artesian spring emerges, it is also known as ....
12. In unsteady flow towards wells equation, $\frac{\partial v}{\partial t} = -\frac{\partial h}{\partial t}$ S.A., S is ....
13. .... refers to the downward movement of water through soil and rock layers, eventually reaching water table.
14. The amount of water that a material can hold in its pore spaces is known as ....
15. The process by which groundwater is replenished, usually through rainfall is, known as ....

ANSWERS

Multiple Choice Questions

1. d	**2.** d	**3.** b	**4.** b	**5.** d	**6.** d	**7.** d	**8.** c	**9.** d
10. c	**11.** d	**12.** b	**13.** b	**14.** c	**15.** d	**16.** a	**17.** b	**18.** c
19. a	**20.** d	**21.** b						

True & False

1.	**2.** F	**3.** F	**4.** F	**5.** T	**6.** F	**7.** F	**8.** T	**9.** T
10. F	**11.** T	**12.** F	**13.** F	**14.** F	**15.** T	**16.** T	**17.** F	

Fill in the Blanks

1. Aquifuge	**2.** Specific Yield	**3.** Radius of Influence
4. Drawdown	**5.** Drawdown	**6.** Transmissibility
7. d_{60} and d_{70}	**8.** Well Development	**9.** Maximum Safe Yield
10. Spring	**11.** Geyser	**12.** Storage Coefficient
13. Percolation	**14.** Porosity	**15.** Infiltration

Section 10

Hydrology

Definitions – 51

Multiple Choice Questions – 52

True & False – 37

Fill in the Blanks – 40

DEFINITIONS

Basin Lag: It is defined as the time interval from the mid-point of rainfall excess to the peak of the unit hydrograph. Physically, lag time represents the mean time of travel of water from all parts of the watershed to the outlet during a given storm.

Catchment: (Drainage Basin or River Basin) is defined as the area drained by a stream or a system of connecting streams such that all the surface runoff originating in this area leaves the area in a concentrated flow through a single outlet.

Condensation: It is the change of state of matter from gas phase into liquid phases and is reverse to evaporation. Condensation normally occurs when vapour is cooled $\frac{\text{and}}{\text{or}}$ compressed to its saturation limit.

Cyclone: It is a storm or system of winds that rotates about a centre of low atmospheric pressure, advances at speeds 30 to 50 $\frac{\text{km}}{\text{hour}}$ or even more, and brings a great deal of rain.

Direct Runoff: It is that part of the runoff which enters the stream immediately after the rainfall.

Drainage Density: It is the ratio of total length of streams of all orders within a basin to its area.

Drizzle: It is the light rain falling in fine droplets of size less than 0.5 mm and intensity less than 1 $\frac{\text{mm}}{\text{hour}}$.

Drought: It is a very long period of abnormally low rainfall. It is a natural disaster.

Effluent Stream: It is a stream which receive groundwater flow.

Ephemeral River: A river that only flows when there is rain or snow melt. The rest of the year there is just a dry river bed with no water.

Evaporation: It is the process in which liquid changes to gaseous state at the free surface, below the boiling point through the transfer of heat energy.

Evapotranspiration: It is the amount of water loss to the atmosphere from a land surface by the processes of transpiration and evaporation. Used as synonym of consumptive use.

Field Capacity: It is the volume of water that the ground can hold.

Flood Routing: It is the technique of determining the flood hydrograph at a section of a river by utilizing the data of flood flow at one or more upstream sections. It uses a conservation of mass approach to route an inflow hydrograph.

Form Factor: It is defined as the ratio of catchment area and square of catchment length.

Glaze: When drizzle comes in contact with cold ground (at 0°C), the water drop freeze to form an ice coating called glare or freezing rain.

Hail: It is precipitation in the form of irregular pellets or lumps of size more than 8 mm. Hails occur in violent thunder storms in which vertical currents are very strong.

Humidity: It is the concentration of water vapour, the gaseous state of water, is generally invisible to the human eye. Normally expressed as percentage.

Hydrograph: It is a plot of flow rate versus time.

Hydrometry: It is the science of monitoring water in natural water resources. It focuses on monitoring the natural precipitation of water, such as rain or snow, and changes in water levels in all kinds of water resources such as lakes, rivers, storage reservoirs and underground water.

Hyetograph: It is a plot of the average intensity of rainfall against the time interval.

Infiltration: It is the flow of water into the ground through the soil surface.

Infiltration Capacity: It is defined as the maximum rate of infiltration. Infiltration capacity decreases as the soil moisture content of soil surface layers increases.

Infiltration Index: It is the assumed constant value of infiltration rate during the storm used in calculations involving floods.

Influent Stream: When water table is below the bed of the stream, the stream water percolates to groundwater. Such stream is called Influent Stream.

Isohyet: A line on a map connecting points having the same amount of **rainfall** in a given period.

Lag: The time difference between the inflow and outflow peak is called lag.

Land Use: It is the term used to describe the human use of land. Land is used for various purposes such as agriculture, forestry, mining, housing, industry, roads etc.

Lapse Rate: It is the **rate** at which an atmospheric variable, normally temperature, in Earth's atmosphere, falls with altitude.

Mass Curve of Rainfall: It is rainfall plot of accumulated precipitation against time, plotted in chronological order.

Monsoon: The season of heavy rain during summer in hot Asian Countries.

Percolation: The infiltrated water first meets the Soil Moisture Deficiency, if any, and thereafter the excess water moves vertically downwards to reach the groundwater table. This vertical movement is called percolation.

Perennial River: These are rivers which flow continuously throughout the year, and are fed by groundwater, snow and glaciers, or by heavy rainfall during monsoon.

Phreatic Surface: In an unconfined aquifer the free surface of the water table is known as phreatic surface. It has constant pressure equal to atmospheric pressure.

ϕ-index (Phi-index): It is the average rainfall above which the rainfall volume is equal to runoff volume.

Potential Evapotranspiration (PET): When sufficient moisture is always available to completely meet the needs of vegetation fully covering the area, the resulting evapotranspiration is called Potential Evapotranspiration.

Precipitation: It denotes all forms of water that reach the earth from the atmosphere. The usual forms are rainfall, snowfall, hail, frost and dew.

Probable Mamimum Precipitation (PMP): It is defined as the greatest or extreme rainfall for a given duration that is physically possible over a basin.

Slope: For a given stretch of the stream, the ratio of horizontal distance between the two end parts of the stretch to the difference in elevation between two points gives the slope of the stretch.

Stream Order: The pattern of branches that unite to form main stream leaving the catchment. The smallest stream to start network is designated as 'order 1', two channels of order 1, produce 'order 2' and so on.

Synthetic Unit Hydrograph (SUH): For ungauged river basins unit hydrographs are derived by relating the selected basin characteristics to the typical unit hydrograph shape. The resulting hydrograph is known as Synthetic Unit Hydrograph.

Time of Concentration (t_c): The time taken for a drop of water from the farthest part of catchment to reach the outlet is designated as time of concentration.

Transpiration: It is the process by which water leaves the body of a living plant and reaches the atmosphere as water vapour.

Tropical Cyclones (Hurricanes or Typhoons): These are rapidly rotating storm that begins over tropical oceans. Next to Earthquakes, they are most dangerous natural hazards.

Unit Hydrograph: It is a hydrograph of surface runoff that would result at a given point in a stream from unit rainfall excess occurring in unit time uniformly over the catchment area above that given point.

Vapour Pressure: the partial pressure exerted by the water vapour is called vapour pressure. The more vapour present in the air, the higher is the vapour pressure.

Water Harvesting: It is defined as the process of collecting and concentrating runoff water from a runoff area into a run-on area, where the collected water is either directly applied to the cropping area or used for other purposes.

Water Year: It is 12-month period to begin and end in a relatively dry season. (Countries have different water years).

Watershed: It is an area of land that drains all the streams and rainfall to a common outlet such as the outflow of reservoir, mouth of bay, or any point along a stream channel. (It is also called drainage basin or catchment).

Wetlands: These are water bearing bodies which help in maintaining freshwater ecology and as well as groundwater recharge. Examples are marshes, lagoons, swamps, bogs and mires.

W-index: It is refined version of ϕ-index, the initial losses are separated from the total abstractions, the average value of thus obtained is called W-index.

MULTIPLE CHOICE QUESTIONS

1. Average Rainfall in India is about:
   (a) 4000 mm (b) 250 mm (c) 1200 mm (d) 11500 mm
2. Lapse rate is related to:
   (a) Humidity (b) Rainfall (c) Temperature (d) Atmospheric Pressure
3. The wind speed is measured by?
   (a) Speedometer (b) Altimeter (c) Wind blower (d) Anemometer

4. Percentage of water available for human use on earth:
   (a) 2.5 (b) 2.75 (c) 1.75 (d) less than 1.0
5. In India Monsoon hits Kerala State on:
   (a) 1st June (b) 14th June (c) 15th September (d) 7th July
6. Which of the following Rain Gauges is non-automatic?
   (a)Simon's (b) Tipping Bucket (c) Float (d) Weighing Bucket
7. According to the I.S. Code 4987-1968 what should be the density of Raingauges in Hilly areas:
   (a) 520 Km^2 per gauge (b) 260-390 Km^2
   (c) 130 Km^2 (d) 900 Km^2
8. Lines of equal Rainfall are called
   (a) Isobars (b) Isoclines (c) Isomer (d) Isohyet
9. The probability P_{ro} of an event being equalled or exceeded is given by Weibull as
   (a) $P_{ro} = \frac{m}{N}$ (b) $\frac{2m-1}{2N}$ (c) $\frac{m}{N+1}$ (d) $\frac{N}{m}$
10. The diameter of Indian standard Evaporation pan is,
    (a) 1000 mm (b) 1220 mm (c) 1020 mm (d) 1425 mm
11. In formula R = kP, the runoff coefficient for urban residential area (single houses) is about
    (a) 0.05 to 0.1 (b) 0.3 to 0.5 (c) 0.7 to 0.85 (d) 0.6 to 0.75
12. In a single peaked Hydrograph the relation between Rising Limb and Recession Limb is,
    (a) Recession curve is longer (b) Rising Limb is longer
    (c) Both are identical (d) There is no relation
13. In the Direct Runoff formula, n= $0.36\frac{\Sigma 0 \times t}{A}$ cm. Area A is in the unit of?
    (a) Hectare (b) Km^2 (c) m^2 (d) Acre
14. Who developed the theory of Synthetic unit hydrograph?
    (a) V.T. Chow (b) Darcy (c) Snyder (d) Penman
15. In Dicken's Formula for Flood discharge, $Q = CA^n$, n is equal to
    (a) $\frac{3}{4}$ (b) $\frac{2}{3}$ (c) $\frac{1}{2}$ (d) $\frac{5}{6}$
16. Unit Hydrograph theory was made by
    (a) Harton (b) V.T. Chow (c) L.K. Sherman (d) A.K. Biswas
17. In India, rainfall is generally recorded at
    (a) 12 Noon (b) 10 AM (c) 8 PM (d) 8 AM
18. Hydrograph is the graphical representation of
    (a) Groundwater flow (b) Rainfall
    (c) Surface Runoff (d) River flow
19. For controlling evaporation chemical film of the chemical is most suitable
    (a) Methyl alcohol (b) Ethyl alcohol (c) Cetyl alcohol (d) Butyl alcohol

20. Amount of water required to meet Evapotranspiration needs of crop during its full growth is called
(a) Consumptive use (b) Effective rainfall
(c) Conjunctive use (d) None of these

21. Unit Hydrograph of a specific duration can be obtained by dividing Direct Runoff hydrograph ordinates by
(a) Direct Runoff volume (b) Period of storm
(c) Total Rainfall volume (d) Base flow volume

22. Probability (P) and Recurrence Interval (T) are related as
(a) $PT^2 = 1$ (b) $\frac{P}{T} = 1$ (c) $\frac{P^2}{T} = 1$ (d) $P \times T = 1$

23. 1 Km^3 is equal to how many litres?
(a) 10^{12} (b) 10^9 (c) 10^6 (d) 10^3 litres

24. According of Falkan Mark, water stress occurs when water availability per person per year is less than
(a) 1000 m^3 (b) 1500 m^3 (c) 2000 m^3 (d) 550 m^3

25. Which state has made rooftop rainwater harvesting structure compulsory to all the houses is
(a) Tamil Nadu (b) Andhra (c) Karnataka (d) Kerala

26. The river which has no base flow is called
(a) Perennial river (b) Ephemeral river (c) Seasonal river (d) Flooded river

27. Orographic Rain Occurs due to lifting of air mass by
(a) Pressure difference (b) Mountain barriers
(c) Cyclones (d) Turbulence

28. Pitot Tube is to measure
(a) Pressure (b) Flow velocity (c) Depth of water (d) Pressure difference

29. According to WWF, there will be acute shortage of water in India by the year
(a) 2025 (b) 2035 (c) 2050 (d) 2090

30. Horton's equation for infiltration capacity is
(a) $f = f_0 + (f_0 - f_c)e^{-kt}$ (b) $f = f_t - (f_0 - f_c)e^{-kt}$
(c) $f = f_c . (f_c - f_0)e^{kt}$ (d) $f = f_c + (f_0 - f_c)e^{-kt}$

31. To determine mean rainfall in a catchment the accurate method is
(a) Isohyetal Method (b) Thiessen's Method
(c) Arithmatic Method (d) Ad hoc Method

32. Hydrology is the science which deals with
(a) Rain water (b) Surface + Groundwater
(c) River water (d) Ocean water

33. Rain Simulators are used for the determination of
(a) Evaporation (b) Infiltration capacity
(c) Runoff (d) Precipitation

34. Phytometer is used to measure

(a) Evaporation (b) Interception (c) Infiltration (d) Sun's radiation

35. Pressure exerted by a fully saturated air is called

(a) Partial pressure (b) Vapour pressure
(c) Total pressure (d) Saturation vapour pressure

36. If the dew point is below 0° C,

(a) Dew will form (b) Frost will be formed
(c) Vapour will form (d) Rainfall-occurs

37. What kind of energy Rain holds relative to Ocean?

(a) Potential (b) Kinetic (c) Electrical (d) Chemical

38. Water cycle is driven by

(a) Clouds (b) Sun (c) Planets (d) Oceans

39. The starting point of water (Hydrological) cycle is

(a) Ocean (b) Clouds
(c) Planets (d) It has no start or end point

40. Water released from the leaves is called

(a) Evapo-transpiration (b) Evaporation
(c) Transpiration (d) Condensation

41. When water vapour turns back in liquid form it is called

(a) Evaporation (b) Condensation (c) Precipitation (d) Transpiration

42. Direct Runoff is made up of

(a) Overland flow only (b) Rainfall + Evaporation
(c) Surface runoff + Infiltration
(d) Surface Runoff + Interflow + Channel precipitation

43. Base Flow denotes

(a) Delayed groundwater flow (b) Delayed groundwater flow + Snowmelt
(c) Delayed groundwater + Interflow (d) Annual minimum flow in stream

44. Flow duration curve is a plot between

(a) Accumulated flow against time
(b) Discharge against time in chronological order
(c) Base flow against time
(d) Stream flow against percentage of times the flow is equalled or exceeded

45. The flow-mass curve is an integral curve of

(a) Hyetograph (b) Hydrograph (c) Flow-duration (d) S-curve

46. How many unit hydrographs can be drawn for a given basin

(a) Only 1 (b) 10 (c) 100 (d) Infinite

47. The runoff as ordinate against the corresponding percentage of time as abscissa gives
(a) Unit Hydrograph (b) Hydrograph (c) Flow duration (d) Mass-curve

48. The tropical cyclones of the Bay of Bengal are called
(a) Hurricane (b) Willy Willy (c) Depressions (d) Typhoons

49. In India monsoon first enters in the state
(a) West Bengal (b) Gujrat (c) Kerala (d) Goa

50. Cyclone represents Low pressure in the
(a) Outer surface (b) Centre
(c) All over (d) At half the diameter

51. A cloud burst is
(a) 50 mm or over rain per hour
(b) Thunderstorm with normal rain
(c) Rapid condensation of very high clouds
(d) Sudden and copious rainfall over a small area of very short duration

52. Flood Routing is mainly to evaluate the
(a) Outflow hydrograph (b) Volume of water
(c) Speed of water (d) Time of Peak

TRUE OR FALSE QUESTIONS

1. The total amount of water on earth remains the same. (T or F)

2. Tank and Lysimeter method is used to measure Evapotranspiration (ET) or consumptive use. (T or F)

3. Penman's Method is based on Energy Budget and Mass Transfer approach to evaluate Evapotanspiration. (T or F)

4. Transpiration is associated with photosynthesis of plants. (T or F)

5. The unit of dew point is $\frac{N}{m^2}$. (T or F)

6. The dew point is the temperature at which humidity is 100%. (T or F)

7. Normal daily temperature is taken as average daily mean temperature for specific 100 years period. (T or F)

8. The maximum rainfall recorded is in Cherrapunji (T or F)

9. For rainfall to occur condensation Nuclii is essential. (T or F)

10. In India monsoon period starts in the month of September. (T or F)

11. In a tipping bucket type raingauge, the capacity of each bucket is 0.25 cm. (T or F)

12. Microwave radar Rainfall measurement gives accurate rainfall information. (T or F)
13. In computing average rainfall over a basin the most accurate method is Thissen Polygon method. (T or F)
14. Isohyets are lines of equal rainfall in a basin. (T or F)
15. In a mass curve of rainfall the horizontal portion shows no rainfall. (T or F)
16. The recurrence interval (T) and return period are same. (T or F)
17. To check the consistency of records, double mass curve method is used. (T or F)
18. Probable Maximum Precipitation (PMP) is the rainfall over a basin which would produce the flood with high risk of being exceeded. (T or F)
19. Dalton gave law of Evaporation as $E = C\,(e_f - e_a)$, where C is a function wind speed. (T or F)
20. Normally Evaportanspiration is more than Evaporation + transpiration from water body. (T or F)
21. Runoff is maximum at the beginning of rainfall. (T or F)
22. The Direct Runoff (DR) includes surface runoff, sub-surface runoff and base flow. (T or F)
23. In hilly areas the Runoff is more than in flat areas (T or F)
24. Rational method is used to find average flow for small catchments. (T or F)
25. IS : 5973-1970 Evaporation Pan's evaporation is more than the Lake Evaporation. (T or F)
26. Water Budget or water balance method gives accurate result of estimation of evaporation from lake or reservoir. (T or F)
27. Single tube infltrometer gives better result than the double ring infiltrometer. (T or F)
28. ϕ-index is defined as the average rate of rainfall **below** which the rainfall volume is equal to runoff volume. (T or F)
29. W-index is less than ϕ-index (T or F)
30. If in a storm Total Rainfall (P) is 158 mm and Direct Runoff is 88 mm. The storm lasts for 7 Hours, the ϕ-index is $10\,\frac{\text{mm}}{\text{Hour}}$. (T or F)
31. A unit hydrograph represents 1 cm or 1 mm of runoff. (T or F)
32. Synthetic Unit Hydrograph is for ungauged basins only. (T or F)
33. In Dicken's formula ($Q = CA^{3/4}$), C has maximum value for Northern India. (T or F)
34. The most popular formula for Annual Flood Series $\left(P = \frac{m}{N+1}\right)$ is given by Gumble. (T or F)
35. Flood Probability Plot on Gumbles paper has Discharge Vs Recurrence Interval on a semi-log paper. (T or F)
36. Area covered with vegetation has less infiltration compared to bare land. (T or F)
37. A heytograph is a bar graph showing the Rainfall Intensity Vs Time. (T or F)

FILL IN THE BLANKS

1. Percent of Earth's Water in Oceans ....
2. Water is said to be .... solvent.
3. The temperature at which air just gets saturated is called ....
4. Wind speed is measured by ....
5. Average annual rainfall over India is about .... mm.
6. In India maximum rainfall occurred at ....
7. IS code 4987-1968 recommends one rain gauge for how much area in predominantly hilly regions ....
8. What percentage should be of recording rain gauges ....
9. Optimum Number of Rain gauges N formula is $N = \left[\frac{C_V}{p}\right]^x$. The value of x is ....
10. Among various method to calculate mean rainfall, the most accurate method is ....
11. The plot of cumulative depth of rainfall against time is known as ....
12. Rainfall is classified as Heavy if intensity is more than .... $\frac{mm}{h}$.
13. If (P) is the probability of occurrence, the recurrence interval (T) will be ....
14. Full form of PMP is ....
15. Line joining places of equal rainfall are called ....
16. The line joining equal rainfall of particular duration and a particular frequency is called ....
17. The rainwater held by vegetation, buildings etc is called ....
18. Dalton's Law of evaporation formula is ....
19. The IS: 5793 – 1970 Evaporimeter is modified form of, ...
20. Single tube inflometer has diameter .... cm.
21. Runoff consists of Direct runoff plus ....
22. The ratio of the total length of all streams of any catchment divided by its area is called, ...
23. In the Runoff, Precipitation relation R = kP, the runoff coefficient k for concrete pavement is about ....
24. The average rainfall above which rainfall volume equals the runoff volume is called ....
25. The travel time of a water particle from hydraulically most remote point in the basin to outflow location is called ....
26. Unit Hydrograph theory was propagated by ....
27. The curve obtained by adding together a number of unit Hydrographs is called ....

28. A unit hydrograph produced by a very short unit duration (tending to zero) is called ....
29. The basin lag (t_L) in a Synthetic Unit Hydrograph is given by the formula $t_L = C_t (L_{ca} . L)^x$ where, x is equal to ....
30. The probability of occurrence of an event expressed as percent is known as ....
31. The rational formula for Flood Discharge is Q = CIA, The (I) in this formula is ....
32. Among plotting position formulae California, Hazen, Weibul and Gumble which is considered better formula ....
33. The Rational formula provides good result for small catchment areas upto .... km^2.
34. Reservoir capacity for a specific yield is found by .... curve.
35. A flood control reservoir with gates at the spillway is known as ....
36. Modification of an inflow hydrograph is done by ....
37. Panman Method is theoretical method to evaluate ....
38. Windex is always .... than ϕ_{index}.
39. A single peaked hydrograph contains. Rising Limb, Peak element and ....
40. The total area of a basin is called ....

ANSWERS

Multiple Choice Questions

1. c	**2.** c	**3.** d	**4.** d	**5.** a	**6.** a	**7.** c	**8.** d	**9.** c
10. b	**11.** b	**12.** a	**13.** b	**14.** c	**15.** a	**16.** c	**17.** d	**18.** d
19. c	**20.** a	**21.** a	**22.** d	**23.** a	**24.** a	**25.** a	**26.** b	**27.** b
28. b	**29.** c	**30.** d	**31.** a	**32.** b	**33.** b	**34.** c	**35.** d	**36.** b
37. a	**38.** b	**39.** d	**40.** c	**41.** b	**42.** d	**43.** a	**44.** d	**45.** b
46. d	**47.** c	**48.** d	**49.** c	**50.** b	**51.** d	**52.** a		

True & False

1. T	**2.** T	**3.** T	**4.** T
5. F °C	**6.** T	**7.** F 30 Yrs.	**8.** F Mawsynram (11, 872 mm)
9. T	**10.** F Hybe	**11.** F 0.25 mm to 0.5 mm	
12. F 0.5	**13.** F Isoheytal	**14.** T	**15.** T
16. T	**17.** T	**18.** F No Risk	**19.** T
20. F	**21.** F	**22.** F No Base Flow	**23.** T
24. F	**25.** T	**26.** F	**27.** F
28. F Above	**29.** T	**30.** T	**31.** T
32. T	**33.** F Minimum	**34.** F Weibull	**35.** T
36. F	**37.** T		

Fill in the Blanks

1. 97%	**2.** Universal	**3.** Dew Point	**4.** Anemometer
5. 1170 mm	**6.** Mawsynram	**7.** 130 km^2	**8.** 10%
9. Two (2)	**10.** Isohyetal Method	**11.** Mass curve	**12.** $7.5 \frac{mm}{Hr}$
13. $\frac{1}{P}$	**14.** Probable Maximum Precipitation		**15.** Isohyet
16. Isopluvial lines	**17.** Interception	**18.** $E = C (e_s - e_a)$	**19.** USWB Class A pan
20. 30 cm	**21.** Base Flow	**22.** Drainage Density	**23.** 0.83
24. ϕ-index	**25.** Time of Concentration		**26.** L.K. Sharman
27. S-curve	**28.** Instantaneous	**29.** 0.3	**30.** Frequency
31. Rainfall Intensit	**32.** Gumble	**33.** 50 m^2	**34.** Mass
35. Detention Basin	**36.** Flood Routing	**37.** Evapotranspiration	
38. Less	**39.** Recession Curve	**40.** Catchment Area	

SECTION 11

Irrigation

Definitions – 48

Multiple Choice Questions – 46

True & False – 38

Fill in the Blanks – 49

DEFINITIONS

Actual Evaportranspiration (AET): The real evaportanspiration occurring in a specific situation is called Actual Evaaportanspiration.

Alluvial Channel: These are excavated in alluvial soils which carry lot of silt alongwith water. According to prevailing velocity the bed and sides may either scour or silting may occur.

Aqueduct: When a natural drain (river-small or big) cross each other cross drainage (CD) is to be provided. When the canal flows over the drain (like a bridge) it is called aqueduct.

Available Moisture: The difference in water content of the soil between field capacity and permanent wilting point is known as available water or available moisture.

Barrage: The function of barrage is also to raise water level to facilitate entry of water into the canal so that water can flow under gravity. The barrage is totally controlled by gates.

Base Period (B): It refers to the whole period of cultivation from the time when irrigation water is first applied for preparation of ground for planting the crop, to its last watering before the crop is harvested.

Border Strip Method of Irrigation: In this farm is divided into a series of long parallel strips called 'borders', which are separated by low ridges. This method is used for closed growing crops like wheat and not for rice.

Canal Falls: These are solid masonry or concrete structures constructed on the canal if the natural ground slope is steeper than the designed canal bed slope. When a canal taking off from its head is in cutting, due to slope difference it soon has to be made in entire filling, which is expensive and results in heavy seepage loss.

Canal Outlets: It is a small structure which admits water from distributing channel to a water course or field channel. While canal network is to be maintained by the Government, the field channel maintenance is with the farmer, outlet is the connecting medium for the two.

Capacity Factor: It is the ratio of the mean discharge or water supplied to the full supply of canal.

Consumptive Use (Evapotranspiration): It is the depth of water used by a crop during its growth by evaporation from the field and transpiration by plants.

Culturable Commanded Areas (CCA): It is the area in which crop can be grown. CCA = GCA – Unculturable areas in GCA.

Drip Irrigation (Trickle Irrigation): In this water is applied in the form of drops directly near the base of the plant through drip nozzles operating at low pressure. The technique is also known as 'feeding bottle technique'. Lot of water is saved through this method.

Duty (D) and Delta (Δ): Duty is the area of agricultural field that can be irrigated with the unit volume of irrigated water. Delta is the total depth of water required by a crop during the entire period of crop is in the field.

Effective Rainfall (Re): It is that part of the precipitation falling during the growing period of a crop that is available to meet the evapotranspiration needs of the crop.

Field Capacity: It is the moisture content of the soil after free drainage has removed most of the gravity water. It depends upon capillary tension in soil and porosity.

Flow Irrigation: It is the type of irrigation in which the supply of irrigation water available is at such a level that it is supplied on the land by gravity flow.

Free Flooding Method: It consists of dividing the entire field to be irrigated into small strips by a number of field channels (or laterals). Water enters at the higher end flows towards the lower end. It is suitable in steep slopes and permeable soils.

Frequency of Irrigation: The frequency of irrigation (or irrigation interval) is so decided that the average moisture content in soil is close to optimum and at each irrigation the soil moisture content is brought to the field capacity. Alternatively, the frequency of irrigation can be decided so as to satisfy the daily consumptive use which varies with stage of plant growth.

Furrow Method: A furrow consists of a narrow ditch between two rows of plants. It is used for row crops like maize, sugarcane, cotton, tobacco, potatoes etc. The size and shape of furrows depend upon type of crop.

Gross Commanded Area (GCA): It is defined as the gross irrigable area which can be commanded by canal irrigation. It includes unculturable areas also like ponds, barren lands, roads etc.

Head-Regulator and Cross-Regulator: These structures are meant to regulate the supplies of the off-taking channel and Parent Channel. The cross regulator raises the water level in the main canal to feed the off-take channel.

Intensity of Irrigation: It is defined as the percentage of the Culturable Commanded Area (CCA) proposed to be irrigated annually. The yearly intensity of irrigation may be obtained by adding the intensity of irrigation for all the crop seasons. This may be more than 100% also.

Kharif Crops (Monsoon Crops): These are sown during June-July and harvested in autumn September-October. These crops are: rice, maize, millets, pulses, groundnut etc.

Kor Period, Kor Watering, and Kor Depth: Crops require maximum water during first watering after crops have grown few centimeter. The first watering is known as Kor Watering. The part of the base period in which Kor watering is needed is called Kor period and the depth of water applied during this period is called Kor depth.

Level Crossing: It is an arrangement provided to regulate the flow of water through the drainage and irrigation canal when they cross each other approximately at the same bed level.

Lift Irrigation: It is practised when the water-supply available is at a low level than the field. In such case water has to be lifted by some mechanical device. Irrigation from wells is an example of lift irrigation.

Lining of Canals: When there is excessive loss of water from a canal by seepage, its bottom and sides are covered (lined) by impervious material like-Brick lining, Concrete lining, Shotcrete lining, Stone lining and Asphaltic lining.

Lysimeter: It is the device, typically a tank or container filled with soil and crop grown on it used to measure actual evaportanspiration.

Ogee Fall: It has gradual convex and concave curves, with an aim to provide smooth transition and to reduce effects of impact and turbulence. There are other types of falls used according to situation. These include-Rapid fall, Notch fall, Glacis type and Sarda type fall.

Paleo Irrigation: It is the watering prior to the sowing. As the soil is very dry and watering makes it soft which helps in ploughing and sowing of crops.

Permanent Wilting Point (Wilting Coefficient): It is the water content at which plants can no longer extract sufficient water from the soil for their growth.

Potential Evaportanspiration (PET): It indicates the water loss which will occur if sufficient moisture is always available to completely meet the needs of plants and evaporation from the soil.

Rabi Crops (Winter Crops): These are sown in autumn (October) and are harvested in spring (March). The crops that fall under this category are : gram, wheat, barley, peas, mustard, potato, tobacco etc.

Relation Between Duty and Delta (D and Δ): The relationship can be easily derived as,

$$\Delta = 8.64 \frac{B}{D} \text{ in meters}$$

where,

D = Duty in $\frac{\text{hactares}}{\text{cumec}}$

Δ = Total depth of water applied (m)

B = Base period in days.

(1 Ha = 10^4 m^2 and 1 cumec day = 8.64 Ha – m)

Saturation Capacity: The amount of water required to fill up the pore spaces in soil particles by replacing all the air held in pores. Also known as Maximum Holding Capacity.

Silt Excluder: It is a device by which silt is taken out from water entering the canal. It is located in the bed in front of the head regulator. The silt laden water in the bottom layers is diverted and top water is led into the canal. It is a preventive measure.

Silt Extractor or Silt Ejector: The silt which has already entered the canal is taken away and thrown out. It is a curative measure.

Sprinkler Irrigation: It consists of applying water in the form of spray as in the ordinary rainfall. Special mechanism of rotating nozzle is used. Method is suitable for unprepared land and irregular topography.

Superpassage: It is the structure in which drainage is passed over the canal. It is reverse of an aqueduct.

Water Application Efficiency (η_a): It is the ratio of the quantity of water stored in the root zone of crops (ws) to the quantity of water delivered to the field (wf) $\eta_a = \frac{w_s}{w_f} \times 100$.

Water Conveyance Efficiency (η_c): It is the ratio of water delivered to the farm (wf) and water diverted from reservoir of river (wr) $\eta_c = \frac{w_f}{w_r} \times 100$.

Water Distribution Efficiency (η_d): It evaluates the degree to which water is uniformly distributed throughout the root zone. The more uniformly the water is distributed, the better will be the crop response. It is determined from the equation $\eta_d = \left(1 - \frac{y}{d}\right) \times 100$.

where,

d = Average depth of water stored during irrigation.

y = Average numerical deviation in depth of water stored from average depth stored.

η_d = Provides a measure for comparing various irrigation methods of water application.

Water Use Efficiency (η_u): It is the ratio of water beneficially used (wu) including leaching water to the quantity of water delivered (wd) $\eta_u = \frac{W_u}{W_d} \times 100$.

Waterlogging: An agricultural land is said to be waterlogged when water table reaches the root zone of the plants. At this stage fertility and productivity are affected adversely.

Weir: It is a solid construction across the river for raising the water level so that water can be easily diverted to the canal.

Wild Flooding (Uncontrolled Flooding): In this method water is spread or flooded on flat land, without much control or prior preparation. It is wasteful use of water.

Zaid Crops: These are grown between Kharif and Rabi Crops (March-June). They require warm and dry weather. Main Zaid crops are: seasonal fruits, vegetables and fodder crops.

MULTIPLE CHOICE QUESTIONS

1. Rajasthan Canal is between States, Rajasthan and:
   (a) Punjab (b) Haryana (c) U.P. (d) Gujrat
2. Contour farming is suitable for areas:
   (a) Hilly (b) Plains (c) Plateau (d) Coastal
3. The water in soil which cannot move by Gravity or Capillary forces:
   (a) Superfluous water (b) Hygroscopic water
   (c) Clear water (d) Confined water
4. Which of the following is not a cash crop:
   (a) Coffee (b) Sugar (c) Cotton (d) Pulses
5. Average Delta (Δ) for Rice crop in mm is:
   (a) 900 (b) 1200 (c) 300 (d) 750
6. The crops which do not require water are called:
   (a) Wet crops (b Dry crops (c) Garden crops (d) Perennial crops
7. The relation between Duty, Delta and Base Period is:
   (a) $\Delta = 8.64 \frac{B}{D}$ (b) $\Delta = 8.64 \frac{D}{B}$ (c) $\Delta = 8.64\, D \times B$ (d) $\Delta = \frac{D}{B}$
8. 1 cumec-day is equal to:
   (a) 10^4 m^2 (b) 86.4 ha.m (c) 8.64 ha.m (d) 864 ha.m
9. Which of the following is finest grained soil?
   (a) Fine sand (b) Silt (c) Clay (d) Silty clay
10. If for a crop Base Period is 320 days, Duty 800 $\frac{\text{ha}}{\text{cumec}}$, then Δ is:
   (a) 0.576 m (b) 1.152 m (c) 0.72 m (d) 3.456 m

11. In India which of the following crops gives maximum Yield per hectare:

(a) Maize (b) Sugarcane (c) Wheat (d) Rice

12. Lane reduced the horizontal creep given by Bligh by a factor,

(a) $\frac{1}{2}$ (b) $\frac{2}{3}$ (c) $\frac{1}{3}$ (d) $\frac{3}{5}$

13. The exit gradient permitted for Fine Sand is

(a) $\frac{1}{6}$ to $\frac{1}{7}$ (b) $\frac{1}{5}$ to $\frac{1}{6}$ (c) $\frac{1}{4}$ to $\frac{1}{5}$ (d) $\frac{1}{3}$ to $\frac{1}{4}$

14. According to IS 6933-1973 the afflux in case of weirs and barrages on alluvial rivers in upper reaches is,

(a) 1 m (b) 0.3 m (c) 2.25 m (d) 1.5 m

15. According to Khosla the most effective piles are,

(a) Upstream pile (b) Short inner piles (c) End piles (d) Deep inner piles

16. The launching apron on downstream of impervious floor of $\frac{\text{weir}}{\text{barrage}}$ should be multiple of downstream pile depth (d_2).

(a) d_2 to 1.5 d_2 (b) 1.5 d_2 to 2.5 d_2 (c) 2.5 d_2 to 3.0 d_2 (d) 3.5 d_2 to 4.0 d_2

17. Silt extractor or silt ejector is provided at

(a) Upstream of the river (b) In the still pool

(c) Just below canal head regulator (d) a little downstream of canal

18. Which of following canals is the last in the system

(a) Minor distributory (b) Branch canal (c) Main canal (d) Water course

19. Which is the disadvantage of an inundation canal

(a) Low cost (b) Less water logging

(c) Manurial value (d) Less duty

20. In a modular outlet the discharge depends upon

(a) Difference of water level in distributary and water course

(b) Difference of water level in distributary

(c) Is independent

(d) Difference of water level in water course

21. The depth discharge relationship remains unaffected in which fall?

(a) Trapezoidal (b) Ogee (c) Straight Glacis (d) Rapids

22. The canal that acts as an irrigation canal as well as feeder canal is known as,

(a) Feeder canal (b) Permanent canal (c) Carrier canal (d) Inundation canal

23. The canal which is farmer's responsibility is,

(a) Contour canal (b) Watershed (c) Field channel (d) Side slope canal

24. In Kennedy's design of irrigation channels which of the following equations in not used

(a) Kutter's Equation (b) Continuity Equation

(c) Kennedy's Equation (d) Darcy's Equation

25. Lacey's silt factor for standard silt is,
(a) 1.25 (b) 0.85 (c) 1.0 (d) 0.5 to 0.7

26. In silt factor formula, $f = 1.76\sqrt{m_d}$, m_d is,
(a) D_{15} (b) D_{85} (c) D_{50} (d) D_{35} in mm.

27. On flatland the channel alignment used is
(a) Side slope (b) Watershed (c) Contour (d) Field channel

28. Which of the following canal alignment needs maximum Cross-Drainage works
(a) Contour canal (b) Side slope canal (c) Watershed canal (d) Field channel

29. In a Navigation canal the velocity of flowing water should be
(a) High (b) Medium (c) Small (d) Very High

30. In an alluvial channel the bed material's average size is 0.5 mm, the value of Lacey's silt factor 'f' will be?
(a) 1.02 (b) 1.24 (c) 1.10 (d) 0.95

31. When a canal is wholly in filling Berm width according to IS-7112-1973 should be
(a) 3D (b) 2D
(c) 1.5 D (d) 2.5 D, D is Full supply depth

32. According to USBR the free board in a canal is in proportion to
(a) D (b) $D^{\frac{1}{2}}$
(c) $D^{\frac{1}{3}}$ (d) $D^{\frac{1}{4}}$, D is Full supply depth

33. Water logging (rise of water table in root zone) occurs, due to
(a) Over irrigation (b) Canal seepage
(c) Lack of surface drainage (d) All these

34. In upper Bari Doab Canal System average canal loss was estimated around
(a) 20% (b) 30% (c) 10% (d) 25%

35. In a lined canal maximum velocity can occur for lining material
(a) Stone (b) Cement concrete
(c) Asphaltic concrete (d) Brick

36. The Sarda Type Falls were developed in Indian state of?
(a) U.P. (b) Punjab (c) M.P. (d) Bihar

37. In a canal discharge intensity is defined as
(a) $\frac{Q}{D}$ (b) $\frac{Q}{B}$ (c) $Q \times D$ (d) $Q \times B$

38. In an aqueduct Canal is
(a) Above drainage (b) Below drainage
(c) Both same level (d) Can either be above or below the drainage

39. In canal transitions the length of expansion transition is
(a) Less than contraction transition (b) Equal to contraction transition
(c) More than contraction transition (d) Depends upon velocity

40. Saturation of soil with water may result in
(a) Alkalinity in soil (b) Soil erosion (c) Soil fertility (d) Swamps

41. Overhead Irrigation is also known as
(a) Canal (b) Sprinkler (c) Tank (d) Drip

42. Sprinkler irrigation is practiced in arid and semi-arid regions as,
(a) It is cheap and easy to install (b) It can be used for all crops
(c) It reduces water loss (d) It requires simple machinery

43. The oldest and cheapest means of irrigation is by
(a) Tube-wells (b) Dug wells (c) Sprinklers (d) Canals

44. Why we need to conserve water, choose the incorrect answer
(a) 71% of Earth is covered with water (b) Our water resources are polluted
(c) Less rainfall due to deforestation (d) Increase in demand

45. Bamboo drip irrigation is popular in
(a) Mizoram (b) Manipur (c) Meghalaya (d) Assam

46. Irrigation globally uses
(a) About half of human water resources (b) Less than 10% of water available
(c) Over 70% of water used (d) Only small amount of human water resources

TRUE OR FALSE QUESTIONS

1. River Water is the only source for irrigation. (T or F)
2. In drip irrigation water is applied close to the plant root. (T or F)
3. Surface and sprinkler irrigation involves wetting the whole soil. (T or F)
4. Drip Irrigation is not suitable for most soils. (T or F)
5. Irrigation canals are generally aligned along ridge line. (T or F)
6. In canals syphons, the flow is like pipe flow. (T or F)
7. The main function of a diversion headwork is to store water. (T or F)
8. The intensity of irrigation is defined as percentage of Cultural Commanded Area (CCA) to be irrigated annually (T or F)
9. In relation of Delta (Δ), Duty (D) and Base Period (B), Δ is inversely proportional to D. (T or F)
10. A fall in canal bed is generally given when Designed bed slope exceeds the ground slope. (T or F)
11. According to Khosla theory the Exit Gradient depends upon $\alpha = \frac{b}{d}$ Ratio. (T or F)

12. Capillary moisture held in soil pores against gravity is utilized by plants. (T or F)
13. If Irrigation efficiency is 80%, conveyance loss 20% and actual depth of watering is 16 cm, the depth of water required at outlet will be 25 cm. (T or F)
14. Scour depth as given by Lacey during flood is $0.47\left(\frac{Q}{f}\right)^{2/3}$. (T or F)
15. The useful moisture of soil is equal to its field capacity. (T or F)
16. Lacey's regime velocity is proportional to $R^{2/3}\,S^{1/2}$. (T or F)
17. The state of soil when plants do not get water for their use is called Permanent Wilting Point. (T or F)
18. In Montague type fall a circular glacis is provided. (T or F)
19. A land is said to be water-logged if water table reaches the root zone of the crop. (T or F)
20. Lacey's silt factor (f) is proportional to square root of average particle size. (T or F)
21. In Kennedy's equation for critical velocity, $V_0 = 0.55\ m\ D^{0.64}$ (m is average particle size). (T or F)
22. In a crop over undulating sandy field, the best method of irrigation is sprinkler Irrigation. (T or F)
23. In a concrete canal the flow velocity should not be more than $5.0\ \frac{m}{s}$. (T or F)
24. For cereal crops the method of Irrigation adopted is furrow method. (T or F)
25. In a cross drainage work if River flows above canal level, it is called an aqueduct. (T or F)
26. Uncontrolled flooding method of irrigation is adopted where there is scarcity of water. (T or F)
27. In border strip method the length of the strip is given more in clayey soil than is sandy soil. (T or F)
28. Furrow method of irrigation is good for paddy crop. (T or F)
29. Contour farming is suitable in plateau area. (T or F)
30. In drip irrigation there is lot of saving of water. (T or F)
31. Water with high conductivity is suitable for all types of crops. (T or F)
32. Low sodium water is good for crops. (T or F)
33. High Boron content is good for crops. (T or F)
34. Average Δ for Rice is about three time Δ of wheat. (T or F)
35. 1 Hectare is equal to 100 m^2. (T or F)
36. The water conveyance efficiency is the ratio of water diverted from river (w_r) and water supplied to field (w_f), $\eta_c = c$. (T or F)
37. Fertility of soil can be maintained by resorting to crop rotation. (T or F)
38. Pressure of 1 bar is equal to $100\ \frac{N}{m^2}$. (T or F)

FILL IN THE BLANKS

1. In Bligh's creep theory no distinction is made between .... and .... creep.
2. Lane's weighted creep theory say that .... creep is more effective.
3. Method of 'Independent Variable' is proposed by ....
4. For fish to move both ways the device is known as ....
5. In a weir divide wall is given to crate a ....
6. Canal Headwork is provided mainly to control ....
7. Silt extractor is provided in the ....
8. The canal which yield a net revenue is called .... canal.
9. The main canal which can command area on both sides is called the ....
10. A head regulator on inundation canal is generally known as ....
11. The irrigable area under Bandhara irrigation scheme is known as ....
12. Upper Bari Doab system is located in ....
13. For upper Bari Doab system Kennedy found non-silting non scouring velocity (critical velocity) V_0 as ....
14. For wider use of his equation Kennedy later introduced to take care of silt grade a factor (m), known as ....
15. For finding mean velocity Kennedy used .... Equation.
16. Relation between Chezy's (C) and Manming's (N) is ....
17. Units of Chezy's coefficient is ....
18. For normal alluvial soils in India Manning's (N) value adopted is around ....
19. For a Trapezoidal channel with bed width B, depth D and side slope $\frac{1}{2}$ H : (V), the area of cross section is ....
20. "Dimensions, width, depth and slope of a regime channel to carry a given discharge with a given silt charge all are fixed by nature", This theory was given by ....
21. In a regime channel silt .... and silt .... are constant.
22. Lacey's Perimeter-Discharge (P – Q) relation is ....
23. To avoid excessive length of canal in filling we provide ....
24. When a canal and natural stream cross each other we require the structures called ....
25. In an aqueduct canal is taken .... the drainage.
26. Presently Percentage of agricultural area is under irrigation in India ....
27. In India the percentage of GDP spent on Agriculture is ....
28. Contour farming is mainly practiced in .... areas.

29. The method of irrigation nicknamed as 'feeding bottle' technique is actually ....
30. Sarda canal system is in the Indian state of ....
31. While contracting width of canal the expansion transition is .... in length than contraction transition.
32. In drip irrigation water is applied at .... of the plant.
33. The greatest advantage of irrigation is increased ....
34. For crops like Maize, Cotton, Potatoes etc, the method of irrigation most suitable is ....
35. Available moisture in soil is between field capacity and ....
36. The eight month crop is ....
37. The twelve month crop is ....
38. Irrigation capacity of one unit of water is called ....
39. Water Depth required by a crop during entire period the crop is in the field is called its ....
40. If a crop required 10 watering of 8.5 cm each during crops life, its Delta will be ....
41. The period between first and last watering is called ....
42. 1 Cumec – day = ... Hectare – Meters.
43. If for a crop Base period is 110 days and $\Delta = 68$ cm, the Duty will be ....
44. The first watering of crop is known as ....
45. The ratio of mean supply to full supply of canal is called ....
46. First watering before sowing of crop is known as ....
47. The total loss of water from a cropped farm is called the ....
48. Evapo-transpiration is also known as ....
49. When nature of crop sown in a particular field is changed year after year it is known as ....

ANSWERS

Multiple Choice Questions

1. a	**2.** a	**3.** b	**4.** d	**5.** b	**6.** b	**7.** a	**8.** c	**9.** c
10. d	**11.** b	**12.** c	**13.** a	**14.** a	**15.** c	**16.** b	**17.** d	**18.** d
19. d	**20.** c	**21.** a	**22.** b	**23.** c	**24.** d	**25.** c	**26.** c	**27.** b
28. a	**29.** c	**30.** b	**31.** a	**32.** b	**33.** d	**34.** c	**35.** b	**36.** a
37. b	**38.** a	**39.** c	**40.** d	**41.** b	**42.** c	**43.** b	**44.** a	**45.** c
46. c								

True & False

1. F **2.** T **3.** T **4.** T

5. T **6.** T **7.** F To divert **8.** T

9. T **10.** F Ground slope exceeds **11.** T

12. T **13.** T **14.** F Power $\frac{1}{3}$

15. F Field Capacity-Wilting Point **16.** F $S^{1/3}$ **17.** T

18. F Parabolic **19.** T **20.** T

21. F m depends on size of particles (coarse m > 1) **22.** T

23. F **24.** F Check Method **25.** F Super passage **26.** F Abundance

27. T **28.** F Flood and Drip Irrigation **29.** F Hilly area

30. T **31.** F Unsuitable **32.** T **33.** F

34. T **35.** F 10,000 m^2 **36.** F $\frac{w_f}{w_r} \times 100$ **37.** T

38. F 1 Lakh $\frac{N}{m^2}$ or Pa

Fill in the Blanks

1. Horizontal and Vertical **2.** Vertical **3.** A.N. Khosla

4. Fish Ladder **5.** Still Pool **6.** Discharge **7.** Canal

8. Productive **9.** Ridge Canal **10.** Flood Regulator **11.** Thal

12. Pakistan Punjab **13.** $0.55\ D^{0.64}$ **14.** Critical Velocity Ratio (CVR)

15. Kutter's **16.** $C = \frac{1}{N} R^{1/6}$ **17.** $\frac{m^{1/2}}{s}$ **18.** 0.02

19. $BD + \frac{D^2}{2}$ **20.** Lacey **21.** Grade or Charge **22.** P = 4.733 Q

23. Falls **24.** Cross-Drainage Works (CD-works) **25.** Above or Over

26. 48.8% **27.** 15% **28.** Hilly **29.** Drip Irrigation

30. Uttar Pradesh (U.P) **31.** More **32.** Near the roots

33. Food Production **34.** Furrow Method **35.** Wilting Point **36.** Cotton

37. Sugarcane **38.** Duty **39.** Delta (Δ) **40.** 85 cm

41. Base Period **42.** 8.64 Ha – m **43.** 1400 Ha – m **44.** Kor Watering

45. Capacity Factor **46.** Paleo **47.** (ET) Evapo-Transpiration

48. Consumptive used $\frac{m}{s}$ **49.** Crop Rotation

SECTION 12

Materials

Definitions – 41

Multiple Choice Questions – 102

True & False – 102

Fill in the Blanks – 108

DEFINITIONS

Acrylic Paint: These are popular as they are durable and fast-drying. They are composed of pigment suspended with an acrylic polymer emulsion, silicon, defoamers, oils etc. These are typically applied to interior wall surfaces. It resists cracking and fading. It is non-toxic and has eco-friendly formulation.

Bitumen: It is an immensely viscous constituent of petroleum. Depending upon its exact composition it can be a sticky, black liquid or an apparently solid mass that behaves as liquid over very large time scales. It is mostly used in road construction and roofing. Its molecular formula is $C_n H_{2n+2}$, where n represents number of carbon atoms in the hydrocarbon chain. Its chemical name is **Polycyclic Aromatic Hydrocarbon**.

Brass and Bronze: Brass is an alloy of copper and zinc. In proportions, it can be varied to achieve different mechanical and electrical, acoustic and chemical properties, but copper typically has the larger proportion, generally 66% and 34% zinc. Similar to brass is **bronze**, a copper ally that contains **tin** instead of zinc. Brass is used in door knobs, drawer pulls and sculptures due to its gold like appearance. Bronze due to timbre property, is used in making bells, cymbals and some musical instruments.

Bulking of Sand: The increase in the volume of sand due to an increase in moisture content is known as the bulking of sand. A film of water is created around the sand particles which forces the particle to get aside from each other and thus the volume increases.

Bullet Proof Glass (Ballistic Glass): It is a strong and optically transparent material particularly resistant to penetration to projectiles. However, like any other material, it is not completely impenetrable. It is made of combination of two or more types of glass, one hard and other soft. Softer layer makes it elastic, so that it can flex instead of shatter.

Bull's Trench Kiln: It is a type of kiln oval in shape used to make bricks in continuous operation. It is primary technology used in India for brick production, accounting for about 70% of the Countries total brick production. A bull's trench kiln is typically 50 m to 100 m long and 1.5 m to 2.5 m deep underground. It is divided into 8 to 12 sections for brick production. Green bricks are added from one end and fired bricks are removed from the other end.

Distemper: It is a type of whitewash made of chalk, pigment and water. This paint is less durable. To make it work, it is bound with an animal based glue like casein. The paint is not applied thickly, or it can lead to uneven finish and cracking or peeling. Painting has to be done on dry surface only.

Eco-Sand: It is high-quality manufactured sand of cubical-shaped fine aggregates that is ideal for ready-mixed concreting, and pre-cast concrete applications. By using Eco-sand, environmentally destructive sand mining is avoided altogether, contributing to sustainable construction.

Enamel Paint: It is a type of paint that is made from a combination of varnish and inorganic pigment. It is hard, glossy, smooth, and of durable finish. Usually it is oil based but, there are some water based also. It is durable as it is weather resistant. It can be used on variety of surfaces.

English Bond: In brick masonry it has one course of stretcher and a course of header above it, i.e. it has two alternating courses of stretchers and headers. Headers are laid centered on the stretchers in course below and each alternate row is vertically aligned.

Final Setting Time: The final setting time is defined as the time taken to reach the cement paste to become into a hardened mass. In construction, **initial setting time** should not be too late. Normally, initial setting time is 30-45 minutes and final setting time is below 10 Hours.

Fire Resistant Material: A fire resistant material is one which is designed to resist burning and withstand heat. Fire resistant material is that which prevents or delays the passage of excessive heat and fire.

First Class Bricks: These are of high quality and uniform size, shape and colour. They have sharp edges, are free from cracks and other defects, and produce a clear **ringing sound** when struck. They are used for load-bearing structures and for exposed walls. Compressive strength of these bricks is about $\frac{10N}{mm^2}$.

Flemish Bond: It is another traditional pattern where stretchers and headers are laid alternately in a single course. Fleming bond is attractive aesthetically, but is weaker than **english bond** for load bearing walls. Often this type of bond is used for two-brick thick walls.

Frog in Bricks: Frog is an indentation or depression on the top face of a brick made with the object of forming a key for mortars. It is kept on the top face while constructing a wall in order to fill mortar properly in it. In modern bricks **frogs** are between 10 mm to 20 mm deep. It may also display manufactures logo or year of manufacture.

Gneiss: It is a metamorphic rock that is commonly found and is formed when igneous or sedimentary rocks are subjected to high temperatures and pressures. Gneiss is made up of feldspar, quartz, mica, amphibole and pyroxene. It displays a well developed foliation-and a poorly developed schistosity and cleavage.

Granite: It is a coarse-or medium grained intrusive igneous rock that is rich in **quartz** and **felspar**. It is most common plutonic rock of Earth's crust formed by cooling of **magma**. The compressive strength of granite is 1050 to 1358 $\frac{kg}{cm^2}$.

Gypsum: It is Calcium Sulphate ($Ca\ SO_4$). Refined gypsum in the anhydrite form (no water) is 29.4% Calcium (Ca) and 23.5% Sulphar (S). Usually, gypsum has water associated in the molecular structure ($Ca\ SO_4\ 2H_2O$) and is approximately 23.3% (Ca and 18.5% (S). (Plaster of Paris POP).

Hardness of Material: Hardness is a measure of material's ability to resist plastic deformation. In other words, it is a measure of how resistant material is to **denting** and **scratching**. Diamond for example, is a very hard material.

Hoffman Kiln: It is a type of kiln used to make bricks and other ceramic products. In this kiln, a fire moves continuously through a closed circuit of bricks, which are stacked in a circular, elliptical or rectangular shape. The fire is moved forward by a chimney or fan. It has higher efficiency. It gives uniform burning of bricks.

Le Chatelier Apparatus: It is a tool to determine the **Soundness** of cement and the potential expansion of concrete. It typically consists of a split brass mould, a base plate and a glass plate. The brass mould is split lengthwise, allowing easy separation after the test. A gradual scale on the glass plate helps measure the expansion accurately.

Marbel: It is basically limestone that is more or less crystallized by metamorphism, that ranges from granular to compact in texture, that is capable of taking a high polish, and that is used especially in architecture and sculpture. Its compressive strength is 827 to 1020 $\frac{kg}{cm^2}$.

Modular Brick: It is a type of building block that is made from clay, concrete or other suitable materials. The actual size as per BIS recommendation of modular brick is 190 × 90 × 90 mm with mortar, the size is 200 × 100 × 100 mm. Average compressive strength is about $\frac{15N}{mm^2}$.

Plastic of Paris (POP): Is a fine white powder that hardens when mixed with water. It is quick-setting gypsum plaster. The chemical formula of Plaster of Paris is Ca $SO_4 \frac{1}{2} H_2O$. It is fire resistant. Used in buildings to avoid fire hazards. It is used in making fine artwork for decoration and making orthopaedic cast.

Plywood: It is a composite material manufactured from thin layers or **plies** of wood **veneer** that have been stacked and glued together with the grains of adjacent layers arranged at right angles or at a wide angle. It has high impact resistance. The most used plywood is of 12.5 mm $\left(\text{approx} \frac{1}{2} \text{inch}\right)$. These are used both as structural and non-structural material. It has multiple uses.

Polyvinyl Chloride (PVC): It is a synthetic plastic which is resistant to oxidation and degradation. It is also flame retardant and has a good electrical insulation properties. PVC can be rigid or flexible and it can be machined, welded and heat formed. It is used in many industries including construction, electronics, medicine (medical packaging), food beverages etc.

Porcelain: It is a hard fine-grained ceramic material made by heating raw material generally including **kaolinite** in a kiln to a temperature between 1,200 and 1,400° C. It is sonorous, non-porous and usually translucent and white ceramic ware that consists essentially of kaoline, quartz and feldspathic rock. It is used in 'pottery' industry and for decorative purpose.

Portland Cement: It is the product obtained by pulverizing clinker, consisting of hydraulic calcium silicates to which some calcium sulphate has usually been provided as an interground addition. The strength and other properties of concrete are mainly derived from the hydration of tricalcium and dicalcium silicates.

Pozzolana: It is finely divided siliceous and aluminous material that reacts chemically with slaked lime at ordinary temperature and in the presence of moisture to form a strong slow-hardening cement. Addition of pozzolana to ordinary Portland cement increases shrinkage but reduces permeability of the concrete. It also increases workability. It is used in marine and hydraulic structures.

Pug-mill: It is a machine in which materials such as clay and water are mixed, blended or kneaded into a desired consistency, with minimum air-pockets and lumps. The materials are mixed into a plastic state in order to make clay fit for next process of preparing bricks.

Putty: It is a soft, pliable material that can be moulded into different shapes. It is often used to fill holes or gaps, and it can also be used as a sealant. Putty is typically made from a mixture of clay, chalk and linseed oil. It is also used to fasten glass in window frames.

Quick Lime: It is an alternate name for the chemical compound known as **Calcium Oxide** (CaO), burnt lime, or caustic lime. Under standard conditions for temperature and pressure, quicklime is known to exist as a white crystalline solid. It is used in the manufacture of iron and steel, and paper and pulp and many more.

Silica: Silica or Silicon dioxide (SiO_2) commonly found in nature as quartz. Sometimes silica is the major constituent of sand. It is used in many commercial products like bricks, glass and ceramics, plaster, concrete etc.

Slate: It is a fine grained, foliated, homogenous, metamorphic rock derived from original shale-type sedimentary rock composed of clay or volcanic ash through low-grade regional metamorphism. Slate is particularly suitable as a roofing material as it has very low water absorbing quality.

Soda-lime Glass: Also called **Soda-lime-silica** glass, used for windowpanes, glass containers like bottles and jars, food items etc. It amounts about 90% of manufactured glass. Soda-lime glass is relatively inexpensive, **chemically stable,** reasonably hard, and extremely workable. Also, it is ideal for glass recycling.

Stainless Steel: It is a corrosion-resistant alloy of iron, chromium and, in some cases, nickle and other metals. Completely and infinitely recyclable, stainless steel is termed as a **green material.** It is used in doors and windows and many domestic gadgets.

Stucco Paints: Stucco or **render** is a construction material made of aggregates, a binder and water. Stucco is applied wet and hardness to a very dense solid. It is used as a decorative coating for walls and ceilings, exterior walls and as a sculptural and artistic material in architecture.

Surkhi: It is fine powdered artificial pozzolonic material made from burnt clay, brickbats or burnt bricks. It is used in mortars and concrete. It is a substitute for sand in mortar and concrete. It can also be used with lime to make mortars. It adds strength to mortar mix, water resistant and reduces cracking in concrete. It has higher shrinkage.

Tree's Age: The number of rings on a stump indicates the number of Years the tree has lived. There are rings of darker and lighter bands; one year of growth is composed of both a dark and a light band. The rings also tell about the environmental conditions for a particular year. Thinner rings show colder or dryer year and thicker one show better growing conditions.

Tri-Calcium Silicate: Tri-calcium **silicate** and Tri-calcium **Aluminate** are responsible for initial strength of cement and **Di-calcium** silicate is responsible for ultimate strength of cement. The setting and hardening of cement after the addition of water is due to hydration of these compounds.

Wrought Iron: It is an iron alloy with a very low carbon content (less than 0.05%) in contrast to that of **cast iron** (2.1% to 4.5%). It is a semi-fused mass of iron with fibrous **slag** inclusion, which gives it wood-like appearance. It is tough and corrosion resistant and can be forge welded. It is used in fencing railings and grills.

MULTIPLE CHOICE QUESTIONS

1. Which is the hardest material of the following?

(a) Steel (b) Silver (c) Diamond (d) Gold

2. Strain has the units of

(a) $\frac{N}{m}$ (b) $\frac{kg}{m}$ (c) $\frac{N}{m^2}$ (d) $\frac{m}{m}$

3. Which of the following has more fire resisting characteristics?

(a) Marble (b) Lime stone

(c) Granite (d) Compact sand stone

4. The rocks which are formed due to cooling of magma at a considerable depth from earth surface are called
   (a) Plutonic (b) Hypabyssal (c) Volcanic (d) Igneous
5. Portland cement was developed by
   (a) James Parker (b) Joseph Aspdin (c) Vicat (d) Puntkii
6. In typical oxide composition of general-purpose Portland cement the percentage of silica is (about)
   (a) 65% (b) 21% (c) 6% (d) 3%
7. Percentage of carbon content in steel varies from
   (a) 0.25 to 1.5 (b) 2 to 3.5 (c) 3 to 4.5 (d) 2.5 to 3.75
8. Soda lime glass is combination of
   (a) Copper Silicate + Potassium Silicate (b) Sodium Silicate + Calcium Silicate
   (c) Aluminum Silicate + Potassium Silicate (d) Aluminum Silicate + Calcium Silicate
9. Gypsum consists of
   (a) H_2S and CO_2 (b) $CaSO_4$ and H_2O (c) Lime and H_2O (d) CO_2 + Ca
10. Stucco paints are suitable for
   (a) Stone masonry (b) Brick walls (c) Both (a) and (b) (d) Neither a nor b
11. The minimum compressive strength of 2nd class brick should be
   (a) $75 \frac{kg}{cm^2}$ (b) $90 \frac{kg}{cm^2}$ (c) $100 \frac{kg}{cm^2}$ (d) $120 \frac{kg}{cm^2}$
12. The minimum compressive strength of 1st class brick should be
   (a) $75 \frac{kg}{cm^2}$ (b) $90 \frac{kg}{cm^2}$ (c) $100 \frac{kg}{cm^2}$ (d) $120 \frac{kg}{cm^2}$
13. Number of bricks required for 1 m^3 of brick masonry
   (a) 550 (b) 400 (c) 450 (d) 500
14. Addition of pozzolana to ordinary portland cement increases
   (a) Bleeding (b) Shrinkage
   (c) Permeability (d) Heat of hydration
15. Which of the following metamorphic rocks have the more weather resistance characteristics?
   (a) Marble (b) Quartzite (c) Slate (d) Limestone
16. Initial setting time for asbestos cement products should not be less than
   (a) 90 minutes (b) 60 minutes (c) 75 minutes (d) 30 minutes
17. As per IS-specifications, the final setting time for ordinary Portland cement is
   (a) 30 minutes (b) 1 hour (c) 6 hours (d) 10 hours
18. The Lime and Sand proportion in brick construction mortar has to be
   (a) 1 : 8 (b) 1 : 2 (c) 1 : 4 (d) 1 : 6
19. Excess of Silica in brick earth results in
   (a) Cracking and Warping of bricks (b) Loss of Cohesion
   (c) Enhancing the impermeability of bricks (d) None of these

20. Quick lime is

(i) Slow in setting (ii) Rapid in Stacking (iii) Good in strength

(a) Only (i) (b) Only (ii) (c) (i) and (ii) (d) (ii) and (iii)

21. Bulking of sand is caused due to

(a) Surface moisture (b) Air voids (c) Viscosity (d) Clay content

22. Early attainment of strength in Rapid Hardening Cement is mainly due to

(a) Gypsum (b) Fine grinding

(c) Tri-calcium silicate (d) Tri-calcium Aluminates

23. Gneiss is obtained from

(a) Igneous rocks (b) Metamorphic rocks

(c) Sedimentary rocks (d) Sedimentary metamorphic rocks

24. Gypsum consists of

(a) $H_2S + CO_2$ (b) $CaSO_4 + H_2O$ (c) Lime + H_2O (d) CO_2 + Ca

25. The basic test for samples of engineering materials is

(a) Tensile test (b) Compression test (c) Bending test (d) Hardness test

26. Eco-sand is rich in

(a) Silica (b) Alumina (c) Ferrous (d) Oxides

27. Quick lime is

(a) Calcium Carbonate (b) Calcium Oxide

(c) Calcium Hydroxide (d) None of these

28. The normal consistency of ordinary portland cement is approximately

(a) 10% (b) 20% (c) 30% (d) 40%

29. The ultimate tensile strength of structural mild steel is about

(a) $160 \frac{N}{mm^2}$ (b) $260 \frac{N}{mm^2}$ (c) $520 \frac{N}{mm^2}$ (d) $420 \frac{N}{mm^2}$

30. A good building stone should not absorb water more than

(a) 5% (b) 10% (c) 15% (d) 20%

31. Which of the following trees yields hard wood?

(a) Deodar (b) Chir (c) Shishum (d) Pine

32. In which direction the strength of timber is maximum

(a) Parallel to grains (b) At 45° to grains

(c) Perpendicular to grains (d) Same in all directions

33. The moisture content in a well seasoned timber is

(a) 4 to 6% (b) 10 to 12% (c) 15 to 20% (d) 100%

34. The age of a tree can be known by examining

(a) Cambium layer (b) Annular rings

(c) Medullary rays (d) Heart wood

35. Plywood is made by binding together thin layers of wood in such a way that the angle between grains of any layer to grains of adjacent layer is

(a) 0° (b) 30° (c) 45° (d) 90°

36. A I-class brick when immersed in cold water for 24 hours should not absorb water more than

(a) 15% (b) 20% (c) 22% (d) 25%

37. Crushing strength of a First Class brick should not be less than

(a) $3.5\frac{N}{mm^2}$ (b) $7.0\frac{N}{mm^2}$ (c) $10.5\frac{N}{mm^2}$ (d) $14.0\frac{N}{mm^2}$

38. Hydraulic lime is obtained by

(a) Burning of Lime Stone (b) Burning of Kankar

(c) Adding water to Quick lime (d) Calcination of pure clay

39. Paints with lead base are suitable for painting of

(a) Wood work (b) Iron work

(c) Both wood & Iron work (d) Any surface

40. The vehicle used in case of enamel paints is usually

(a) Linseed oil (b) Water (c) Varnish (d) Any of these

41. Distemper is used to coat

(a) External concrete surfaces

(b) Wood work

(c) Interior surfaces not exposed to weather

(d) Compound walls

42. Putty is made up of

(a) White lead and turpentine (b) Red lead + linseed oil

(c) Chalk powder + raw linseed oil (d) Zinc oxide + boiled linseed oil

43. The approximate ratio of strength of 15 cm × 30 cm concrete cylinder to that of 15 cm cube of same concrete is

(a) 1.25 (b) 1.00 (c) 0.85 (d) 0.50

44. The ratio of Young's modulus of high tensile steel to that of mild steel is about

(a) 0.5 (b) 1.0 (c) 1.5 (d) 2.0

45. Stainless steel resists corrosion due to

(a) Carbon (b) Sulphur (c) Vanadium (d) Chromium

46. Wrought iron is used for

(a) Structural works in beams (b) Columns & struts

(c) Small size water pipes (d) Large pipes

47. The normal period for curing of lime mortar is

(a) 1 day (b) 3 days (c) 7 days (d) 10 days

48. Glazing is used to make earthware

(a) Impervious (b) Porous (c) Hard (d) Soft

49. Pug mill is used for
    (a) Preparation of clay (b) Moulding of clay
    (c) Drying of bricks (d) Burning of bricks
50. Normal size of modular brick is (in mm)
    (a) $190 \times 90 \times 90$ (b) $190 \times 190 \times 90$ (c) $200 \times 100 \times 100$ (d) $200 \times 200 \times 100$
51. In a good brick clay and silt content must be atleast
    (a) 25% (b) 30% (c) 40% (d) 50%
52. For underwater construction of structures the type of lime used is
    (a) Hydraulic (b) Fat (c) Quick (d) Pure lime
53. Polyvinyl Chloride (PVC) is a material
    (a) Thermosetting (b) Thermoplastic (c) Elastoplastic (d) Rigid-plastic
54. Before testing setting time of cement following test should be done
    (a) Soundness (b) Strength (c) Fineness (d) Consistency
55. Surkhi is added to lime mortar to
    (a) Prevent shrinkage (b) Decrease setting time
    (c) Increase bulk (d) Impart hydraulicity
56. The specific gravity of a good building stone should be greater than
    (a) 1.5 (b) 1.7 (c) 2.2 (d) 2.7
57. In arches, stratified stones are placed such that their planes are
    (a) Parallel (b) Perpendicular (c) Radial (d) Anyways
58. The tendency of a stone is to split along
    (a) Texture (b) Fracture (c) Cleavage (d) Structure
59. Which of these stone is best suited for construction of piers and abutments of a Railway bridge?
    (a) Granite (b) Sand stone (c) Lime stone (d) Quartize
60. What type of rock is marble?
    (a) Metamorphic (b) Argillaceous (c) Calcerious (d) Siliceous
61. As per Indian standards for bricks minimum acceptable compressive strength of any class of burnt clay bricks in dry state is
    (a) 10 MPa (b) 0.75 MPa (c) 5.0 MPa (d) 3.5 MPa
62. Bull's Trench kiln is used to manufacture
    (a) Lime (b) Cement (c) Bricks (d) Tiles
63. The Le Chatelier apparatus is used to determine
    (a) Compressive strength of cement (b) Fineness of cement
    (c) Setting time (d) Soundness
64. The most important purpose of 'frog' in a brick is to
    (a) Emboss manufacturer's name (b) Reduce brick's weight
    (c) Form keyed joint (d) Improve insulation

65. Bricks are burnt at a temperature range of
(a) 500° to 700° C (b) 700° to 900° C (c) 900° to 1200° C (d) 1200° to 1500° C

66. The 'frog' of the brick is kept at
(a) Bottom face (b) Top face (c) Shorter side (d) Longer side

67. In a brick masonry the bond produced by laying alternate headers and stretchers in each course is known as
(a) English bond (b) Double Flemish bond
(c) Zigzag bond (d) Double Flemish bond

68. Minimum thickness of wall where single Flemish bond can be used is
(a) Half brick thick (b) One brick thick
(c) One and a half brick thick (d) One brick thick

69. The type of bond provided in brick masonry for carrying heavy loads is
(a) Single Flemish (b) Double Flemish
(c) English (d) Zigzag

70. Number of vertical joints in a stretcher course is x times the number of joints in the header course, where x is
(a) $\frac{1}{2}$ (b) 1 (c) 2 (d) $\frac{1}{4}$

71. The process of mixing clay, water and other ingredients to make brick is known as
(a) Kneading (b) Moulding (c) Pugging (d) Wetting

72. The internal size of mould used in making brick compared to fully burnt brick is
(a) Equal (b) Smaller (c) Greater (d) Much bigger

73. The slump recommended for mass concrete is
(a) 25 to 50 mm (b) 50 to 100 mm (c) 100 to 125 mm (d) 125 to 150 mm

74. Soundness test of cement determines
(a) Quality of free lime (b) Ultimate strength
(c) Durability (d) Initial setting

75. The important constituent of an oil paint is
(a) Thinner (b) Vehicle (c) Pigment (d) All these

76. In stone masonry, stones (stratified rocks) are so placed that the direction of pressure to the plane of bedding is
(a) Right Angle (b) 45° (c) 60° (d) Parallel

77. Chemically, marble is a rock
(a) Metamorphic (b) Calcareous (c) Silliceous (d) Argillaceous

78. Geologically, marble is a type of rock
(a) Igneous (b) Metamorphic (c) Sedimentary (d) Stratified

79. Commonly used colour pigments in paints, is
(a) Ambers (b) Carbon black (c) Iron oxide (d) All these

80. Which of the following is the purest form of Iron?
(a) Cast iron (b) Wrought iron (c) Mild steel (d) High carbon steel

81. The base material for distemper is
(a) Lime putty (b) Lime (c) Chalk (d) Lime cement

82. The amount of water used for one kg of distemper is
(a) 0.2 liter (b) 0.4 liter (c) 0.6 liter (d) 0.8 liter

83. Excess of alumina in brick earth makes the brick
(a) Impermeable (b) Brittle and weak
(c) Losse cohesion (d) To crack and warp on drying

84. For preparing porcelains, the clay should be
(a) Pure (b) Tenacious (c) Plastic (d) All these

85. For testing compressive and tensile strength of cement, the mortar is made by mixing cement and standard sand in the ratio of
(a) 1 : 2 (b) 1 : 3 (c) 1 : 4 (d) 1 : 6

86. Bullet proof glass is made of thick glass sheet sandwiched by a layer of
(a) Steel (b) Stainless steel
(c) High test plastic (d) Chromium plate

87. The cracks caused by shrinkage of exterior surface of wood exposed to atmosphere, are called
(a) Radial shakes (b) Heart shakes (c) Twisted fibre (d) Wind cracks

88. Plaster of Paris is obtained by calcining
(a) Bauxite (b) Gypsum (c) Limestone (d) Any of these

89. For making sports good, timber preferred is
(a) Mulberry (b) Mahogany (c) Sal (d) Deodar

90. The quantity of Nickel in INVAR, is
(a) 12% (b) 24% (c) 30% (d) 36%

91. The ultimate tensile strength of structural mild steel is about. ($\frac{N}{mm^2}$)
(a) 160 (b) 260 (c) 420 (d) 520

92. Generally wooden moulds are made from
(a) Ply wood (b) Shishum (c) Deodar (d) Teak

93. Bitumen is generally obtained from
(a) Organic material (b) Synthetic material
(c) Petroleum product (d) Coal

94. The most commonly used base for timber painting, is
(a) Red lead (b) Zinc white (c) White lead (d) Titanium white

95. A badly mixed cement concrete results in
(a) Segregation (b) Honey combing (c) Bleeding (d) All these

96. To give brilliant finish, the type of varnish used, is

(a) Water varnish (b) Spirit varnish
(c) Turpentine varnish (d) Oil varnish

97. Compared to Mild Steel, cast Iron has

(i) High compressive strength (ii) High Tensile strength
(iii) Low compressive strength (iv) Low tensile strength

(a) (i) and (ii) (b) (ii) and (iii) (c) (iii) and (iv) (d) (i) and (iv)

98. The ingredient of the brick earth which enables the brick to retain its shape

(a) Alumina (b) Silica (c) Iron (d) Magnesia

99. Strength of concrete primarily depends on

(a) Quality of water (b) Quality of aggregate
(c) Quality of cement (d) Water-cement ratio

100. Seasoning of timber is essentially to remove

(a) Knots (b) Sap (c) Twisted fibre (d) Roughness

101. Brass is an alloy of

(a) Cu + Zn (b) Zn + Pb (c) Sn + Ag (d) Zn + Ni

102. Which of the following impurities in water is destructing

(a) Calcium Chloride (b) Lead Nitrate (c) Alkalies (d) Algae

TRUE OR FALSE QUESTIONS

1. Plywood has same tensile strength in all direction. (T or F)
2. The stretcher bond in brick masonry is used only when wall thickness is 280 mm. (T or F)
3. For massive concrete structures the most suitable cement is, Low Head Cement. (T or F)
4. Wooden moulds for bricks are made from Ply Wood. (T or F)
5. Pressure and stress have same units (T or F)
6. Strain $\in = \dfrac{\text{Change in length}}{\text{Original length}}$. (T or F)
7. Units of strain is m. (T or F)
8. According to Hookes Law, within elastic limits stress is directly proportional to strain. (T or F)
9. The total strain energy stored in the body is generally known as resilience. (T or F)
10. If a free body is heated it experiences thermal stress. (T or F)
11. Shear $\dfrac{\text{Stress}}{\text{Shear}}$ strain is called Rigidity modulus. (T or F)
12. Resilience and Proof Resilience are same thing. (T or F)
13. The higher the Young's modulus softer is the material. (T or F)

14. Young's Modulus has same units as pressure. (T or F)
15. Young's Modulus is defined as $\frac{\text{Stress}}{\text{Stain}}$. (T or F)
16. Poisson's Ratio is a dimensionless quantity. (T or F)
17. Rubber is more elastic than steel. (T or F)
18. The dimensional formula for Young's Modulus is $[MLT^{-2}]$. (T or F)
19. The strain right angle to the direction of applied load is called longitudinal strain. (T or F)
20. For most of the materials Poisson's Ratio lies between 0 and 0.5. (T or F)
21. Bending Moment is maximum when shear force changes its sign. (T or F)
22. When shear Force is zero, bending moment is also zero. (T or F)
23. In case of equal like principal stresses the diameter of Mohr's Circle is zero. (T or F)
24. The value of Young's Modulus (E) in simple bending is more in compression than tension. (T or F)
25. If E is Young's Modulus, k is Bulk Modulus and ν is Poisson's Ratio then $E = 3k(1 + 2\nu)$. (T or F)
26. Bulk Modulus $k = \frac{\text{Direct stress}}{\text{Volumetric strain}}$. (T or F)
27. A concentrated load on a beam acts only on supports. (T or F)
28. On a beam sagging and hogging Bending Moments are same. (T or F)
29. In a simply supported beam subjected to uniformly distributed load of w over the full length L, the maximum Bending Moment is $\frac{wL^2}{8}$. (T or F)
30. Principal plane has zero shear stress. (T or F)
31. For a solid circular shaft of Diameter D the Moment of Inertia is $\frac{\pi D^4}{32}$. (T or F)
32. Hookes Law is valid only within Elastic Limits. (T or F)
33. In a cantilever beam one end is fixed and other end is simply supported. (T or F)
34. When two equal forces acting in the opposite direction acting on a body, it causes torsion. (T or F)
35. Polar modulus of Torsional section modulus of a solid circular rod is $\frac{\pi}{16} d^3$, where d is diameter of rod. (T or F)
36. The shear stress induced in a shaft due to applied torque is zero at the circumference. (T or F)
37. Torque per unit twist of a shaft is called. Stiffness of shaft. (T or F)
38 Macaulay's method is used to find strength of beam. (T or F)
39. In a simply supported beam deflection is maximum where slope is also maximum. (T or F)
40. In a beam when Bending Moment changes its sign, the point is known as Point of Contraflexure. (T or F)
41. Factor of safety is the Ratio of ultimate stress to permissible stress. (T or F)
42. Change in temperature does not induce stress in any body. (T or F)
43. In the Neutral, layer of a beam subjected to simple bending, there is neither tension nor compression. (T or F)

44. Power transmitted by shaft rotating at (n) RPM and Torque T (N – m) is $P = \frac{2\pi\, nT}{60}$. (T or F)

45. Polar Modulus of a hollow shaft is $Z_p = \frac{\pi(D_0^2 - D_1^2)}{16 D_0}$, D_0 = Outside diameter, D_1 = Inside diameter. (T or F)

46. A spring when loaded stores energy. (T or F)

47. The solid length of a spring is the distance between top and bottom coil. (T or F)

48. Solid Length of a spring is given be L = n × d (n) is number of coils and (d) diameter of the wire. (T or F)

49. Stiffness of a spring is given by $\delta = W \times \delta$. (T or F)
W = Load and δ = deflection

50. In a simply supported beam with point load W at the centre, deflection Y_C is given by (T or F)

$$Y_C = -\frac{WL^2}{16\,EI}$$

51. In a simply supported beam carrying a point load (W) at the centre the slope at the support is given by (T or F)

$$\theta_C = \frac{WL^2}{48\,EI}$$

52. For maximum deflection in a beam $\frac{dy}{dx}$ must be zero. (T or F)

53. Struts are vertical elements of structure subjected to axial compressive load. (T or F)

54. A column is a vertical element of a structure subjected to compressive load and fixed at ends or hinged. (T or F)

55. When lime ($CaCO_3$) is burnt to redness in contact with air, quick lime is formed? (T or F)

56. Process of chemical combination of quick lime with water is known as Calcination. (T or F)

57. Curing of concrete is not required always. (T or F)

58. The ratio of weights of partially compacted concrete to fully compacted concrete is called Compaction Factor. (T or F)

59. Slump test is required to check the workability of concrete. (T or F)

60. For M-25 concrete the slump value should be between 25 mm to 60 mm. (T or F)

61. Hardened plain cement concrete is known as artificial or cast stone. (T or F)

62. Artificial stones are weak in strength. (T or F)

63. Shell lime is the purest form of lime. (T or F)

64. According to IS 712-1984 lime is qualified into 6 categories. (T or F)

65. Gypsum is added in manufacture of cement to reduce its setting time. (T or F)

66. Alkalies, oxides of K and Na add strength to cement. (T or F)

67. The strength of concrete depends upon water-cement ratio. (T or F)

68. Green concrete is environmentally friendly. (T or F)

69. Slump test is used to test the workability of concrete. (T or F)

70. Structural wood is called Timber. (T or F)
71. The proportion of lime and sand in mortar for brick construction is 1 : 4. (T or F)
72. Stainless steel resists corrosion due to carbon content. (T or F)
73. Asphalt is obtained from Bitumen distilation. (T or F)
74. Normal curing period for lime mortar is 3 days (T or F)
75. Glazing is used to make earthwares impervious. (T or F)
76. Pug mill is used for preparation of clay. (T or F)
77. Crushing strength of the first class brick should not be less than 14 $\frac{N}{mm^2}$. (T or F)
78. For construction of Structures under water Hydraulic Lime should be used. (T or F)
79. Compared to mild steel cast iron has low compressive strength. (T or F)
80. For good building stone, specific gravity should be more than 2.7. (T or F)
81. In arches stratified stones are placed such that their planes are radial. (T or F)
82. Chemically marble is Metamorphic rock. (T or F)
83. Geologically marble is Metamorphic rock. (T or F)
84. Fibre boards cannot be used for furniture. (T or F)
85. The process of mixing clay, water etc. to make brick is known as moulding. (T or F)
86. For massive structures like dams low heat cement should be used. (T or F)
87. The standard size of masonry bricks is 20 cm × 10 cm × 10 cm. (T or F)
88. Soft Steel is obtained by Bessemer Process. (T or F)
89. A Queen closure is a brick with half the breadth. (T or F)
90. Quick lime is used for white washing. (T or F)
91. Quick lime is calcium carbonate. (T or F)
92. Soundness of cement is determined by Le-Chatelier's device. (T or F)
93. According to IS 399-1963 the weight of timber is specified at 8% moisture content. (T or F)
94. Alumina helps brick to retain its shape. (T or F)
95. The main ingredient of Portland cement are Lime and Silica. (T or F)
96. The trunk of a tree after cutting all branches is called log. (T or F)
97. Minimum thickness of wall where Flemish Bond cab be used is two brick thick. (T or F)
98. Brick masonry meant for carrying heavy loads must use English Bond. (T or F)
99. For lining of furnaces Over-burnt bricks are suitable. (T or F)
100. Age of any tree can be known by examining Annular rings. (T or F)
101. The fire clay contains pure oxide of iron. (T or F)
102. Alcohol does not react with concrete. (T or F)

FILL IN THE BLANKS

1. Lime obtained from calcinations of pure lime stone is called .... lime.
2. Purest form of lime is .... lime.
3. Strength of concrete depends upon .... Ratio.
4. In mortar, the binding materials is ....
5. Initial setting of cement is caused due to Tri-calcium ....
6. Turpentine oil is used in paints as a ....
7. In small size water pipes .... Iron is used.
8. The timber having maximum resistance against white ants is obtained from ....
9. Lime concrete is generally used for flooring at .... level.
10. Gypsum consists of water and ....
11. Plastics are compounds of Oxygen, Nitrogen and ....
12. Number of bricks required for two cubic meter of brick masonry will be ....
13. For First-class bricks compressive strength should not be loss than .... $\frac{kg}{cm^2}$.
14 For Second-class bricks compressive strength should not be less than .... $\frac{kg}{cm^2}$.
15. Gneiss is obtained from sedimentary .... rocks.
16. For preparation of clay .... mill is used.
17. Clay and silt content in a good brick earth must not be less than .... %.
18. When struck Second-class bricks produce a .... sound.
19. The coefficient of hardness of stones used in road work should be greater than ....
20. Normal consistency of ordinary Portland cement is about .... %.
21. Rock preferred for roofing is .... type.
22. Paint which consists of powdered chalk, pigments and water is called ....
23. Angle between grains of two adjacent layers in plywood should be ....
24. Jhumb bricks are .... burnt.
25. Lacquer is spirit ....
26. Plastic asphalt is a mixture of cement and ....
27. Wooden moulds are made (preferably) of .... wood.
28. Plaster of Paris is obtained by calcination of ....
29. Invar contains about .... % nickel.
30. Seasoning of timber is done for decreasing .... content.

31. The initial setting time of lime-pozzolana, is .... hours.
32. Alumina in brick earth provides mainly ....
33. Crushing strength of a good building stone should be more than .... M Fa.
34. The amount of water used for one kg of distemper should be .... Litre.
35. Expansion joints in masonry walls are provided in walls of length more than .... m.
36. Mortar joint in masonry which is normal to the face of wall is known as .... joint.
37. The number of steps in a flight normally should not be less than ....
38. The maximum number of steps in a flight should generally be not more than ....
39. Minimum width of landing should be equal to .... of stairs.
40. Half-turn stairs are also called .... stairs.
41. The outer projection of tread is known as ....
42. The window provided on the sloping side of a pitched roof is called .... window.
43. The opening left in a flat roof for lighting purpose is called ....
44. Putty is made up of Linseed oil and ....
45. The compacting factor test for cement concrete determines its ....
46. The approximate ratio between the 7 days and 28 days strength is about ....
47. For complete hydration of cement the water-cement ratio should be more than 0.35 but less than ....
48. Seasoning of timber is done to remove .... from the timber.
49. To prepare 1.0 m^3 of 1 : 2 : 4 concrete (By volume), the coarse aggregate required will be ....
50. The most commonly used retarder in cement is ....
51. Marble contains mostly Lime and ....
52. A sample of cement is said to be sound when it does not contain free ....
53. Loss on Ignition in Portland cement shall not be more than .... %.
54. Portland Cement is manufactured by burning in a kiln limestone and ....
55. By calcining gypsum is obtained ....
56. Initial setting time of concrete ceases at temperature .... °C.
57. In case of concrete transported by pumping, the slump should be between 75 mm to .... mm.
58. If slump of concrete mix is 75 mm, its workability is considered to be ....
59. When cement paste rises to the surface of concrete it is called .... of concrete?
60. The tensile strength of concrete is about .... % of its compressive strength.
61. As per Indian standards concrete is designated into .... grades.
62. For determination of compressive strength of concrete minimum number of cubes required in a sample is ....

63. The Unit weight of RCC is generally taken as .... $\frac{KN}{m^3}$.
64. Light weight aggregate is produced by sintering ....
65. Light-weight concrete is used for .... walls.
66. Efflorescence is caused due to an excess of ....
67. .... in brick earth enables brick to retain its shape.
68. The position of brick cut to form angles other than 90° in plan is known as .... closer.
69. Magnesite bricks are suitable for use in ....
70. Sir Abel's process makes timber more .... resistant.
71. On application of external stress on timbers it behave like an .... material.
72. The timber's strength is maximum when load is applied .... to grain.
73. Slate is formed by metamorphic action on ....
74. The tool jumper is used for .... of stones.
75. The soft wood is obtained from Chir and .... trees.
76. The tensile strength of plywood is .... M Pa.
77. The percentage of carbon content in mild steel is less than ....
78. As compared to stretcher course, the thickness of joints in header course should be ....
79. To transmit heavy load through steel columns the type of foundation is ....
80. Total settlement for isolated foundations on clayey soils should be limited to .... mm.
81. The minimum depth of foundation in clayey soil is .... m.
82. The maximum total settlement for Raft foundation on clayey soils should be limited to .... mm.
83. The slenderness ratio for masonry walls should not be more than ....
84. The expansion and shrinkage of plywoods are very low because plies are placed at angles .... to each other.
85. King closures are related to .... masonry.
86. The optimum number of revolutions over which concrete is required to be mixed in a concrete mixer is ....
87. The compound used for initial setting and early gain of strength of Portland cement is ....
88. Low heat cement contains lower percentage of ....
89. A gauged mortar is obtained by adding sand and .... to cement.
90. Grading of aggregate in concrete mix results in reduction in ....
91. The Portland cement has maximum amount of ....
92. A good Portland cement should not have specific surface (cm^2/g) less than ....
93. Moorum is decomposed ....

94. Granite is mainly composed of felspar, mica and ....

95. The compressive strength of felspar is .... than Quartz.

96. The crushing strength of stone depends upon its texture as well as its ....

97. The compressive strength of perforated bricks should not be less than ($\frac{MN}{m^2}$) ....

98. The dolomite bricks are basic .... bricks.

99. The clinker is formed at a temperature of .... °C.

100. For initial setting time, final setting time and Normal consistency of cement .... apparatus is used.

101. The shrinkage of concrete .... its bond strength.

102. As per Indian standard the ideal temperature for curing is .... °C.

103. The liquid medium used in Enamel paints is ....

104. Commonly used solvent in oil paints is either Naptha or ....

105. For cement paints commonly used cement variety is .... cement.

106. The best primer used for structural steel work is .... lead.

107. Bronze paint has high .... property.

108. Asbestos paint is highly resistant to ....

ANSWERS

Multiple Choice Questions

1. c	2. d	3. d	4. a	5. b	6. b	7. a	8. b	9. b
10. c	11. a	12. c	13. d	14. b	15. b	16. a	17. d	18. b
19. b	20. c	21. a	22. b	23. d	24. b	25. a	26. a	27. b
28. c	29. d	30. a	31. c	32. a	33. b	34. b	35. d	36. b
37. c	38. b	39. a	40. c	41. c	42. c	43. c	44. b	45. d
46. c	47. c	48. a	49. a	50. a	51. d	52. a	53. b	54. d
55. d	56. d	57. c	58. c	59. a	60. c	61. d	62. c	63. d
64. c	65. c	66. b	67. b	68. c	69. c	70. a	71. a	72. c
73. a	74. a	75. d	76. a	77. b	78. b	79. d	80. b	81. c
82. c	83. d	84. d	85. b	86. c	87. d	88. b	89. a	90. d
91. c	92. b	93. c	94. c	95. b	96. b	97. d	98. b	99. d
100. b	101. a	102. b						

True & False

1. T	2. F 90 mm	3. T	4. F Sheeshan
5. T	6. T	7. F unitless	8. T
9. T	10. F Not in Free bodies		11. T
12. F	13. F Stiffer	14. T	15. T
16. T	17. F	18. F $M L^{-1} T^{-2}$	19. F Lateral
20. T	21. T	22. F Maximum or Minimum	
23. T	24. F	25. T	26. T
27. F	28. F	29. T	30. T
31. T	32. T	33. F Free	34. T
35. T	36. F	37. T	38. F Deflection
39. F Zero	40. T	41. T	42. F Not in Fixed body
43. T	44. T	45. F $\frac{\pi}{16 D_0}(D_0^4 - D_1^4)$	
46. T	47. F Full compressed length		48. T
49. F $\frac{w}{\delta}$	50. F $-\frac{wL^2}{48 EI}$	51. F $-\frac{wL^2}{16 EI}$	52. T
53. F Inclined	54. T	55. T Calcination	56. F Slaking
57. F	58. T	59. T	60. T

61. T
62. F
63. T
64. T
65. F Increase
66. F
67. T
68. F
69. T
70. T
71. F 1 : 2
72. F Chromium
73. F Petroleum
74. F 7 days
75. T
76. T
77. F 10.5 $\frac{N}{mm^2}$
78. T
79. F High
80. T
81. T
82. F Calcareous
83. T
84. F
85. F Kneading
86. T
87. T
88. T
89. T
90. T
91. F Calcium oxide
92. T
93. F 12%
94. F Cilica
95. T
96. T
97. F $1\frac{1}{2}$ Brick
98. T
99. F Refractory
100. T
101. F Hydrated Aluminum silicate
102. T

Fill in the Blanks

1. Quick
2. Shell
3. Water-cement
4. Cement
5. Aluminate
6. Thinner
7. Wrought
8. Sal
9. Ground
10. Calcium Sulphate $CaSO_4$
11. Hydrogen
12. 1000
13. $\frac{100 \text{ kg}}{cm^2}$
14. $\frac{75 \text{ kg}}{cm^2}$
15. Metamorphic
16. Pug
17. 50%
18. Metallic
19. 17 (Seventeen)
20. 30%
21. Slate
22. Distemper
23. 90°
24. Over
25. Varnish
26. Asphalt
27. Shisham
28. Gypsum
29. 36%
30. Moisture
31. 2 (Two)
32. Plasticity
33. 100 M Pa
34. 0.6 Litre
35. 40 m
36. Cross
37. 3 (Three)
38. 12 (Twelve)
39. Width
40. Dog Legged
41. Nosing
42. Dormer
43. Lantern
44. Chalk
45. Workability
46. 2 : 3
47. 0.45
48. Sap
49. 0.85 m^2
50. Gypsum
51. Silica
52. Lime
53. 4%
54. Clay
55. Plaster of Paris
56. 0° C
57. 100 mm
58. Medium
59. Bleeding
60. 10%
61. Seven
62. 3 (Three)
63. $\frac{25 \text{ KN}}{m^3}$
64. Fly ash
65. Partition
66. Alkalies
67. Silica
68. Squint
69. Furnaces
70. Fire
71. Elastic
72. Parallel
73. Shale
74. Quarrying
75. Deodar

76. Ply **77.** 0.25% **78.** Less **79.** Grillage

80. 65 mm **81.** 0.9 m **82.** 100 mm **83.** 20 (Twenty)

84. 90° **85.** Brick **86.** 20 (Twenty) **87.** Tricalcium (C_3S) Silicate

88. Tricalcium Aluminate (C_3A) **89.** Lime **90.** Voids

91. Lime (CaO) **92.** 2250 cm^2/g **93.** Laterite **94.** Quartz

95. Less **96.** Specific Gravity **97.** $\frac{75\ MN}{m^2}$ **98.** Refractory

99. 1500° C **100.** Vicat's **101.** Increases **102.** 27° C

103. Varnish **104.** Turpentine **105.** White **106.** Red

107. Reflective **108.** Fire

SECTION 13

Open Channel Flow

DEFINITIONS

Afflux: It is the net differential water-surface elevation between upstream and downstream sections of a construction to the flow.

Alluvium: It refers to the loose, non-cohesive material (such as sand and silt) transported by suspended in or deposited by water.

Alternate Depths: The two possible depths having the same specific energy are known as alternate depths.

Aspect Ratio: It is defined as the ratio of width of the channel to the depth of flowing fluid. It plays an important role in the formation of secondary currents.

Backwater Curve: In Gradually Varied Flow (GVF) the surface curves which have positive slopes $\left(\frac{dy}{dx} > 0\right)$ are called Backwater Curves.

Bed Forms: When the shear stress on the bed of an alluvial channel due to flow of water is more than critical shear stress (τ_c), the bed becomes dynamic causing deformation of bed and creating various shapes. These shapes are called bed forms (like ripples, dunes etc.)

Bed Load: All the sediment that is transported in a thin layer of the order of two grain diameter next to the bed is classified as bed load.

Best Section (Hydraulically Efficient Section): With slope, roughness coefficient and area of flow fixed, a minimum wetted perimeter section will represent best section. For most common trapezoidal channel, one half of a regular hexagon is the best section.

Celerity or Surge Front Speed: Is the speed at which the surge (or wave) is moving with respect to fluid medium. This means that if there is a surge in the fluid medium it will move at a different speed, the relative speed is known as celerity.

Chezy's Coefficient (C): In Chezy's Equation ($U = C\sqrt{RS_o}$), C is Chezy's coefficient which depends upon the nature of the surface and flow parameters. It has units of $\sqrt{g}$ or $[L^{1/2} T^{-1}]$.

Compound Channel: It is a channel section composed of a main deep portion and one or two flood plains that carry high-water flows. It is also known as a two-stage channel.

Critical Depth: The condition of minimum specific energy is known as critical-flow condition and corresponding depth (y_c) is known as critical depth.

Critical Slope: The slope of a channel which carries a given discharge as a uniform flow at the critical depth is called critical slope (s_c).

Darcy-Weisbach Friction Factor (f): It represents the loss of pressure of a fluid in a pipe due to the interaction between the fluid and pipe boundary. For pipe flows the Darcy-Weisbach equation giving head loss in pipe length L is, $h_f = f\frac{L\,V^2}{2gD}$, where f is friction factor. Friction factor has been evaluated by number of investigators for various flow conditions.

End Contraction: Due to the curvature of streamlines there is a decrease in the width of nappe. The approximate value of end contraction is 0.1 H, where. H is the height over the crest.

Energy Loss (E_L): When supercritical flow depth (y_1) changes into subcritical flow depth (y_2) causing a hydraulic jump, there is much turbulence causing considerable energy loss. It is estimated as EL = $\frac{(y_2 - y_1)^3}{4\,y_1\,y_2}\frac{Nm}{N}$.

Energy Slope: Unlike Uniform Flow, in Non-uniform Flow, such as Gradually Varied Flow the bed slope (S_o), water surface slope (S_w) and energy slope (S_f) will all differ from each other. Considering Manning's Equation applied to G V F, $S_f = \frac{n^2\,V^2}{R^{3/4}}$. The differential equation for energy slope will be $\frac{dH}{dx} = \frac{dZ}{dx} + \frac{dY}{dx} + \frac{d}{dx}\left(\frac{V^2}{2g}\right)$. $\frac{dZ}{dx}$ is bed slope, $\frac{dY}{dx}$ is water surface slope.

Fall Velocity: It is defined by the equation giving equilibrium between gravity force, resistance to the fall of the particle and buoyant force. For spherical particles and Reynold's Number less than 1, fall velocity (w) is given by formula $w = \frac{1}{18\,\mu}\,d^2\,(\gamma_s - \gamma w)$, known as Stokes Law. For higher Reynold's Number $w = \sqrt{gd\,\Delta}$, where $\Delta = \frac{\rho_s - \rho_w}{\rho_w}$.

Ganguillet and Kutter's Formula: For Chezy's 'C' some investigator developed formulae, one of these used in India is Ganguillet and Kutter's formula as follows, $C = \dfrac{23 + \frac{1}{n} + \frac{0.00155}{S_o}}{1 + \left[23 + \frac{0.00155}{S_o}\right]\frac{n}{\sqrt{R}}}$; where (n) is same as Manning's n.

Gradually Varied Flow (GVF): A steady non-uniform flow in a prismatic channel with gradual changes in its water surface elevation is termed as Gradually Varied Flow. Regions of high curvature are excluded in the analysis of this flow.

Gradually Varied Flow Profiles: In gradually varied flow the surface profiles are either backwater curve $\left(\frac{dy}{dx}\text{ is} + \text{ve}\right)$ of drawdown curve $\left(\frac{dy}{dx}\text{ is negative}\right)$.

Hydraulic Grade Line: It is defined as the line which gives the sum of pressure head $\left(\frac{p}{\gamma}\right)$, and datum head. HGL $= \frac{p}{\gamma} + Z$.

Hydraulic Jump: It is the rapid transition from supercritical to sub-critical flow. It is a highly turbulent process resulting in much energy dissipation and air entrainment.

Hydraulic Radius: In an open channel having cross-sectional area A and Wetted Perimeter P, hydraulic radius is defined as $R = \frac{A}{P}$. It plays important role in developing flow equations. Velocity in open channels is directly proportional to $\sqrt{R}$.

Hydraulically Efficient Channel Section: Same as Best Section.

Hydraulically Smooth Surface: When the average height of irregularities (k) is much lesser than the thickness of the laminar sub-layer (δ), then the boundary is called hydradynamically smooth $\frac{k}{\delta} < 0.25$.

Hydrodynamically Rough Surface: When average height of irregularities (k) is much greater than laminar sub-layer (δ) then boundary is designated as rough surface $\frac{k}{\delta} < 6$. (Note : between 0.25 and 6 it is boundary transition condition).

Incipient Motion (Critical Motion): The condition of flow at which the bed particles will just begin to move is known as condition of incipient motion. The bed shear corresponding to incipient motion is known as critical shear stress or critical tractive force.

Isovels: These are the lines (contours) linking points of equal velocity, and as such indicate velocity variation within a given section of the flow field.

Kennedy's Equation: Kennedy proposed a relation between the Velocity (V) and depth (y_0) of a stable channel as, $V = 0.56\, m\, y_o^{0.64}$. In designing stable channel this equation is not in use now.

Kinetic Energy Correction Factor (α): It is defined as the ratio of the Kinetic Energy of flow based on actual velocity across a section to kinetic energy of flow based on average velocity across the same section. $\alpha = 2$ for Laminar Flow, $\alpha = 1.02$ to 1.05 for turbulent flow and $\alpha = 1$ for uniform flow.

Lacey's Regime Channel: Lacey defined a regime channel as one which carries a constant discharge under uniform flow in an unlimited incoherent alluvium having the same characteristics as that transported without changing the bottom slope, shape or size of cross section over a period of time.

Length of the Jump: It is the horizontal distance between the toe of the jump to a section where water level attains maximum depth. Length L_j is taken as 6.1 y_2.

Manning's Formula (Equation): It is a resistance formula for uniform flow in open channels, written as $V = \frac{1}{n} R^{2/3} S^{1/2}$ where *n* is roughness coefficient or Manning's *n*. It can be calculated by Strickler's Formula, $n = \frac{d_{50}^{1/6}}{21.1}$ or $\frac{d_{90}^{1/6}}{26}$, the value of *n* varies between 0.011 for smooth channels to 0.125 for very rough surfaces.

Maximum Velocity: In open channels maximum velocity occurs just below the free surface, at about 0.8 to 0.95 depth from bottom.

Mild Slope: If the normal depth is more than the critical depth, that is the flow is subcritical and slope $S_o < S_c$ is called Mild Slope.

Moody Chart or Moody Diagram: (also Stanton diagram) is a graph in non-dimensional form that relates the Darcy-Weisbach friction factor (f), Reynold's Number (R_e) and surface roughness ($\in$) for fully developed flow in a circular pipe.

Most Efficient Open Channel: The semi circular section has the lowest wetted perimeter, and is, therefore, the section of the highest efficiency. Due to practical limitation in construction and maintenance of these, mostly trapezoidal section is adopted.

Moving Hydraulic Jump: The surges both moving upstream or downstream are often termed as moving hydraulic jumps in view of their similarity to a steady hydraulic jump.

Musknigham Method: It is a popular method of channel routing. It is simple to use and gives reasonably good results in many situations.

Narrow Channel: The channels which are not wide enough to have two-dimensional flow, the resistance of side walls become significant. Channels with $\frac{B}{y_o} < 5$ are classified as narrow channels.

Non-Erodible Channels: Lined channels are non-erodible also called Rigid Boundary Channels. The lining material can be concrete, stone masonry, steel, iron, timber or plastic.

Normal Depth y_o or y_n): When the water surface is parallel to the channel bottom flow is said to be uniform and the water surface is at normal depth.

Ogee Spillway: It represents the shape of the downstream face of the spillway. It conforms to the profile of lower nappe coming out from a well-ventilated sharp crested rectangular weir.

Oscillating Jump: This type of Jump is chracterised by an instability of the high velocity flow in the jump which oscillates in a random manner between the bed and surface. It occurs at Froude Number $2.5 < F_i < 4.5$.

Parshall Flume: It is a fixed hydraulic structure in an open channel that is designed to measure water discharge in the open channel or non-full pipe. It has three main parts, the contraction, throat and expansion part.

Permissible Velocity: While high velocities are permitted in lined canals, in irrigation channels in alluvium high velocities cannot be permitted for fear of erosion of bed. Also very low velocities may cause deposition of sediment and choking of channel. These velocities allowed are known as non-silting, non-scouring velocities.

Prismatic Channel: A channel in which the cross-sectional shape, size and bottom slope are all constant is termed as prismatic channel.

Rapidly Varied Flow: If the curvature in a varied flow is large and depth of flow changes appreciably over short lengths, such a phenomenon is termed as Rapidly Varied Flow (RVF). Examples are flow over spillways and hydraulic jump.

Regime Channel: For alluvial channels Lacey (1930) gave the concept of Regime Channel as one which carries a constant discharge under uniform flow in an unlimited incoherent alluvium having the same characteristics as that transported without changing the bottom slope, shape or size of cross-section over a period of time.

Section Factor: It is one of the geometrical property of channel section. For uniform flow computation it is defined as, $Z = A\,R^{2/3}$ and for critical flow, $Z = \left(\frac{A^3}{T}\right)^{1/2}$.

Sequent Depth: The depths in a hydraulic jump having same Specific Force $\left(\frac{Q^2}{gA} + A\bar{Z}\right)$ are known as sequent depths or conjugate depths. Depths at same specific energy $\left[y + \frac{V^2}{2g}\right]$ are known as alternate depths.

Stilling Basin: Hydraulic Jump is accompanied by lot of energy loss. The downstream portion of the hydraulic structure where the energy dissipation is deliberately allowed to occur so that outgoing stream flows safely to the channel below is known as stilling basin.

Suspended Load: It consists of finer particles which remain in suspension due to turbulence. Significant suspended load occurs if turbulent velocity fluctuations are larger than settling velocity.

Total Load: The sum of the suspended load and bed load is total load. As total load material comes from the boundary of channel, it is sometimes called as "total bed material load".

Transitions: These are short length canal structures that facilitate the needed change in cross-section geometry of main canal. A typical example is fluming of a trapezoidal canal into a rectangular cross section in cross drainage works like aqueduct.

Undular Jump: In this water surface has very small ripples. The sequent depth difference is very small and $1.0 < Fr \leq 1.7$.

Varied Flow: A flow in which the flow properties vary along the channel is termed as varied or non-uniform flow.

Wash Load: It is the finer Fraction of sediment load, that has not originated in the channel bed, and is not likely to deposit.

Wide Rectangular Channel: In a wide rectangular channel $B \gg y$, the hydraulic Radius $R = \frac{A}{B + 2y}$, $B + 2y \approx B$, hence $R = \frac{By}{B}$ or Hydraulic Radius can be assumed to be equal to depth of flow. In such channel there is increase in normal depth as well as discharge.

MULTIPLE CHOICE QUESTIONS

1. In an open channel the driving force is
   (a) Pressure difference
   (b) Kinetic Energy difference
   (c) Gravitational force
   (d) All these

2. For uniform flow the main equation is Chezy's equation $V = C\sqrt{RS_O}$ the units of C are of
   (a) $\sqrt{R}$
   (b) $\sqrt{S_o}$
   (c) V
   (d) $\sqrt{g}$

3. In Manning's Formula, $V = \frac{1}{N} R^{2/3} S_0^{1/2}$ N is related to Chezy's C as
   (a) $C = \frac{R}{n}$
   (b) $\frac{R^{1/6}}{n}$
   (c) $\frac{\sqrt{R}}{n}$
   (d) $\frac{R^2}{n}$

4. The formula, $C = \dfrac{23 + \frac{1}{n} + \frac{0.00155}{S_o}}{1 + \left[23 + \frac{0.00155}{S_o}\right] \frac{n}{\sqrt{R}}}$ is

(a) Kutter's Formula
(b) Bazin's Formula
(c) Strickler's Formula
(d) Kavman's Formula

5. For Kutter's N for Soil CWPC has recommended N value,
(a) 0.02 to 0.03
(b) 0.2 to 0.3
(c) 0.035 to 0.05
(d) 0.05 to 0.08

6. In an open channel the maximum velocity occurs at
(a) Surface
(b) 0.6 depth
(c) 0.8 depth
(d) 0.2 depth

7. A trapezoidal channel of Base width 26 m and depth 1.50 m, has side slopes 1 : 1, the wetted perimeter is about
(a) 39 m
(b) 28 m
(c) 41 m
(d) 30 m

8. The relation between average velocity and surface velocity is $V_a = k\,V_s$ where k is
(a) 0.6 to 0.8
(b) 0.8 to 0.95
(c) 0.7 to 0.9
(d) 1.0

9. Steady flow in open channel will exist when
(a) Flow is uniform
(b) Depth does not change with time
(c) Channel is frictionless
(d) Channel bed is straight

10. One-dimensional method of flow analysis means
(a) Uniform flow
(b) Steady Uniform flow
(c) No variation of parameters in transverse direction
(d) Neglecting variation in longitudinal direction

11. For an-open channel flow to take place between two section
(a) The channel bed must always slope in the direction of flow
(b) The $\frac{u}{s}$ depth must be more than $\frac{d}{s}$ depth
(c) The $\frac{u}{s}$ momentum must be more than $\frac{d}{s}$ momentum
(d) The total energy at u/s end must be larger than the total energy at the d/s section

12. The difference between total head line and piezometric head line represents the
(a) Velocity head
(b) Pressure head
(c) Elevation of channel bed
(d) Depth of flow

13. The momentum equation in x-direction $\Sigma F_x = \rho Q\,(V_{x2} - V_{x1})$ makes the assumption that the flow is
(a) Steady
(b) Uniform
(c) Unsteady
(d) Frictionless

14. The term 'alternate depth' in open channel flow is used to denote the depths having the same
(a) Kinetic energy for a given discharge
(b) Specific force for a given discharge
(c) Specific energy for a given discharge
(d) Total energy for a given discharge

15. In a rectangular channel the alternate depths are 1.0 m and 2.0 m respectively, the specific energy head in (m) is
(a) 3.38
(b) 1.33
(c) 2.33
(d) 3.0

16. At critical depth, the discharge is

(a) Minimum for a given specific energy
(b) Maximum for a given specific force
(c) Maximum for a given specific energy
(d) Minimum for a given specific force

17. Given F_o = Froude Number of flow in an open channel of rectangular cross-section and y_o = Depth of flow and y_c = Critical depth, then $\frac{y_c}{y_0}$ will be

(a) $\sqrt{F_0}$
(b) $F_o^{2/3}$
(c) $F_o^{1/3}$
(d) $F_o^{3/2}$

18. The Froude Number of a flow in a rectangular channel is 0.73. If the depth of flow is 1.50 m, specific energy in (m) is

(a) 1.90
(b) 1.50
(c) 1.73
(d) 0.73

19. The flow in a rectangular channel is sub-critical. If the width is expanded at a certain section, the water surface

(a) At a downstream section will drop
(b) At a downstream section will rise
(c) At a upstream section will rise
(d) At a upstream section will drop

20. In a triangular channel the vertex angle is 60° has a critical depth of 0.25 m , in the discharge in litre per second is

(a) 60
(b) 640
(c) 160
(d) 40

21. For the value of Manning's N, Strickler's formula is

(a) N = cc
(b) $N = \frac{(d_{50})^{1/6}}{21.1}$
(c) $\frac{d_{90}}{26}$
(d) $\frac{d_{50}}{21.1}$

22. The critical depth in a rectangular channel of width B. is given as

(a) $\left[\frac{Q}{Bg}\right]^{1/3}$
(b) $\left[\frac{Q}{2g}\right]^{1/3}$
(c) $\left[\frac{Q^2}{Bg}\right]^{1/3}$
(d) $\left[\frac{Q^2}{B^2 g}\right]^{1/3}$

23. For critical flow in open channels. Froude Number is

(a) < 1
(b) ≥ 1
(c) ≤ 1
(d) $= 1$

24. Critical depth in an open channel can be produced by

(a) Raising bed
(b) Lowering bed
(c) Decreasing width
(d) Increasing width

25. In a non-prismatic channel

(a) Unsteady flow is not possible
(b) Uniform flow is not possible
(c) Flow is always uniform
(d) Flow is not possible

26. Uniform flow is not possible, if the

(a) Friction is large
(b) Fluid is viscous
(c) So ≤ 0
(d) $S_o > 0$

27. The dimensions of Manning's rugosity coefficient N are

(a) $L^{1/6}$
(b) $L^{1/2} T^{-1}$
(c) $L^{-1/3} T$
(d) $L^{-1/3} T^{-1}$

28. In uniform flow there is balance between

(a) Gravity and Frictional forces
(b) Gravity and Inertial forces
(c) Inertial and Frictional forces
(d) Inertial and Viscous forces

29. The following two channels have same bed slope and area of cross-section and made of same material, the ratio of discharge will be

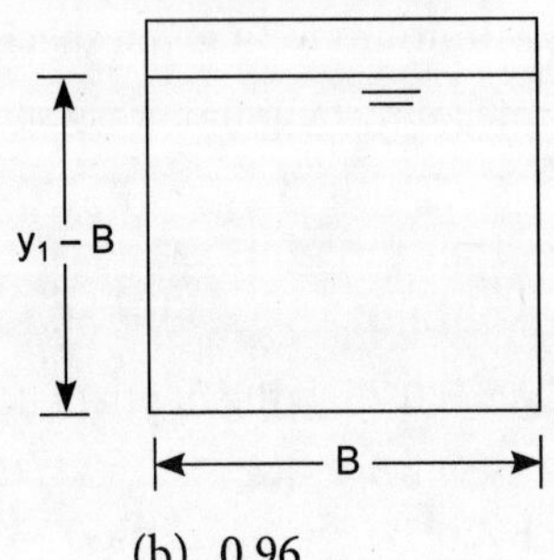

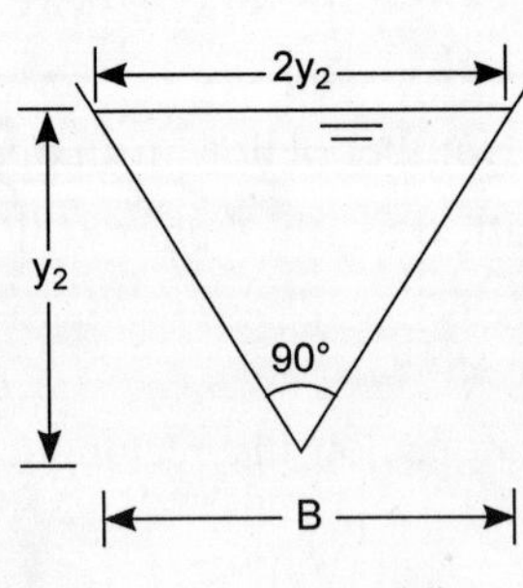

(a) 1.0 (b) 0.96 (c) 0.85 (d) 1.1

30. In a circular channel of diameter D, the maximum discharge will occur when depth (y_0) is equal to

(a) D (b) $\frac{D}{2}$ (c) 0.83 D (d) 0.938 D

31. A triangular section is hydraulically efficient when vertex angle (θ) is

(a) 90° (b) 120° (c) 60° (d) 30°

32. Shear velocity (u^*) is given by

(a) $\sqrt{g\, y_o\, s_o}$ (b) $\sqrt{g\, R\, S_o}$ (c) $g\, R\, S_o$ (d) $g\, y_o\, s_o$

33. For most efficient rectangular channel the condition is

(a) $R = y$ (b) $R = \frac{y}{3}$ (c) $R = \frac{y}{4}$ (d) $R = \frac{y}{2}$

34. Considering Manning's formula $Q = \frac{A}{n} R^{2/3} S_0^{1/2}$, the conveyance K is defined as

(a) $\frac{1}{n} R^{2/3}$ (b) $A\, R^{2/3}$ (c) $\frac{n}{A} R^{2/3}$ (d) $\frac{A}{n} R^{2/3}$

35. In Gradually varied flow as $y \to yc$, $\frac{dy}{dx}$ will tend to

(a) 0 (b) ∞ (c) S_o (d) A finite value

36. The total number of possible types of GVF profiles are

(a) 9 (b) 11 (c) 12 (d) 15

37. When a mild slope channel gets into a milder slope the GVF curve is of type

(a) M_1 (b) M_2 (c) M_3 (d) Any of these

38. A hydraulic jump occurs when there is a break in grade from a

(a) Mild to steep slope (b) Seep to mild slope
(c) Steep to steeper slope (d) Mild to Milder slope

39. In a hydraulic jump the initial and final depths are 0.3 m and 1.50 m, the energy (head) loss will be

(a) 1.92 m (b) 1.50 m (c) 0.96 m (d) 1.2 m

40. The hydraulic jump always occurs from

(a) M_2 to M_1 curve (b) H_3 to H_1 curve
(c) Below normal to above normal depth (d) Below critical to above critical depth

41. In the following trapezoidal channel, if the discharge is 8.0 $\frac{m^3}{s}$ the Froude Number will be

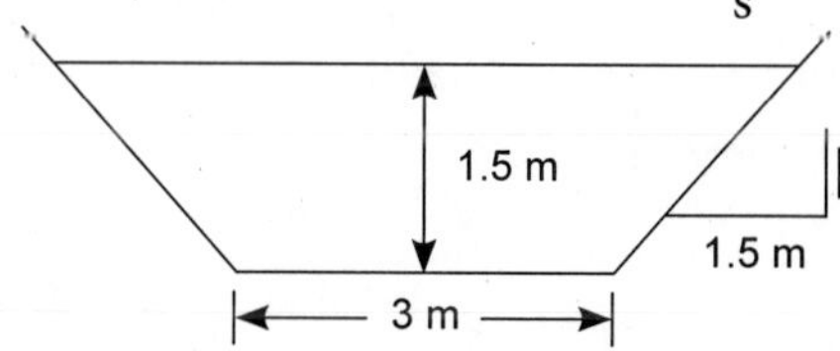

(a) 0.316 (b) 0.265 (c) 0.132 (d) 0.06

42. The critical depth at a section of a rectangular channel is 1.5 m. The specific energy at that section will be

(a) 0.75 m (b) 1.0 m (c) 1.5 m (d) 2.25 m

43. A discharge in a triangular notch and a rectangular suppressed weir both having the same head and C_d are identical when the ratio of water surface width in the V-notch to the length of rectangular weir is

(a) 1.00 (b) 1.5 (c) 1.88 (d) 2.5

44. The most efficient channel of Trapezoidal section as shown here is when

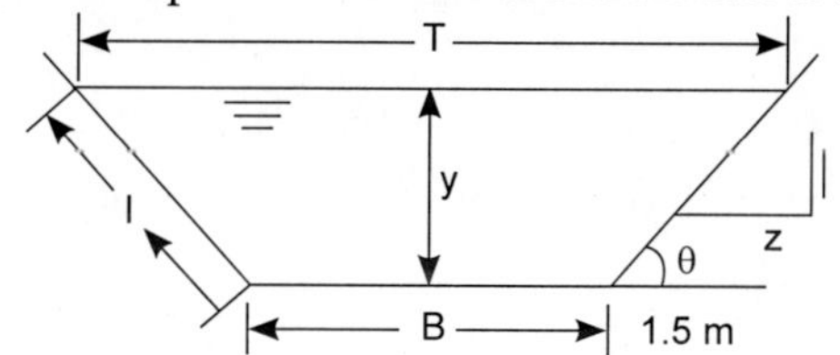

(a) B = 2y (b) T = 2l (c) T = l (d) θ = 30°

45. In any open channel at critical depth, specific energy is

(a) Maximum (b) Any value (c) Minimum (d) Half of total energy

46. For critical depth in any channel, Froude Number $\frac{Q^2T}{gA^3}$ should be

(a) = 1 (b) > 1 (c) < 1 (d) 1.5

47. In a hydraulic jump at the initial and final depths

(a) Energy should be same
(b) Specific energy should be same
(c) Momentum should be same
(d) Specific force should be same

48. The energy loss in hydraulic jump is given by

(a) $\frac{y_2 - y_1}{4\,y_1\,y_2}$ (b) $\frac{(y_2 - y_1)^3}{4\,y_1\,y_2}$ (c) $\frac{(y_2 - y_1)^2}{4\,y_1\,y_2}$ (d) $\frac{y_2 - y_1}{y_1\,y_2}$

49. In the given channel if $S_0 = \frac{1}{100}$ Manning's N = 0.016. The discharge will be

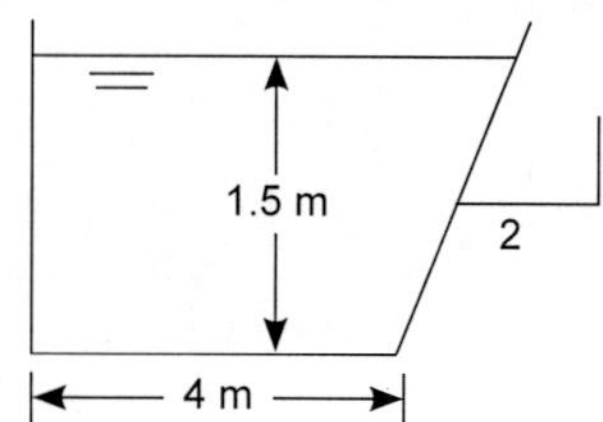

(a) $20.5 \frac{m^3}{s}$ (b) $15.55 \frac{m^3}{s}$ (c) $18.55 \frac{m^3}{s}$ (d) $10.3 \frac{m^3}{s}$

50. If the energy loss in a jump formed in a rectangular channel is 5 N – m/N when Froude Number $F_1 = 5$, the conjugate depths will be

(a) 0.755 m and 5.0 m (b) 0.397 m and 6.0 m

(c) 0.665 m and 5.5 m (d) 0.555 m and 6.2 m

51. The discharge through a circular channel of diameter 4.5 m and N = 0.03, $S_0 = 3 \times 10^{-3}$ at depth y = 2m will be

(a) $20.5 \frac{m^3}{s}$ (b) $23.5 \frac{m^3}{s}$ (c) $25.5 \frac{m^3}{s}$ (d) $29.5 \frac{m^3}{s}$

52. A Cipoletti weir has a side slope of

(a) 1 V to 4 H (b) 1 H to 4 V (c) 1 V to 3 H (d) 1 H to 5 V

53. The discharge equation of a Parshall flume is expressed as

(a) $Q = K\sqrt{H_a}$ (b) KH_a^n (c) KH_1 (d) $K(H_1^{3/2} - H_a^{3/2})$

54. The equation of Gradually varied Flow (GVF) IS

(a) $\frac{dy}{dx} = \frac{S_o - S_e}{1 + F^2}$ (b) $\frac{dy}{dx} = \frac{S_o - S_e}{1 - F^2}$ (c) $\frac{dy}{dx} = \frac{S_o + S_f}{1 + F^2}$ (d) $\frac{dy}{dx} = \frac{S_o - S_e}{1 - F^2}$

(where, S_o = Bottom slop, S_e = Energy line slope and F = is Froude Number)

TRUE OR FALSE QUESTIONS

1. Hydraulic Radius (R) is defined as the ratio of Area and Wetted perimeter. (T or F)
2. Hydraulic Depth (D) is defined as the ratio of Area and Top width at free surface (T). (T or F)
3. The Section Factor is defined as $A\sqrt{R}$. (T or F)
4. Average velocity in an open channel occurs at 0.6 depth is only approximate. (T or F)
5. Maximum velocity in open channel occurs at some point below the surface. (T or F)
6. In Uniform flow slope of Channel bottom (S_o), Water Surface (S_w) and Energy line (S_f) are related as, $S_f > S_w > S_o$. (T or F)
7. Chezy's C and Manning's N are dimensionless. (T or F)
8. Chezy's C is directly proportional to Manning's N. (T or F)
9. For each specific energy there are two depths one sub-critical and other is super critical known as sequent depths. (T or F)
10. The Froude Number range for a steady jump is $4.5 < Fr \leq 9.0$. (T or F)
11. In a hydraulic jump as the initial Froude Number increases energy loss also increases. (T or F)
12. The ratio of initial depth to final depth in a hydraulic jump is $\frac{y_1}{y_2} = \frac{1}{2}[\sqrt{1+8F_1^2} - 1]$. (T or F)

13. In a sharp crested rectangular weir the discharge is proportional to $H^{3/2}$. (T or F)

14. In a backwater curve $\frac{dy}{dx}$ is less than zero (Negative). (T or F)

15. The two alternate depths in a wide rectangular channel are 3.86 m and 1.0 m respectively. The discharge in the channel is, $\frac{31\ m^3}{s}$. (T or F)

16. Primary aim of creating hydraulic jump is to conserve energy. (T or F)

17. Analysis of hydraulic jump is based on energy conservation. (T or F)

18. A hydraulic jump in a rectangular channel has initial Froude Number of 5.0, the Froude Number at the end will be about 0.3. (T or F)

19. A clinging nappe can be restored by providing air vents. (T or F)

20. In a channel transition the contraction portion is longer than the expansion portion. (T or F)

21. Flow in a curved open channel the level of water on outer side is more than in inner side. (T or F)

22. If head H_0 over an overflow spillway is less than the design head H_d, the coefficient of discharge C_0 will be more than design discharge coefficient C_{d0}. (T or F)

23. A gradually varied flow with non-uniform discharge is termed as 'Spatially Varied flow'. (T or F)

24. A moving hydraulic jump is a typical case of unsteady flow. (T or F)

25. In a Gradually varied unsteady flow in open channel flow $\frac{dQ}{dx} = 0.10$, if top width (T) of the channel is 10.0 m $\frac{\partial A}{\partial t}$ is equal to 0.1. (T or F)

26. A mobile boundary channel carries with it, the bed load and suspended load and formation of bed forms. (T or F)

27. The dynamic equation for Gradually varied flow is $\frac{dy}{dx} = \frac{S_f - S_0}{1 - F_r^2}$. (T or F)

28. All surface profile curves with suffix 2 are drawdown curves. (T or F)

FILL IN THE BLANKS

1. For uniform flow $\frac{\partial y}{\partial s}$ is equal to ....

2. In open channel flow $R_e = \frac{VR}{\nu}$, Reynolds Number, should be less than ....

3. Froude Number for open channel sub-critical flow should be .... than 1.

4. For a wide rectangular channel Hydraulic Radius R is equal to .... of flow.

5. Conveyance K for open channel in terms of discharge (Q) and bed slope (S_0) is ....

6. Isovels are the lines of .... velocity.

7. The bed shear in an open channel is given by τ_0 = ...

8. Most efficient cross section of a trapezoidal channel is half of regular ....
9. Specific energy is the energy of flow measured taking datum as .... of the channel.
10. The specific energy diagram is asymptotic to the x-axis and line 45° to the ....
11. In a rectangular channel of width 6 m, the discharge is measured as 15 $\frac{m^3}{s}$, the discharge intensity is ....
12. In a rectangular channel the specific energy for maximum discharge is .... of critical depth.
13. In a rectangular channel of width 5 m, if the discharge is 10 $\frac{m^3}{s}$, the critical depth will be .... m.
14. The flow in a rectangular channel is sub-critical, if a small hump is created at some section, the water surface will .... at that section.
15. A rectangular channel 3 m wide carrying water at 1.85 $\frac{m^3}{s}$ at a depth of 0.5 m. A channel contraction is required at a section. The greatest allowable contraction width without altering the upstream flow is ....
16. The specific force expression is $F = \frac{Q^2}{gA}$ + ...
17. For an oscillating jump the Froude Number range is ....
18. The energy loss in a choppy or strong jump can be upto .... %.
19. The length of hydraulic jump is 5 to 7 times its ....
20. Flow over a spillway is in the category of .... Varied Flow.
21. When Normal depth $y_o > y_c$, the slope is categorized as .... slope.
22. The section of open channel where a definite relationship exists between the discharge (Q) and depth (y) is known as .... point.
23. The flow of water in a wash basin is an example of .... vortex.
24. Venturi flume is used to measure .... of liquid.
25. When a cylindrical vessel containing liquid is revolved, the surface of liquid takes the shape of a ....
26. Before passage of a surge, the depth and velocity are 1.8 m and 3.72 $\frac{m}{s}$ and after passing of surge these are 0.6 m and 7.46 $\frac{m}{s}$ respectively, the speed of surge is .... $\frac{m}{s}$.

ANSWERS

Multiple Choice Questions

1. c 2. d 3. b 4. a 5. a 6. d 7. d 8. b 9. b
10. c 11. d 12. a 13. a 14. c 15. c 16. c 17. b 18. a
19. a 20. d 21. b 22. d 23. d 24. a, c 25. b 26. c 27. c
28. a 29. b 30. d 31. a 32. b 33. d 34. d 35. b 36. c
37. a 38. b 39. c 40. d 41. a 42. d 43. d 44. b 45. c
46. a 47. d 48. b 49. b 50. a 51. d 52. a 53. b 54. b

True & False

1. T 2. T 3. $A\sqrt{D}$ 4. T
5. T 6. F $S_f = S_w = S_0$ 7. F 8. F Inversely
9. F Alternate 10. T 11. T 12. F (F_2^2)
13. T 14. F > 0 (positive) 15. T 16. F To Dissipate Energy
17. F Momentum Conservation 18. T 19. T
20. F 21. T 22. F $C_0 < C_{d0}$ 23. T
24. T 25. F (– 0.1) 26. T 27. F $S_0 - S_f$
28. T

Fill in the Blanks

1. Zero 2. 500 3. Less 4. Depth
5. $K = \frac{Q}{S_0}$ 6. Equal 7. γRS_0 8. Hexagon
9. Bed 10. x-axis 11. $\frac{2.5\ m^3}{s/m}$ 12. $\frac{3}{2\,y_c}$
13. 0.742 m 14. Drop 15. 2.47 m 16. A Z
17. 2.5 to 4.5 18. 80 to 85 19. Height 20. Rapidly
21. Mild 22. Control 23. Free 24. Discharge
25. Parabola 26. $+1.8\ \frac{m}{s}$

SECTION 14

Railway Engineering

Definitions – 37

Multiple Choice Questions – 105

True & False – 51

Fill in the Blanks – 52

DEFINITIONS

Buffer Stop: (Or Bumper, Bumper block) is a device to prevent railway vehicles from going past the end of physical section of track.

Cant in Railway Track: Also called superelevation or Cross-slope is the rate of change in elevation or height between two rails. The cant is measured as the level difference between the rail heads. Limiting value of cant is given by $\frac{G}{10}$ where G is the gauge.

Central Standard Trial (CST) Sleepers: These are most widely used sleepers in the Indian Railways. The CST-9 sleeper consists of two triangular inverted pots on either side of the rail seat, a central plate with a projected **keel**, and a box on the top of the plate. Cast Iron is used for CST-series of sleepers.

Colour Light Signals: Seamaphore signals are being replaced by high intensity beam colour light signals both during day and night. In normal position green light indicates, **proceed** and red light indicates **stop** or **danger**.

Composite Sleeper Index: It is the term used in Civil Engineering to define the strength and hardness of composite material used in sleepers. CSI = $\frac{(S + 10\,N)}{20}$ where S is strength and H is hardness.

Concrete Sleepers: These are railway track sleepers made of steel reinforced concrete. With the characteristic of water resistant, sun resistant and corrosion resistance, concrete sleepers are widely used all over the world.

Dog Spike: It is a fastening which is used for fixing rails with wooden sleepers. It is a large nail with an offset head that is used to secure rails and base plates to sleepers.

Driving Wheels: The powered wheels under the locomotive are called **driving wheels**. These are initially cast or forged then heat-treated to have a specific hardness.

Fish Plate: These are specially rolled section used for joining the rails end to end with the help of fish bolts. The section looks like a fish. It helps in transferring load from one rail to the next. It also allows expansion and contraction due to change in temperature.

Flag Station: A railroad station where trains stop only when a flag or other signal is displayed or when passengers are to be discharged. Also called a **flag stop**.

Indian Railways: From a very modest beginning in 1853, when the first train steamed off from **Mumbai to Thane**, a distance of 34 km, Indian Railway have grown into a vast network of 7,308 stations spread over a route length of 68,043 km. with a fleet of 13,215 locomotives, 74,744 passenger service vehicles, 10,103 other coaches and 3,18,896 wagons. The network runs multi-gauge operations extending over 68,043 route kilometers. 94% of rail routes are now electrified.

Ladder Track: It is a type of railway track in which the track is laid on longitudinal supports with transverse connectors holding the two rails at the correct gauge distance.

Mono-Block Sleepers: These are Pre-stressed-concrete (PSC) sleepers. These days pre-tensioned type are mostly used against post-tensioned sleepers. They have a trapezoidal section.

Monorail: It is a railway system in which the track consists of a single elevated rail. The entire set of beams, bogies, wheels etc. is more complex and expensive. Requires lot of maintenance. It is not very popular. The longest monorail (96.5 km) is in China. India has one monorail system (19.5 km)- Mumbai.

Number of Wheels: Most locomotives in India have 6 Axles and 12 Wheels. Some have 4 Axles and 8 Wheels. Diesal and electric locomotive have 6 to 12 Wheels. Each bogie is 4 or 6 Wheeled.

Outer Signal: It is a stop signal on a railway line that is located outside the station and is the first stop a train encounters before entering the station. A train is stopped at outer signal, if another train is already at the station.

Point and Crossings: These are special arrangement provided on railway track to facilitate trains to be diverted from one track to another. Combination of points and crossings are also known as **turn out**.

Rail Chair: It is installed between the rail sleepers and the rails, fixed with screws, and locked with a steel key. It helps in distributing load from train to the rail to the sleepers to achieve uniform force. They are often used for double headed or bull headed rails.

Rail Gauges: The gauge of a railway track is defined as the clear minimum perpendicular distance between the inner faces of the two straight parallel rails. In India 3-gauges are used- Broad Gauge (BG) : 1676 mm, Meter Gauge: 1000 mm and Narrow Gauge : 610 mm. For Metro rails standard Gauge : 1435 mm is used.

Rail Track Ballast: It is component of track which provides levelled bed or support for the railway sleepers. It transfers the load from sleepers to sub-grade and distributes the load uniformly on sub-grade. It holds the track in place and facilitates drainage.

Rails Life: The rails have a lift of 800 gross million tonnes, which works out to 12 to 13 years under Indian traffic conditions. Corrosion, however, may reduce its life even to 50%.

Railway Platform: It is an area alongside a railway track providing convenient access to trains. All stations have platforms, with large stations having multiple platforms. Traditional platforms are elevated relative to track but often lower than the train floor. In India there are two standard heights of platforms (200 mm and 760 mm). The minimum length recommended for the passenger platform in the case of all gauges is 180 m. Desirable length for BG is 305 m.

Scissors Crossover: It is a type of railway track where two tracks cross each other at a level and the rails of one track overlap the rails of other track. It is made up of diamond-shaped metal pieces that are placed in the intersection of the two tracks.

Semaphore Signals: This type signals display their different indications to train drivers by changing the angle of inclination of a pivoted **arm**. It shows three aspects: arm up-**clear**, horizontal-**stop**, arm diagonal (low) **Approach.**

Shunting Signals: It controls low speed movement where provision of main signal is not appropriate, such as movement in siding. It does not necessarily imply that line ahead is clear of vehicles.

Sleeper Density: It is the number of sleepers used per rail length. It is also expressed in number of sleepers per kilometer of track. Sleeper density depends upon maximum permissible speed and traffic density.

Speed Limit of Trains in India: The maximum speed limit of Indian Railways varies depending upon the category of train and the track condition. For passenger trains, the, the maximum speed limit is usually 110 $\frac{\text{Km}}{\text{Hour}}$ on BG tracks and 100 $\frac{\text{Km}}{\text{Hour}}$ on meter gauge tracks. However, some trains are allowed speed of 160 $\frac{\text{Km}}{\text{Hour}}$. For freight trains speed limit varies between 65 to 75 $\frac{\text{Km}}{\text{Hour}}$. For narrow gauge speed limit is 130 $\frac{\text{Km}}{\text{Hour}}$.

Stop Signals: A stop signal has a red arm with a white strip towards the left hand end. It shows a red light when on and a green light when off. Trains are not supposed to pass stop signal when on except under specially authorised conditions.

Stresses on Rail Sleepers: The sleepers are subjected to a large number of forces such as dead and live load and dynamic components of track.

Superelevation (or Cant): It is the difference in height between the outer and the inner rail on a curve. The inner rail is taken as the reference rail and is normally kept its original level. The inner rail is also known as the **gradient rail**. Superelevation ensures better load distribution on both rails, neutralizes the effect of lateral forces and smooth movement.

Terminal (Station): In railway terminology it refers to a station at the end of a railway line, where trains can only enter to exit in one direction only because tracks end there. Essentially it is the final stop on a line where trains cannot go further ahead. Here trains have to reverse direction to leave.

Track Modulus: (or Track Foundation Modulus) is defined as the supporting force per unit length of rail per unit deflection. Or, it is an index of measure of resistance due to deformation.

Transition Curves: These are used to enable a smooth transition between straight and curved railway segments. The Railway Board has decided that on Indian Railways transition curves will normally be laid in the shape of a cubic parabola. It is said to be the cure for uncomfortable ride as it helps in managing centrifugal and centripetal forces on the vehicle.

Triangles in Railway Line: In railway track a triangle is where 3 lines meet at junction and where trains can proceed from a line to either of the two. There are two types of triangles : A triangle junction where 3 lines meet and a turning triangle where one leg is a short dead end.

Turn Table: It is a device for turning railway rolling stock, usually locomotives so that they can be moved back in the direction from which they came. These are useful where there is lack of sufficient space and also for economic consideration.

Warner Signal: A Warner signal is to warn of an approach to a stop signal further ahead, or to advise a driver of the condition of the block section being entered. As such, it is a permissive signal and may be passed or must reduce speed.

Wheel Burns: Heat is generated and depression is caused due to movement on adverse gradient. Heat is also generated due to application of brakes.

MULTIPLE CHOICE QUESTIONS

1. The first passenger train in India was introduced in
   (a) 1851 (b) 1853 (c) 1853 (d) 1815
2. The Railway has what degree of freedom for its movement?
   (a) One (b) Two (c) Three (d) Four
3. Track Modulus is defined as
   (a) $\frac{\text{Load}}{\text{unit}}$ length of sleeper
   (b) $\frac{\text{Load}}{\text{unit}}$ length of sleeper to produce depression in rail
   (c) $\frac{\text{Load}}{\text{unit}}$ length of rail to produce depression in sleeper
   (d) $\frac{\text{Load}}{\text{unit}}$ length of rail to produce unit $\frac{\text{depression}}{\text{deflection}}$ in track
4. Which of the following causes stresses in sleepers?
   (a) Eccentric vertical loads (b) Contact shear stress between wheel and rail
   (c) Lateral deflection of sleepers (d) Track components
5. There are how many types of rail sections?
   (a) 2 (b) 3 (c) 4 (d) 5
6. How does the depth of ballast cushion affect the rail-section?
   (a) Higher the depth bigger the rail section (b) Less depth, bigger is section
   (c) Less depth, smaller section (d) Both depth and section same
7. What must be done to wooden sleepers before use?
   (a) Seasoning (b) Washing (c) Painting (d) Hydrating
8. In a shunting signal if the Red Band is inclined at 45°, it indicates
   (a) Stop (b) Proceed
   (c) Proceed cautiously (d) None of these
9. The normal size of Ballast used for points and crossings is
   (a) 25 mm (b) 40 mm (c) 50 mm (d) 10 mm
10. The standard length of Rail for Broad and Meter Gauge are
    (a) 12 m and 12 m (b) 12 m and 13 m (c) 13 m and 12 m (d) 13 m and 13 m
11. Gauge is the distance between
    (a) Centre to centre of rails (b) Running faces of rails
    (c) Outer faces of rails (d) None of these
12. The spike used for fixing chairs of bull-headed rails to wooden sleepers is
    (a) Dog spike (b) Rail screw (c) Elastic spike (d) Round spike

13. Super elevation on curves is provided by means of
(a) Cant-board (b) Straight edge (c) Spirit level (d) All these

14. The device provided to prevent the vehicles from moving beyond the end of rail at terminal is called
(a) Turn table (b) Buffer stops (c) Triangles (d) Scotch blocks

15. The sleepers in the form of two bowls placed under each rail and connected together bar, are known as
(a) Pot sleeper (b) Box sleepers (c) Plate sleepers (d) Duplex sleepers

16. Fish bolts are made of
(a) Cast iron (b) Low carbon steel (c) High carbon steel (d) Stainless steel

17. A triangle is used for?
(a) Diverting trains from the main line to branch line
(b) Crossing over between parallel tracks
(c) Changing the direction of the engine
(d) all the above

18. The resistance of the train is due to?
(a) Speed (b) Gradient (c) Curves (d) All these

19. A good sleeper should be such that?
(a) The rails can be easily fixed and taken out from the sleeper without moving them
(b) It can provide sufficient bearing area for the rail
(c) If can provide sufficient effective bearing area on ballast
(d) All the above

20. An advance starter signal is used for?
(a) Shunting (b) Goods train (c) Loco-sheds (d) All these

21. A welded rail joint is generally?
(a) Supported on a sleeper (b) Suspended
(c) Supported on a metal plate (d) None of these

22. Which of the following sleepers provide best elasticity of tract?
(a) Wooden sleeper (b) Cast iron sleeper (c) Steel slipper (d) RCC sleepers

23. Creep causes?
(a) Opening of rail joints (b) Distortion of points and crossing
(c) Buckling of track (d) All of these

24. Rail chairs are used to fix
(a) Flat footed rails (b) Bull headed rails
(c) Double headed rails (d) None of these

25. RCC sleepers are used in railways due to their
(a) Suitability for track circuiting
(b) Capacity to maintain the gauge properly

(c) Heavy weight which improves track modulus
(d) All the above

26. The chief function of providing ballast in the railway track is
(a) To uniformly distribute the load from sleepers over a large area of formation
(b) To hold the sleepers in position
(c) To provide elasticity resilience to the track
(d) All the above

27. Which of the following sleepers have highest service life?
(a) Wooden sleeper (b) Cast Iron sleeper
(c) Steel sleeper (d) Concrete sleepers

28. Minimum packing space provided between two sleepers is between
(a) 10 to 25 cm (b) 25 to 30 cm (c) 30 to 35 cm (d) 35 to 40 cm

29. Dog spikes are used for fixing rail to the
(a) Wooden sleepers (b) CST-9 sleepers (c) Steel trough (d) Concrete sleepers

30. A mono-block sleeper has a section
(a) Square (b) Rectangular (c) Trapezoidal (d) Semi-circular

31. Width of the top portion of a flat footed rail, is
(a) 69.85 mm (b) 63.50 mm (c) 66.77 mm (d) 136.52 mm

32. Rails are bent to correct curvature if the degree of curve, is more than
(a) 1° (b) 2° (c) 3° (d) 4°

33. Minimum depth of ballast prescribed of BG trunk lines of Indian Railways is
(a) 20 cm (b) 15 cm (c) 25 cm (d) 30 cm

34. If G is gauge in m, V is speed of train in KMPH and R is the radius of a curve in m, the equilibrium superelevation, is
(a) $\frac{GV^2}{R}$ (b) $\frac{GV^2}{17\,R}$ (c) c (d) $\frac{GV^2}{130\,R}$

35. On Indian Railway standard length of rails for M.G. track is
(a) 33 feet (10.06 m) (b) 36 feet (10.97 m) (c) 39 feet (11.89 m) (d) 42 feet (12.08 m)

36. The nominal size of ballast used for points and crossing is
(a) 25 mm (b) 40 mm (c) 50 mm (d) 10 mm

37. Rail section is generally designated by its
(a) Total weight (b) Total length
(c) Wight per meter length (d) Area of its cross-section

38. The quantity of stone ballast required per meter tangent length, is
(a) 1.15 m^3 (b) 1.14 m^3 (c) 1.13 m^3 (d) 1.11 m^3

39. Weight and cross-section of rails are decided on
(a) Gauge of track (b) Speed of trains
(c) Spacing of sleepers (d) All these

40. A scissors cross over consists of one diamond and
(a) One turn out (b) Two turn outs (c) Three turn outs (d) Four turn outs

41. The slipping of driving wheels of locomotives on the rail surface causes
(a) Wheel burns (b) Hogging of Rails
(c) Scabbing of Rails (d) Corrugation of rails

42. Wooden sleepers used on the girders of bridges, are generally made of
(a) Sal (b) Chir (c) Teak (d) Deodar

43. The shape of transition curve used by Indian Railways, is
(a) Cubic parabola (b) Spiral
(c) Sine curve (d) Lemniscates of Bernoulli

44. If w is the width of sleepers, S is sleeper spacing and d is depth of ballast then
(a) $d = \frac{(S-w)}{2}$ (b) $d = \frac{(S-w)}{3}$ (c) $d = \frac{(S-w)}{4}$ (d) $d = \frac{(S-w)}{5}$

45. The reception signal is
(i) Outer signal (ii) Home signal (iii) Starter (iv) Advanced starter
The correct answer is
(a) (i) and (ii) (b) (ii) and (iii) (c) (iii) and (iv) (d) (i) and (iv)

46. The standard width of ballast for B.G. Track in Indian Railways, is kept
(a) 3.35 m (b) 3.53 m (c) 2.35 m (d) 2.53 m

47. Which of the following rail failures is caused by loose fish bolts at expansion joints
(a) Crushed head (b) Angular break (c) Split head (d) Transverse fissures

48. Bull headed rails are generally provided on
(a) Points and crossing (b) Straight tangent
(c) Curved tracks (d) Meter gauge tracks

49. The type of bearing plate used in all joints and on curves to give better bearing area to the rail, is
(a) Flat mild steel plate (b) Mild steel canted plate
(c) Cast iron anti-creep plate (d) None of these

50. In a Railway track, permissible gauge with tolerance under loaded condition, is
(a) G + 0.1 mm (b) G + 1.5 mm (c) G – 1.0 mm (d) G – 1.5 mm

51. The station having two or more lines is called, a
(a) Flag station (b) Crossing (c) Terminal (d) Junction

52. The longitudinal movement of the rails in a track is technically known, as
(a) Buckling (b) Creeping (c) Hogging (d) Failure

53. To prevent the change in gauge and creep, the steel sleepers are fixed bolts and
(a) One key (b) Two keys (c) Three keys (d) Four keys

54. The number of dog spikes normally used per rail seat on curved track, is?
(a) One inside and two outside (b) One on either side
(c) Two on either side (d) One outside and two inside

55. For main cities and routes of maximum intensity the type of gauge most suitable, is
(a) Broad (b) Narrow (c) Meter (d) Any of these

56. The platform should be provided distance away from the centreline of the track for broad-gauge
(a) 1.219 m (b) 1.346 m (c) 1.676 m (d) 1.854 m

57. The length of platform compare to the longest train should be
(a) Equal to (b) More than (c) Less than (d) Double

58. The Indian Railway has bee divided into
(a) Six zones (b) Eight zones (c) Twelve zones (d) Sixteen zones

59. The speed of a locomotive, in India, on broad gauge is restricted between
(a) 120 – 140 KMPH (b) 60 – 75 KMPH (c) 75 – 96 KMPH (d) 96 – 120 KMPH

60. The largest dimension of rail section is
(a) Height (b) Foot width (c) Head width (d) Thickness

61. To reduce the wearing of rails, the rails are placed at an
(a) Outward slope of 1 : 30 (b) Inward slope of 1 : 30
(c) Outward slope of 1 : 20 (d) Inward slope 1 : 20

62. The length of each fish plate is
(a) 623.4 mm (b) 514.3 mm (c) 457.2 mm (d) 317.5 mm

63. A Warner signal which is first seen by train driver is known as
(a) Outer (b) Home (c) Disc (d) Routing

64. The joint generally not used on Indian Railway, is
(a) Base joint (b) Supported joint (c) Bridge joint (d) Suspended joint

65. The gauge representing the maximum width and height upto which a railway vehicle is built, is known as
(a) Narrow gauge (b) Loading gauge (c) Broad gauge (d) None of these

66. In India for meter gauge track, maximum height of a loaded goods wagon, is fixed at
(a) 5.87 m (b) 3.4 m (c) 5.32 m (d) 4.72 m

67. The height of the platform above the rail surface for broad gauge tracks should be?
(a) 229 to 406 mm (b) 838 to 982 mm
(c) 305 to 406 mm (d) 762 to 838 mm

68. Creep is greater
(a) On curves (b) In new rails than old rails
(c) Both a and b (d) In heavy rails

69. The maximum superelevation prescribed on narrow gauge of Indian Railway is
(a) 165.1 mm (b) 83.2 mm (c) 76.2 mm (d) 101.6 mm

70. In order to counteract the increased lateral thrust on curves, an extra shoulder provided outside of curves is
(a) 50 mm (b) 100 mm (c) 150 mm (d) 200 mm

71. The choice of gauge depends upon

(a) Volume and nature of traffic (b) Speed of train
(c) Physical features of country (d) All these

72. The length of platform for broad gauge should not be less than

(a) 100 m (b) 200 m (c) 300 m (d) 400 m

73. For broad gauge main lines with maximum traffic loads, the rail section provided should be

(a) 29.8 to 37.25 $\frac{kg}{m}$ (b) 44.7 to 56.8 $\frac{kg}{m}$ (c) 49.8 to 52.3 $\frac{kg}{m}$ (d) 49.8 to 56.8 $\frac{kg}{m}$

74. The width of platform, under no circumstances, should be less than?

(a) 2 m (b) 4 m (c) 6 m (d) 8 m

75. The standard width of ballast for meter gauge track on Indian Railways, is

(a) 2.3 m (b) 2.9 m (c) 3.35 m (d) 5.53 m

76. Sand may be used as ballast for

(a) Wooden sleepers (b) Steel sleepers (c) Cast iron sleepers (d) All the above

77. The effective bearing area of all types of sleepers is

(a) 0.40 m^2 (b) 0.42 m^2 (c) 0.44 m^2 (d) 0.46 m^2

78. Packing of ballast is done

(a) Near end of sleeper (b) On the shoulders
(c) Under sleepers (d) Between two rails

79. In a diamond crossing, number of noses are

(a) 2 (b) 3 (c) 4 (d) 6

80. Spacing of sleepers is kept

(a) Uniform throughout the length of rail (b) Closer near joints
(c) Closer in the middle of rails (d) None of these

81. Rail tops of track are placed

(a) Horizontal (b) At inward slope of 1 : 20
(c) At outward slope of 1 : 20 (d) At outward slope of 1 : 30

82. Tensile strength of steel used in rails should into be less than

(a) 450 MPa (b) 500 MPa (c) 700 MPa (d) 850 MPa

83. The slope of embankment for a railway track is generally taken as

(a) 1 : 1 (b) 1 : 5 : 1 (c) 2 : 1 (d) 1 : 2

84. Minimum Composite Sleeper Index (CSI) prescribed on Indian Railway for track sleepers, is

(a) 552 (b) 783 (c) 1352 (d) 1455

85. If S is strength and H is hardness Index of timber at 12 % moisture content, the composite sleeper Index

(a) $\frac{(H + 10\ S)}{20}$ (b) $\frac{(H + 20\ S)}{20}$ (c) $\frac{(10\ H + S)}{20}$ (d) $\frac{(S + 20\ H)}{20}$

86. Coal Ash (Cinder) is used in the initial stages of new construction of a railway for
(a) Wooden sleepers (b) Steel sleepers (c) C.I. sleepers (d) None of these

87. Normally the limiting value of cant is
(a) $\frac{G}{8}$ (b) $\frac{G}{10}$ (c) $\frac{G}{12}$ (d) $\frac{G}{15}$
where G is the Gauge

88. A triangle used for turning the face of the locomotives, consists of
(a) Three turnouts (b) One turnout and two splits
(c) Two turnouts and one split (d) Three splits

89. The size of ballast used on Indian Railways for wooden sleepers is
(a) 25 mm (b) 38 mm (c) 43 mm (d) 50 mm

90. In order to counteract the increased lateral thrust on curves, an extra shoulder provided outside of curves is?
(a) 50 mm (b) 100 mm (c) 150 mm (d) 200 mm

91. When one track is superimposed on the other track, it is known as
(a) Ladder track (b) Double-slip track (c) Gauntlet track (d) None of these

92. The broad gauge is
(a) 0.6096 m wide (b) 0.762 m wide (c) 1.00 m wide (d) 1.676 m wide

93. Generally how many times more the rail can bear a load of its weight per meter length?
(a) 100 times (b) 350 times (c) 460 times (d) 560 times

94. The minimum composite sleeper index for wooden sleepers used over bridge girders, is
(a) 1352 (b) 1455 (c) 1555 (d) 1652

95. The rail seat is given a slope of
(a) 1 in 10 (b) 1 in 20 (c) 1 in 30 (d) 1 in 40

96. The main disadvantage of cast-iron sleeper, is
(a) That they are liable to crack (b) That they are liable to break
(c) That they are liable to get rusted (d) Both a and b

97. In USA rail length is
(a) 20 m (b) 30 m (c) 40 m (d) 50 m

98. Which of the following bolt is not used in rail track?
(a) Fish bolt (b) Dog bolt (c) Eye bolt (d) Rag bolt

99. To hold the adjoining ends of rails in correct horizontal and vertical planes, the rail fastening used are
(a) Fish plates (b) Spikes (c) Anchors (d) Bearing plates

100. George Stephenson succeeded in running the first train of thc world in?
(a) 1825 (b) 1835 (c) 1841 (d) 1853

101. The speed of locomotive, in India, on meter gauge is restricted upto
(a) 60 KMPH (b) 80 KMPH (c) 100 KMPH (d) 120 KMPH

102. Which of the following types of gauges present in Indian Railways?
- (a) Broad Gauge, Meter Gauge and two Narrow Gauges
- (b) Meter Gauge, Narrow Gauge and two Broad Gauges
- (c) Standard Gauge, Meter Gauge and Narrow Gauge
- (d) Broad Gauge, Standard Gauge and Meter Gauge

103. Size of which Gauge of Railway is not correct?
- (a) Broad Gauge 1435 mm
- (b) Meter Gauge 1000 mm
- (c) Narrow Gauge 762 mm
- (d) Narrow Gauge 610 mm

104. Metros and monorails in India use
- (a) Broad Gauge
- (b) Meter Gauge
- (c) Standard Gauge
- (d) Narrow Gauge

105. India's first train from Bori Bunder (Mumbai) to Thane was run on Which Gauge?
- (a) Standard Gauge
- (b) Broad Gauge
- (c) Meter Gauge
- (d) Narrow Gauge

TRUE OR FALSE QUESTIONS

1. The track modulus is not affected by gauges. (T or F)
2. The Broad Gauge in India adopted a compromise width equal to 1.676 m. (T or F)
3. The first Indian Railway was laid in the year 1853. (T or F)
4. Minimum depth of ballast recommended for B.G. Trunk line of Indian Railway is 15 cm. (T or F)
5. Boxing of ballast is done under rails. (T or F)
6. Best ballast size ranges between 2 to 5 cm. (T or F)
7. For flat footed rails no chairs are used to hold rail in position. (T or F)
8. In railways a triangle is used to change the direction of engines through 180°. (T or F)
9. Rails are bent to correct curvature if the degree of curve is more than 6°. (T or F)
10. Rail section first designed on Indian Railway was, flat footed. (T or F)
11. Wooden sleepers used on bridges are made of Teak Wood. (T or F)
12. Maximum wheel base distance provided on Indian B.G. track, is about 5 m. (T or F)
13. The tread of wheels is provided on outward slope is 1 in 20. (T or F)
14. Bearing plates are used to fix flat footed rails to wooden sleepers. (T or F)
15. A welded rail joint is generally suspended. (T or F)
16. The switch generally used for B.G. and M.G. track is articulated. (T or F)
17. An ideal sleeper is made of wood. (T or F)
18. To divert train from one track to another is called a railway point. (T or F)
19. At a rail joint, the ends of adjoining rails are connected with a pair of fish plates with **two** fish bolts. (T or F)

20. The width of the top portion of a flat footed rail is 66.77 mm. (T or F)
21. Minimum packing space provided between two sleepers is 20 to 25 cm. (T or F)
22. No diamond crossing should be flatter than 1 in $8\frac{1}{2}$. (T or F)
23. A scissors cross over consists of one diamond and four turnouts. (T or F)
24. Sleepers act as a non-elastic medium between the rails and ballast. (T or F)
25. Gauge is the distance between centre to centre of rails. (T or F)
26. The side slopes of embankments for a railway track is generally 2 : 1. (T or F)
27. Longitudinal movement of rail is termed as 'Creep'. (T or F)
28. Maximum degree of curvature for Meter Gauge is limited to 30°. (T or F)
29. Dog spikers are used for fixing rail to wooden sleepers. (T or F)
30. Standard size of wooden sleepers for Broad Gauge track is 250 × 26 × 12 cm. (T or F)
31. The limiting value of Cant-gradient for all gauges of Indian Railway is 1 : 1000. (T or F)
32. The limiting value of Cant is given by $\frac{\text{Gauge}}{10}$. (T or F)
33. Switch angle is angle between the gauge face and tongue rail. (T or F)
34. Maximum value of 'throw of switch' for B.G. track is 100 mm. (T or F)
35. Metal sleepers are superior to wooden sleepers in terms of life only. (T or F)
36. Wear of rails is maximum in sharp curves. (T or F)
37. For a sleeper density of (n + 5), the number of sleepers required for constructing a B.G. track of length 650 m will be 975. (T or F)
38. Mostly used method for designating crossing is Right angle method. (T or F)
39. The reception signal is outer as well as home signal. (T or F)
40. A treadle bar is used for setting points and crossings. (T or F)
41. The turnouts mostly used for goods train on Indian Railways is 1 in 20. (T or F)
42. Heel divergence is always **less** than flangeway clearance. (T or F)
43. Traverser is used to transfer locomotives to and from parallel tracks without need of shunting. (T or F)
44. For changing the direction of Engine a Triangle is used. (T or F)
45. The starter signal is the reception signal. (T or F)
46. Yellow light signals indicate 'proceed cautiously'. (T or F)
47. In a shunting signal when the red band is inclined at 45° it indicates 'proceed'. (T or F)
48. **Split head** rail failure is caused by loose fish bolts at expansion joints. (T or F)
49. Rails are fixed to steel sleepers by keys. (T or F)
50. Wear of rails can be controlled by **increasing** the number of rail joints. (T or F)
51. In locomotives coning of wheels are done at a slope 1 in 30. (T or F)

FILL IN THE BLANKS

1. The rail is designated by its ....
2. The Mascot of Indian Railway is .... the elephant.
3. In Broad Gauge mostly .... kg rails are used.
4. The distance between running faces of rails is designated as ....
5. The limiting value of cant gradient for all gauges is 1 in ....
6. Permissible cant deficiency for Broad Gauge line .... mm.
7. For fixing rails on wooden sleepers the spikes used are called .... spikes.
8. A falling gradient followed by a rising gradient in valleys is called .... gradient.
9. The role of super-elevation in railway line is to counteract the effect of .... force.
10. The nominal size of ballast used for points and crossings is .... mm.
11. Width of ballast section for B.G. line is .... m.
12. Ordinary rails are made of high .... steel.
13. To join two rails together .... plate is used.
14. 60 R rails are mostly used in .... Gauge.
15. Number of fish bolts per fish plate is ....
16. The largest dimension of a rail section is its ....
17. Maximum speed of locomotive on Broad Gauge line is .... KMPH.
18. Maximum speed of locomotive on Meter Gauge line is .... KMPH.
19. The rail section used in Indian Railways .... headed type.
20. The first Indian Railway was laid in the year ....
21. On B.G. trunk lines of Indian Railways minimum depth of ballast should be .... cm.
22. An advance starter signal is used for ....
23. To correct curvature rails are bent if the degree of curvature is more than ....
24. The length of platform for B.G. line should not be less than .... m.
25. The width of platform should never be less than .... m.
26. The height of the platform above rail surface should not be less than .... mm.
27. The standard length of rail for Metre Gauge line is .... m.
28. The standard length of rail for Broad Gauge line is .... m.
29. Vertical curves are provided where algebraic difference between grades is equal to or more than .... $\frac{mm}{m}$.
30. The shape of transition curve used by Indian Railways is .... parabola.

31. In a 4-8-2 locomotive the number of driving wheels is ....
32. Device provided to stop the train from moving beyond the end of rail at terminal is called .... stops.
33. Two bowls sleepers under each rail and connected by a bar, are known as .... sleepers.
34. The best elasticity of track is provided by .... sleepers.
35. The height of the centre of arm of a semaphore signal above the ground is .... m.
36. The grade compensation on a 4° curve on a B.G. line is .... %.
37. In a Railway survey scheme the first step is .... survey.
38. The number of sleepers used per rail length on a track is known as sleeper ....
39. Outer signal on a B.G. line is kept at a distance of .... m.
40. Device used for changing the direction of engine is called ....
41. For directing the train from main lines to branch lines arrangement is called .... and ....
42. When two tracks of same gauge cross each other at any angle is known as .... crossing?
43. The first train in India ran between Bombay (Mumbai) and ....
44. The width of narrow gauge line is .... m.
45. The adjustment of rail is required when creep exceeds .... mm.
46. In Indian Railways the permissible limit of rail's wear is .... %.
47. The composite sleeper index is the index of strength and ....
48. The minimum composite sleeper index for wooden sleepers over bridge is ....
49. The total affective area of a pot sleeper is .... m^2.
50. Anti-creepers are known as ....
51. In a shunting signal, if red band is horizontal it indicates ....
52. The super-elevation (S) to be provided on rails follows the formula ....
    (in terms of train speed, Gauge, Radius of curve and acceleration due to gravity)

ANSWERS

Multiple Choice Questions

1. b	2. a	3. d	4. d	5. b	6. b	7. a	8. b	9. a
10. c	11. b	12. d	13. d	14. b	15. a	16. c	17. c	18. d
19. d	20. a	21. b	22. a	23. d	24. c	25. d	26. d	27. d
28. c	29. a	30. c	31. c	32. d	33. c	34. c	35. c	36. a
37. c	38. d	39. d	40. d	41. a	42. c	43. a	44. a	45. a
46. a	47. a	48. a	49. b	50. b	51. d	52. b	53. d	54. a
55. a	56. c	57. b	58. d	59. d	60. a	61. d	62. c	63. a
64. c	65. b	66. b	67. d	68. c	69. c	70. c	71. d	72. c
73. b	74. b	75. a	76. c	77. d	78. c	79. d	80. c	81. b
82. c	83. c	84. b	85. c	86. a	87. b	88. c	89. d	90. c
91. c	92. d	93. d	94. b	95. b	96. d	97. b	98. c	99. a
100. a	101. a	102. a	103. a	104. c	105. b			

True & False

1. F	2. T	3. T	4. F 25 cm
5. F At the rails	6. T	7. T	8. T
9. F (Four) 4°	10. F Double Headed	11. T	12. F 6.096 m
13. T	14. T	15. T	16. F Over-riding
17. T	18. F Turnout	19. F Four	20. T
21. F 30 to 35 cm	22. T	23. T	24. F
25. F Running faces of rails		26. T	27. T
28. F 16°	29. T	30. F 275 25 × 13 cm	
31. F 1 : 720	32. T	33. T	34. F 115 mm
35. T	36. T	37. F 900	38. T
39. T	40. F Interlocking Points & Signals		41. F 1 in $8\frac{1}{2}$
42. F Greater	43. T	44. T	45. F Outer or home
46. T	47. T	48. F Crushed Head	49. T
50. F Decreasing	51. F 1 in 20		

Fill in the Blanks

1. Weight per unit length		2. Bholu	3. 52 kg
4. Gauge	5. 1 in 720	6. 75 mm	7. Dog
8. Momentum	9. Centripetal	10. 25 mm	11. 3.35 m

12. Carbon	**13.** Fish	**14.** Metre	**15.** Four
16. Height	**17.** 120 KMPH	**18.** 80 KMPH	**19.** Bull
20. 1853	**21.** 25 cm	**22.** Shunting	**23.** 4 degrees
24. 300 m	**25.** (Four) 4 m	**26.** 762 mm or 2.5 feet	
27. 12 m	**28.** 13 m	**29.** (Four) $\frac{4\text{ mm}}{\text{m}}$	**30.** Cubic
31. (Eight) 8	**32.** Buffer	**33.** Pot	**34.** Wooden
35. 7.5 m	**36.** 0.16%	**37.** Reconnaissance	**38.** Density
39. 540 m	**40.** Triangle or Turn Table		**41.** Points and Crossings
42. Diamond	**43.** Thane	**44.** 0.762 m	**45.** 150 mm
46. 5%	**47.** Hardness	**48.** 1455	**49.** 0.46 m^2
50. Anchors	**51.** Stop	**52.** $\frac{V^2G}{gR}$	

SECTION 15

R.C. Structures

Definitions – 43

Multiple Choice Questions – 103

True & False – 53

Fill in the Blanks – 60

DEFINITIONS

Balanced Section: It is the section of a RCC beam where most distant concrete fibre in compression and tension in steel attains permissible stresses simultaneously. The percentage of steel included in the balanced section is known as **balanced steel** and the neutral axis as the critical neutral axis.

Creep Strain: It is the time-dependent strain induced by a constant mechanical loading. The strain is a function of stress level, the loading time and the temperature. The **creep modulus** for a specific stress, time and temperature is the stress divided by strain measured after a selected time.

Curing: It is the process of preventing the loss of moisture from the concrete while maintaining a satisfactory temperature regime. By compensating for moisture loss, it will reduce shrinkage in concrete. It also ensures hydration of cement which results in increased durability and weather resisting qualities.

Deep Beams: These are recognised by their relatively small span-to-depth ratio. Hence, **deep beams** are designed for bending moment and checked for shear. Deep beams are structural elements loaded as simple beams in which a significant amount of the load is carried to the supports by a compression force combining the load and the reaction. These are used in Bridge girders, foundation beams, shear walls etc.

Design Yield Stress of Steel: If fy is the characteristic Yield strength of steel, then the design Yield stress of steel according to IS code is related as 0.87 fy. At Yield stress material goes permanent deformation or plastic flow.

Doubly Reinforced Beams: Reinforced concrete sections having steel reinforcement both on tensile as well as compressive faces are known as doubly reinforced section.

Eccentric Load: It refers to a force application that deviates from the direct Centre of Gravity (CG) of a structural. Component. This off-centre application of force can induce bending or flexural stress within the member, even when the load is compressive in nature.

Fineness Modulus of Aggregate: It is an empirical figure obtained by adding the total percentage of sample of an aggregate retained on each of a specified series of **sieves**, dividing the sum by 100. A smaller value indicates a finer aggregate. (FM 2 to 4) and coarse aggregate (FM 6.75 to 8).

Flexural Strength: It is one measure of the tensile strength of concrete. It is a measure of an unreinforced concrete beam or slab to resist failure in bending. The flexural strength is expressed as Modulus of Rupture (MR) in MPa. Flexural MR is about 10 to 20 percent of compressive strength depending upon quality of concrete.

Formwork (Shuttering): It is a temporary structure used as a **mould** to pour the concrete. It is a vertical or horizontal arrangement made to deep concrete in position until the same gains strength and shape.

Grades of Concrete: Grades of concrete are defined by the strength and composition of the concrete, and the minimum strength the concrete should have following 28 days of initial construction. The grade is understood in measurements of MPa, where M stands for **mix** and MPa denoted overall strength. For example M15 (Mix : cement 1, sand 2, aggregates 4, strength : 15 MPa or $\frac{N}{mm^2}$).

High Yielding Strength Deformed (HYSD) Steel Bars: These steel bars are known for their flexibility and durability in the construction industry. The manufacturing of H Y S D steel bars is unique as they are formed at a hot temperature and then they are cold twisted.

Hoop Stress: Also known as Circumferential stress, is a type of stress that acts tangentially to the circumference of circular structure, like water towers. It arises primarily due to external loads such as torsion or bending. In water tanks Hoop tension is resisted by steel alone.

Limit States Design: It is a design method in which the performance of a structure is checked against various limiting conditions at appropriate load levels. The Limit State Design requires the structure to satisfy two principal criteria-the Ultimate Limit State (ULS) and the Serviceability Limit State (SLS).

Linear Elasticity: It is a mathematical model as to how solid objects deform and become internally stressed by prescribed loading conditions. Linear elastic theory describes the relationship among stress, strain, and displacement.

Long and Short Columns: Short columns are those whose slenderness ratio is less than 32 or length to diameter ratio is less than 8. Such columns are always subjected to direct compressive stress only. Long columns have a slenderness ratio more than 120 or length to diameter ratio more than 30.

Maximum Size of Bars in RCC: According to IS 456 : 2000 the diameter of Reinforcing bars shall not exceed One-eighth of the total thickness of the slab.

Minimum Cover in Slabs: The minimum required cover thickness for slab-reinforcement usually ranges between 20 mm and 30 mm depending on the environmental conditions present. For dry climate recommended cover is 20 mm and 30 mm for **wet** locations.

Minimum Percentage of Steel: The minimum percentage of reinforcement is 0.15% of gross area if mild steel bars are used and 0.12% of gross area if the High Yielding Strength Deformed (HYSD) steel is used.

Modular Ratio: In RCC modular ratio is defined as the ratio between modulus of elasticity of steel and modulus of elasticity of concrete $\left(\frac{E_S}{E_C}\right)$. Here ES is the modulus of elasticity of steel (taken as 2,00,000 $\frac{N}{mm^2}$). Concrete does not have definite value for the same as concrete is not a perfectly elastic material.

Neutral Axis: It is an axis in the cross-section of a beam (a member resisting bending) along which there are no longitudinal stresses or strains. All fibres on one side of the neutral axis are in tension, while those on the other side are in compression.

One-way Slab: It is a concrete slab that carries load in one direction and is supported by beams on two sides. The ratio of the longer span to shorter span of a One-way slab is ≥ 2.0.

Partial Factor of Safety: It is a factor greater than unity by which either the loads are multiplied or the resistance are divided to obtain the design value.

Permissible Shear Stress in RCC: It is a function of the percentage of steel reinforcement and the grade of concrete. The higher the percentage of steel reinforcement and the grade of concrete, the higher the permissible shear stress. For example according to IS code, the maximum shear stress which can be sustained by M20 grade of concrete is 0.18 times the square root of its compressive strength, i.e., $\tau_{max} = 0.18\sqrt{f_{ck}}$, where f_{ck} is the characteristic compressive strength of concrete.

Pre-stressing: It is the introduction of a compressive force to the concrete to counteract the stresses that will result from an applied load. It is done either by pre-tensioning or by post-tensioning.

Reinforced Concrete Cantilever Retaining Wall: RCC retaining wall consist of a relatively thin stem and base slab. The base is divided into two parts-the heel and toe. The heel is the part of the base under the back fill. The retaining wall is fixed to the reinforced concrete slab foundation with a **shear key** for sliding resistance.

Segregation in Concrete: It refers to the separation of the constituent materials in freshly mixed concrete. This occurs when the heavier aggregate settles down due to gravity leaving behind the lighter cement and water mixture on top. Segregation can also occur when the concrete mixture is not properly mixed or higher water-cement ratio causing some areas to have more cement or water than others.

Shallow Foundation: It is a type of building foundation that transfers structural load to the Earth very near to surface, rather than to a sub-surface layer or a range of depths as does a deep foundation. Generally shallow foundation (as defined by Terzaghi) is equal to or less than its width.

Shear Stress Distribution in a RCC-Section: The section above the neutral axis is in compression and homogenous as no steel is present in the upper portion, so shear stress distribution is **parabolic** till the neutral axis and thereafter till the depth of reinforcement it is rectangular.

Shrinkage of Concrete: It is defined as the contraction due to loss of moisture. Due to shrinkage of concrete the pre-stress in tendon gets reduced with time. The main causes of shrinkage are – higher water-cement ratio, Higher cement content and increased ambient temperature.

Singly Reinforced Beams: Singly reinforced beams are those which are suitable for simple bending situations where the compression zone of the beam can be adequately managed with concrete alone.

Spacing of Stirrups: As per IS-456 : 2000 the spacing between two successive stirrups in Reinforced concrete beams shall not be less than 100 mm. Also the maximum spacing of shear reinforcement measured along the axis of the member shall not exceed (0.75 d) for vertical stirrups and (d) for inclined stirrups at 45°, where (d) is the effective depth of the section under consideration. In no case shall the spacing exceed 300 mm.

Standard Hook in RCC: A standard hook has an anchorage value equivalent length of 16ϕ. The anchorage value of Standard U-type hook shall be 16 time the diameter of bar.

Steam Curing: It is curing in water vapour. Steam curing is used where early strength gain is needed and where heat is required for hydration, such as in cold weather.

Stirrups: Stirrups are used in beams and **ties** are used in columns. In beams these are provided to resist shear force and hold longitudinal bars. Ties in columns are provided to prevent buckling and also hold longitudinal bars in position.

T-Beam: It is a load-bearing structure of reinforced concrete (or wood, metal) with a capital 'T'-shaped cross section. The top of the T-shaped cross-section serves as a **flange** or compression member. The vertical section or web serves to resist shear. For tensile stresses reinforcement is provided in the bottom of web.

Tensile Strength Vs Compressive Strength of Concrete: According to IS : 456 – 2000, the tensile strength of concrete is 0.7 times the square root of the characteristic compressive strength of the concrete cube. Approximately Tensile strength is almost $\frac{1}{10}$ of the compressive strength.

TMT-Bars: Thermo Mechanically Treated bars are high-strength reinforcement bars having a tough outer core and a soft inner core. They are strong, corrosion-resistant, fire-resistant, Earthquake proof and cost-effective.

TOR: It is the short name for "Toristeg Steel Corporation of Luxembourg". It is Cold Twisted Deformed Steel (CTD) and is a form of HYSD steel.

Twisted Steel Bars: These are also known as **Prestressing Tendons.** These are comprised of different wires almost of wires spun together into a strand. They carry high tensile strength. These are mainly used in braces of scaffolds for Prestressing Reinforced Cement Concrete.

Two-way Slab: It is a concrete slab that is supported by beams on all four sides and can bend in two directions. It is designed to resist bending in both the longer and shorter directions and can handle larger spans and heavier loads than one-way slab. It has aspect ratio of longer to shorter theoretical span ≤ 2.0.

Under Reinforced Section: In such sections steel reinforcement reaches the Yield strain at stresses lower than the stress at which the concrete fails. Therefore, all singly reinforced RCC structures must be designed as under-reinforced sections to give adequate warring before the sections fail.

Workability of Concrete: It is a broad and subjective term describing how **easily** freshly mixed concrete can be mixed, placed, consolidated, and finished with minimum loss of homogeneity, avoid segregation.

MULTIPLE CHOICE QUESTIONS

1. A twisted bar has more yield stress than ordinary MS bar by?

(a) 10% (b) 20% (c) 35% (d) 50%

2. In a simply supported RCC beam the reinforcement is placed

(a) Near bottom (b) At Neutral axis (c) Near top (d) Anywhere

3. As per IS : 456-2000, the number of grades of concrete mixes is

(a) 3 (b) 4 (c) 5 (d) 7

4. Modular Ratio is the Ratio of

(a) Young's Modulus of steel to that of concrete

(b) Young's Modulus of concrete to that of steel

(c) Load share of steel to that of concrete

(d) Load share of concrete to that of steel

5. In the limit-state approach the spacing of main reinforcement control primarily

(a) Collapse (b) Cracking (c) Deflection (d) Durability

6. The lever arm in a single reinforced beam is equal to

(a) $\frac{d-n}{3}$ (b) $\frac{2d-n}{3}$ (c) $\frac{3d-n}{3}$ (d) $\frac{4d-n}{3}$

where, d = Distance between top to beam and centre of steel bars and n = Depth of Neutral axis below top of beam

7. The section in which concrete is not fully stressed its permissible value while stress in steel reaches to maximum value, is called
   (a) Under-reinforced (b) Over-reinforced
   (c) Critical section (d) Balanced section
8. The neutral axis of a balanced section is called
   (a) Balanced Neutral axis (b) Critical Neutral axis
   (c) Equivalent Neutral axis (d) Perfect Neutral axis
9. The deep beams are designed for
   (a) Shear Force only (b) Bending Moment only
   (c) Both shear and Bending (d) Bearing
10. In a RCC beam the tension zone lies at the
    (a) Top (b) Middle (c) Side (d) Bottom
11. The minimum cement content required for R.C.C. work under extreme environmental condition is, (as per IS 456-2000)
    (a) $300 \frac{kg}{m^3}$ (b) $320 \frac{kg}{m^3}$ (c) $340 \frac{kg}{m^3}$ (d) $360 \frac{kg}{m^3}$
12. The span to effective depth ratio shall not be more than .... for cantilever $\frac{\text{beam}}{\text{slab}}$ upto spam of 10 m
    (a) 7 (b) 20 (c) 26 (d) 48
13. The maximum size of the bars to be used in slab as per IS 456-2000 is for thinness of slab (T) is
    (a) $\frac{T}{8}$ (b) $\frac{T}{5}$ (c) $\frac{T}{4}$ (d) $\frac{T}{3}$
14. The maximum area of Tension reinforcement required for the beam shall not exceed
    (a) $0.01\ b \times D$ (b) $0.02\ b \times D$ (c) $0.12\ b \times D$ (d) $0.04\ bD$
15. In case of 2-way slab, the limiting deflection of the slab is
    (a) Primarily a function long span (b) Primarily a function of short span
    (c) Independent of short or long spans (d) Depends on both
16. The maximum percentage of Moment redistribution allowed in RCC beams is
    (a) 10% (b) 20% (c) 30% (d) 40%
17. In a RCC slab 75 mm thick, the maximum size of Reinforcement bars that can be used are of diameter
    (a) 12 mm (b) 10 mm (c) 8 mm (d) 6 mm
18. RCC structure design method is based on which linear elastic theory?
    (a) Ultimate load (b) Limit state (c) Kanis (d) Working stress
19. The section in which concrete is not fully stressed to its permissible value when stress in steel reaches its maximum value, is called
    (a) Under Reinforced (b) Over Reinforced
    (c) Critical (d) Balanced

20. The main reinforcement of a R.S. slab consists of 10 mm bars at 10 cm spacing. If 12 mm bars are used, what should be new spacing?
(a) 12 cm (b) 14 cm (c) 14.4 cm (d) 16 cm

21. Doubly reinforced beams are recommended when
(a) The depth of the beam is restricted (b) The breadth of the beam is restricted
(c) Both depth and breadth are restricted (d) The shear is high

22. Permissible shear stress in concrete is a function of
(a) Grade of concrete (b) Grade of steel
(c) Percentage of steel reinforcement (d) a and c

23. Deep beams are designed for
(a) Shear Force only (b) Bending Moment only
(c) Both SF and BM (d) Bearing

24. The minimum reinforcement using mild steel in slabs should not be less than
(a) 0.10% (b) 0.12% (c) 0.15% (d) 0.2%

25. Spacing of shear stirrups in a rectangular RC beam is
(a) Kept constant throughout the span (b) Decreased towards the centre of the beam
(c) Increased at the end (d) Increased towards the centre of the span

26. Normally, the tensile strength of concrete is about .... of its compressive strength
(a) 10 to 15% (b) 15 to 20% (c) 20 to 25% (d) 25 to 30%

27. Tension capacity of steel is greater than combined compression of steel and concrete in the .... beam
(a) Over reinforced (b) Under reinforced (c) Singly reinforced (d) Doubly reinforced

28. In a balanced section, the moment of resistance obtained from compressive force will be .... the moment obtained from tensile force
(a) Greater than (b) Less than (c) Equal to (d) None of these

29. In a RC beam, the shear stress distribution above the neutral axis follows a
(a) Straight line (b) Circular curve (c) Parabolic curve (d) None of these

30. Permissible bending tensile stress in High Yield strength deformed bars of grade 415 in a beam is $\left(\frac{N}{mm^2}\right)$.
(a) 190 (b) 230 (c) 140 (d) 250

31. The main reason for providing number of bars at the support in a simply supported beam is to resist
(a) Compressive stress (b) Shear stress
(c) Bond stress (d) Tensile stress

32. When shear stress exceeds the permissible limit in a slab, then it is reduced by
(a) Increasing the depth` (b) Providing shear reinforcement
(c) Using high strength steel (d) Using more thinner rods

33. Maximum percentage reinforcement in case of slabs is limited to

(a) 2% (b) 4% (c) 6% (d) 3%

34. According to Whitney's theory ultimate strain of concrete is assumed to be

(a) 0.03% (b) 0.1% (c) 0.3% (d) 3%

35. According to Whitney's Theory, depth of stress block for balanced section of a concrete beam is limited to

(a) 0.43 d (b) 0.537 d (c) 0.68 d (d) 0.85 d

36. Ratio of permissible stress in direct compression and bending compression is

(a) Less than 1 (b) 1 to 1.5 (c) 1.5 to 2 (d) More than 2

37. According to IS : 456-1978, the maximum reinforcement in a column is

(a) 2% (b) 4% (c) 6% (d) 8%

38. The minimum cover in a slab should neither be less than the diameter of bar nor less than

(a) 10 mm (b) 15 mm (c) 25 mm (d) 13 mm

39. The imposed load on staircase in Residential buildings as per IS 875 is

(a) $5\frac{KN}{m^2}$ (b) $3\frac{KN}{m^2}$ (c) $1.5\frac{KN}{m^2}$ (d) $1.3\frac{KN}{m^2}$

40. For an axially loaded RCC column of size 400 mm × 400 mm with unsupported length of 5 m, the minimum eccentricity to be considered is

(a) 15.6 mm (b) 20.5 mm (c) 23.3 mm (d) 30.6 mm

41. As per IS code, the minimum grade of concrete for the design of a pre-tensioned concrete structure is

(a) M 20 (b) M 25 (c) M 30 (d) M 40

42. The flexural strength of M 25 concrete as per IS-456 is

(a) 3.5 MPa (b) 5.5 MPa (c) 21.5 MPa (d) 30.0 MPa

43. The minimum diameter of longitudinal bars in a column is

(a) 6 mm (b) 8 mm (c) 12 mm (d) 16 mm

44. According to IS 456 : 1978 the maximum compressive stress in concrete for design purpose in taken as

(a) $0.37\ f_{ck}$ (b) $0.416\ f_{ck}$ (c) $0.446\ f_{ck}$ (d) $0.67\ f_{ck}$

(where f_{ck} represents the strength of concrete cube in MP_a)

45. The design Yield stress of steel according to IS 456-1978 is

(a) $0.37\ f_y$ (b) $0.57\ f_y$ (c) $0.67\ f_y$ (d) $0.87\ f_y$

(where f_y is Yield strength)

46. Minimum thickness of load bearing RCC wall should be

(a) 50 mm (b) 100 mm (c) 150 mm (d) 200 mm

47. In reinforced concrete footing on soil, the minimum thickness at the edge should not be less than

(a) 100 mm (b) 150 mm (c) 200 mm (d) 250 mm

48. The recommended value of modular ratio for reinforced brick work is

(a) 18 (b) 30 (c) 40 (d) 58

49. Creep strain is caused due to
(a) Dead load only (b) Live load only
(c) Both dead and live load (d) None of these

50. Due to shrinkage stresses a simply supported beam having reinforcement only at bottom tends to
(a) Deflects downward (b) Deflects upward
(c) Can deflect both ways (d) None of these

51. In a spherical dome the hoop stress due to concentrated load at the crown is
(a) Compressive everywhere (b) Tensile everywhere
(c) Partly compressive and partly tensile (d) Zero

52. Normally prestressed wires are arranged in the
(a) Upper part of the beam (b) Lower part of the beam
(c) In the middle (d) Anywhere

53. Ratio of compressive to tensile strength of concrete
(a) Increases with age (b) Decreases with age
(c) Remains constant (d) None of these

54. Which one of the following set of values give the minimum clear cover (in mm) for the main reinforcements in the slab, beam, column and footing respectively
(a) 20, 25, 30, 75 (b) 5, 15, 25, 50 (c) 15, 25, 40, 75 (d) None of these

55. A reinforced concrete wall carrying vertical loads is generally designed as per recommendations given for columns. The ratio of minimum reinforcement in vertical and horizontal direction is
(a) 2 : 1 (b) 5 : 3 (c) 1 : 1 (d) 3 : 5

56. The property of the ingredients to separate from each other while placing concrete is called
(a) Segregation (b) Compaction (c) Shrinkage (d) Balking

57. Workability of concrete is directly proportional to
(a) Aggregate cement ratio (b) Time of transit
(c) Grading of the aggregate (d) All these

58. As compared to ordinary portland cement use of pozzuolanic cement
(a) Reduces workability (b) Increases bleeding
(c) Increases shrinkage (d) Increases strength

59. Percentage of voids in cement is approximately
(a) 25% (b) 40% (c) 60% (d) 80%

60. For better workability of concrete, the best shape of aggregate is
(a) Rounded (b) Elongated (c) Angular (d) None of these

61. In a random sampling procedure for cube strength of concrete, one sample has X specimens. These are tested at 28 days and average strength is considered as test result, provided the variation in strength is not more than ± Y percent of average. The value of X and Y as per IS 456 : 2000, are
(a) 4 and 10 (b) 3 and 10 (c) 4 and 15 (d) 3 and 15

62. Unit weight of RCC is $\left(\frac{kg}{m^3}\right)$.

(a) 2200 (b) 2300 (c) 2400 (d) 2500

63. As per IS 456 : 2000, the unit weight of plain cement concrete should be taken as in $\left(\frac{kg}{m^3}\right)$.

(a) 2500 (b) 2400 (c) 2300 (d) 2200

64. M20 concrete mix is approximately

(a) 1 : 3 : 6 (b) 1 : 1 : 2 (c) 1 : 1 . 5 : 3 (d) 1 : 2 : 4

65. For a bar in tension, a standard hook has an anchorage value equivalent to a straight length of

(a) 8 ϕ (b) 12 ϕ

(c) 16 ϕ (d) 24 ϕ, where ϕ is dia of hook

66. A beam curved in plan is designed for

(a) BM and SF (b) BM & Torsion (c) SF & Torsion (d) BM, SF & Torsion

67. The initial and final setting times for ordinary portland cement are approximately related as

(a) T = 530 + t (b) T = 270 + t (c) T = 90 + 1.2 t (d) T = 530 – 1.2 t

where, T is find setting time and + is initial setting time, both in minutes.

68. The diameter of needle of Vicat's apparatus for the determination of Initial setting time is

(a) 0.5 mm (b) 1 mm (c) 5 mm (d) 10 mm

69. Minimum Water-cement Ratio required for complete hydration of cement is

(a) 0.23 (b) 0.36 (c) 0.45 (d) 0.60

70. Increase in fineness modulus of aggregate indicates

(a) Finer grading (b) Coarser grading (c) Gap grading (d) None of these

71. If a beam size 350 mm × 600 mm is subjected to torsional moment to 50 KN – m, then equivalent shear will be

(a) 129 KN (b) 229 KN (c) 329 KN (d) Zero

72. The effective length (as per IS 456 : 2000) for RC cantilever beam of depth (d) as shown below is

L

(a) $L + \frac{d}{2}$ (b) L + d (c) L + 2d (d) L

73. The minimum number of longitudinal bars provided in a RC column of circular cross section?

(a) 4 (b) 5 (c) 6 (d) 8

74. Foundation rests on

(a) Base of foundation (b) Sub-grade

(c) Foundation soil (d) All these

75. The anchorage value of standard hook of Reinforcement bar is

(a) 4 D (b) 8 D (c) 12 D (d) 16 D

76. If HYSD bars are used instead of mild steel bars in a beam, the bond strength

(a) Does not change (b) Increases (c) Decreases (d) Becomes zero

77. As per IS : 456 the deflection in RC beams is controlled by

(a) Using large aspect ratio (b) Small modular ratio

(c) Controlling $\frac{\text{Span}}{\text{Depth}}$ ratio (d) Moderating water – cement ratio

78. A simply supported RC beam of clear span 5 m and support width 300 mm is shown. The effective spam of the bean as per IS : 456 is

400 mm

250 mm

(a) 5300 mm (b) 5400 mm (c) 5200 mm (d) 5150 mm

79. For a continuous slab of 3 m × 3.5 m size minimum overall depth of slab to satisfy vertical deflection limit is

(a) 5 cm (b) 7.5 cm (c) 10 cm (d) 12 cm

80. Flexural collapse in an over-reinforced beam is due to

(a) Compressive failure (b) Tensile failure

(c) Yielding of steel (d) None of these

81. Nominal cover for tensile reinforcement in a RCC beam is measured from .... to nearest outer surface

(a) Centroid of Tensile reinforcement (b) Outer face of longitudinal tensile reinforcement

(c) Centre of transverse reinforcement (d) Outer surface of transverse reinforcement

82. For M 20 grade concrete, the maximum shear stress shall not exceed

(a) $1.6 \frac{N}{mm^2}$ (b) $1.9 \frac{N}{mm^2}$ (c) $2.8 \frac{N}{mm^2}$ (d) $2.2 \frac{N}{mm^2}$

83. For a RCC section the shape of the shear stress diagram is

(a) Wholly parabolic

(b) Wholly rectangular

(c) Above Neutral axis parabolic and below rectangular

(d) Above Neutral axis rectangular and below parabolic

84. Lateral ties in RC columns are provided to resist

(a) Bending Moment (b) Shear

(c) Buckling of longitudinal steel bars (d) Both Bending Moment and shear

85. The minimum nominal percentage of longitudinal reinforcement to be provided in a concrete pedestal as per IS code
(a) 0.4 (b) 0.2 (c) 0.15 (d) 0.1

86. In case of hand mixing of concrete, extra cement to be added is
(a) 5% (b) 10% (c) 15% (d) 20%

87. The factor of safety for
(a) Steel and concrete are same (b) For steel is lower
(c) For steel is higher (d) None of these

88. High strength steel used in pre-stressed concrete can bear how much maximum strain?
(a) 2% (b) 3% (c) 4% (d) 5%

89. For walls, columns and vertical faces of all structures, the form work is removed after
(a) 1 to 2 days (b) 3 days (c) 7 days (d) 14 days

90. For design of retaining walls, the minimum factor of safety against overturning is taken as
(a) 2 (b) 1.4 (c) 1.5 (d) 3.0

91. The minimum grade of RCC around sea coast construction should be
(a) M – 35 (b) M – 30 (c) M – 25 (d) M – 20

92. The cover of longitudinal reinforcement bar in a beam subjected to sea-spray should not be less than
(a) 30 mm (b) 70 mm (c) 75 mm (d) 80 mm

93. For durability of RCC, the water-cement ratio compared to plain concrete is
(a) More (b) Less (c) Equal (d) None of these

94. The characteristic strength of concrete is defined as that compressive strength below which not more than x% result fall
(a) x = 5% (b) x = 10% (c) x = 2% (d) None of these

95. As the span of bridge increases, how the impact factor vary?
(a) Decreases
(b) Increases
(c) Remains constant
(d) Increases upto critical value of span, then decreases

96. In pre-stressed concrete
(a) Forces of Tension and compression change but lever arm remains same
(b) Forces remain same but lever arm changes with Moment
(c) All the three change (d) None of these change

97. The coefficient of shrinkage for high grade concrete for pre-tensioned work is
(a) 3×10^{-4} (b) 3×10^{-2} (c) $\dfrac{3 \times 10^{-2}}{\log (T + 2)}$ (d) $\dfrac{3 \times 10^{-4}}{\log (T + 2)}$

98. A pre-stressed concrete beam has a cross-section with following data
$A = 46{,}400\ mm^2$, $I = 75.8 \times 10^7\ mm^4$, $Y_{bottom} = 244\ mm$, $Y_{top} = 156\ mm$

It is subjected to a pre-stressing force at an eccentricity 'e' so as to have zero stress at top fibre, the 'e' should be

(a) 66.66 mm (b) 66.95 mm (c) 104.72 mm (d) 133.33 mm

99. Consider the singly R.C. beam as below

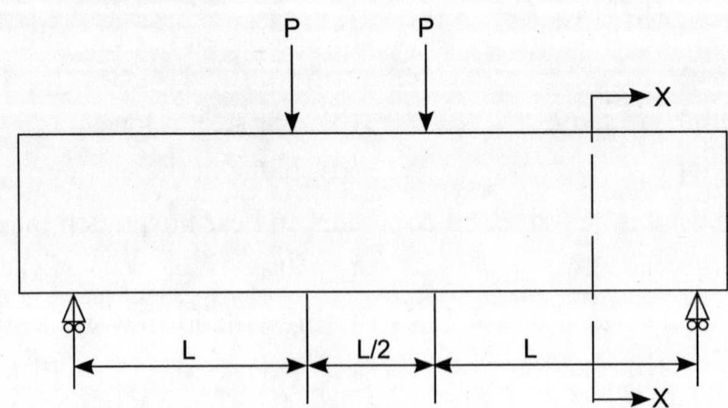

At cross-section X – X, choose the correct option.

(a) Variation strain is linear and of stress non-linear
(b) Variation of stress in linear and of strain non-linear
(c) Variation of both linear
(d) Variation of both non-linear

100. An isolated T-beam is used as a walkway. The beam is simply supported with effective span of 6 m. The effective width of the flange for the shown cross-section is

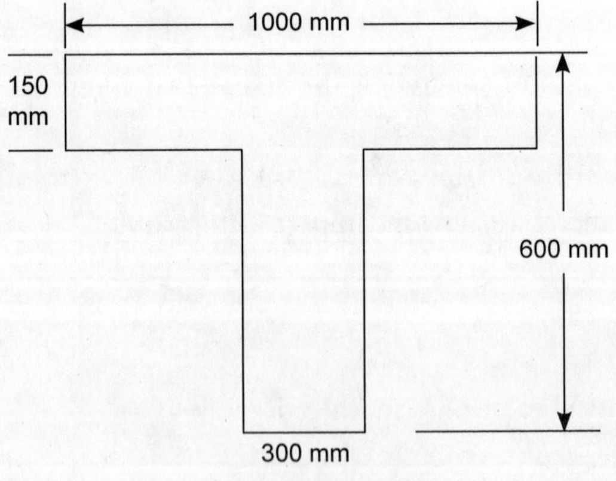

(a) 900 mm (b) 1000 mm
(c) 1259 mm (d) 2200 mm

101. Steam curing is not recommended for use with

(a) Ordinary Portland cement (b) Rapid hardening cement
(c) High alumina cement (d) Nonc of these

102. 1% of voids in a concrete mix would reduce its strength by about

(a) 5% (b) 10%
(c) 15% (d) 20%

103. According to Terzaghi, a foundation is said to be shallow if depth is

(a) Equal to its width
(b) Less than its width
(c) More than its width
(d) Either (a) or (b)

TRUE OR FALSE QUESTIONS

1. The compressive strength is only about 10 to 15% of its tensile strength. (T or F)
2. In a cantilever, the tensile zone is above the neutral axis. (T or F)
3. The stress in concrete increases in direct proportion to increase in strain. (T or F)
4. When load on a beam is increased, the tensile stress below the neutral axis will also increase. (T or F)
5. Under normal conditions the tensile stress in concrete will be more than the permissible stress. (T or F)
6. The actual Neutral axis of an under-reinforced section is below the critical neutral axis of a balanced section. (T or F)
7. The reinforcement in concrete beams takes care of Tensile stresses. (T or F)
8. As the percentage of steel in a beam increases, the depth of Neutral axis also increase. (T or F)
9. As per IS 456-2000, the value of maximum shear stress for M 40 and above concrete will be 1.9 $\frac{N}{mm^2}$. (T or F)
10. Shearing stresses in the beam are not caused by the variation of bending moments along the span. (T or F)
11. In a singly reinforced beam, if the stress in concrete reaches its allowable limit later than the steel reaches its permissible value, the beam section is said to be under-reinforced. (T or F)
12. In a cantilever beam carrying gravity load, main reinforcement is provided above the Neutral axis. (T or F)
13. In a singly reinforced beam, the effective depth is measured from the compression edge to centre of tensile reinforcement. (T or F)
14. The moment of resistance of an under-reinforced section is computed on the basis of compressive force developed in concrete. (T or F)
15. In a singly reinforced concrete beam, as the load increases only steel bars resist tension. (T or F)
16. The maximum shear stress in a rectangular beam is 1.15 times the average shear stress. (T or F)
17. Minimum grade of concrete for RCC work is M 20. (T or F)
18. Tensile strength of concrete in flexure is known as Modulus of Rupture. (T or F)
19. As per IS 456 : 2000 short term static modulus of Elasticity in $\frac{N}{mm^2}$ is EC = 5000 f_{ck}. (T or F)
20. Deep beams are designed for Bending Moments only. (T or F)

21. Shear reinforcement is provided in the form of horizontal bars. (T or F)
22. According to IS 456 : 2000 the spacing of stirrups shall not exceed a distance equal to the lever arm of the resisting moment. (T or F)
23. The longitudinal shearing stresses acting on the surface between steel and concrete are called compressive stresses. (T or F)
24. Deformed bars can be used without hooks provided anchorage requirements are adequately met with. (T or F)
25. For the reinforcement bar of diameter (ϕ), the lap length in compression should not be less than 58 ϕ. (T or F)
26. The hooks and other means of anchoring steel bar do not prevent the initial slipping of RC structure. (T or F)
27. The term 'bond' is used to describe the means by which crack between steel and concrete is prevented. (T or F)
28. As per IS code 456-2000, the permissible value of bond strees for M-15 grade concrete is limited to 1.5 $\frac{N}{mm^2}$. (T or F)
29. When the diameter of a reinforcement bar is ϕ, the anchorage value of the hook alone is equal to 8ϕ. (T or F)
30. When an inclined or horizontal member is carrying mainly axial loads, it is called a strut. (T or F)
31. The Neutral axis in a T-beam may fall within the flange when the stab is comparatively thicker. (T or F)
32. The moment of resistance of the T-beam is founded by multiplying the total tension and the lever arm. (T or F)
33. A long column is one whose ratio of effective length to its least lateral dimension exceeds 20. (T or F)
34. The effective lateral support to a column is given by transverse reinforcement. (T or F)
35. In a slab, the transverse reinforcement is provided at 45° to the span of slab. (T or F)
36. The Ratio of span to overall depth for a continuous slab spanning in one direction should not exceed 35. (T or F)
37. The distribution reinforcement is also called transverse reinforcement. (T or F)
38. The shrinkage in concrete slab causes tension cracks. (T or F)
39. For a continuous slab of two equal spans the maximum bending moment near the centre of each span is (+$\frac{\omega L^2}{8}$, L = Effective span) (T or F)
40. The reinforcement in short span is placed above the reinforcement in the long span. (T or F)
41. The maximum BM and deflection for a two way slab is much smaller than that of a one-way slab. (T or F)
42. For a slab spanning in two directions the ratio of span to depth of slab should not exceed 35. (T or F)

43. The part of a flat slab around the column, which is of greater thickness than the rest of the slab is known as column head ? (T or F)

44. The total thickness of flat slab should never be less than 20 cm. (T or F)

45. The drops are provided in flat slabs to resist shear. (T or F)

46. A RC slab built monolithically with the supporting columns and reinforced in two directions without any beams is called a flat slab. (T or F)

47. The theory of doubly reinforced section is based on the same assumptions as for singly reinforced section. (T or F)

48. The slab forms the compression flange of T-beam. (T or F)

49. The intensity of compressive stress just above the neutral axis is of very large magnitude. (T or F)

50. If the area of tensile steel reinforcement is doubled, the moment of resistance is also doubled. (T or F)

51. At the centre of beam, the shearing stresses are practically negligible. (T or F)

52. When lever arm of a RC beam is Z and width of beam is B, then the maximum shear stress in a beam subjected to shear Force F will be equal to $F \times \frac{B}{Z}$? (T or F)

53. In case of deformed bars, the value of bond stress for various grades of concrete is greater by about 20% than plain bars. (T or F)

FILL IN THE BLANKS

1. The neutral axis of a balanced section is called .... neutral axis.
2. Stiffness of a simply supported beam is satisfied if the ratio of span to depth does not exceed ....
3. A continuous beam is deemed to be 'deep beam' when the ratio of effective span (l) to overall depth is less than ....
4. Half of the main steel in a simply supported slab is bent up near the support at a distance, span divided by .... from the centre of slab.
5. The slab is designed as one way if the ratio of long span to short span is greater than ....
6. The Flexural tensile strength of M 25 grade of concrete in c as per IS 456 : 2000, is ....
7. For M 25 concrete with creep coefficient of 1.5, the long-term static modulus of elasticity in MPa as per IS : 456-2000 is, ...
8. As per the provision of IS 456 : 2000 the (short term) modulus of elasticity of M 25 grade concrete $\left(\frac{N}{mm^2}\right)$ can be assumed as ....

9. In designing for steel column of Fe 250 grade, a base plate resting on concrete pedestal of M 20 grade, the bearing strength of concrete $\left(\frac{N}{mm^2}\right)$ in limit state method design as per IS 456 : 2000 is ....

10. In the reinforced concrete beam shown made of M 25 concrete and Fe 500 steel. Total Cross-sectional area of steel 942 mm². As per limiting state Design of IS 456 : 2000, the design moment capacity in KN–m of the beam section is ....

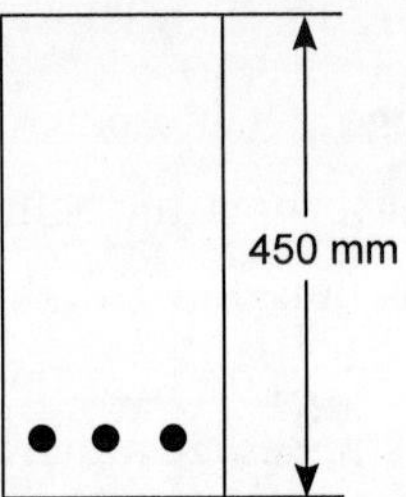

11. The maximum area of reinforcement in a RC beam for compression and tension to be provided is .... %.

12. Adoptable maximum spacing between vertical stirrups in an RCC beam of rectangular cross-section having effective depth D is .... % of D.

13. The side face reinforcement, if needed in a T-beam will be .... % of the web area.

14. Minimum cover to the main steel bars in footing should be .... mm.

15. Among slab, beam, column and footing, minimum clear cover to main steel is given in case of ....

16. When the depth of actual neutral axis in a beam is more than the depth of critical neutral axis, the beam is called .... reinforced beam.

17. In a RC slab the spacing between main bars should not exceed .... times its affective depth.

18. Due to shrinkage stresses a simply supported beam having reinforcement at the bottom only tends to deflect ....

19. A deep continuous beam with effective depth of 500 mm and cover 50 mm has effective length of .... mm.

20. The ratio of permissible shear stress in limit state method of design and working stress method of design is ....

21. If RC retaining wall is not safe against sliding a shear .... is to be provided.

22. The bond strength between steel and concrete is due to adhesion and ....

23. When a RC beam fails in bond its bond strength can be increased by using .... bars in more numbers.

24. The ratio of the maximum and minimum percentage of longitudinal reinforcement that can be provided in a column is ....

25. In very short column (Pedastal), the factor by which the effective length should not exceed the least lateral dimension is .... ?

26. The load carrying capacity of a column designed by working stress method is 500 KN, the ultimate, collapse load of the column is ....

27. According to IS : 456 the slenderness ratio for a short column should not exceed ....

28. In a combined footing the resultant of column loads passes through the CG of footing slab such that the net soil pressure obtained is ....

29. In a RC footing on soil, the thickness at the edge should not be less than .... cm.

30. For footing minimum cover as per IS : 456 should be .... mm.

31. The purpose of reinforcement in pre-stressed concrete is to impart initial .... stress in concrete.

32. High carbon content in steel causes .... in tensile strength.

33. The magnitude of force induced due to earth quake depends upon the .... of the structure.

34. The modulus of rupture of concrete gives the direct .... strength of the concrete.

35. With increase in rate of loading during testing compressive strength of concrete ....

36. Minimum grade of concrete to be used in RCC IS ....

37. In a Balanced section maximum stresses in concrete and steel reach simultaneously and are equal to ....

38. For RC exposed to sulphate attacks minimum grade required is ....

39. The rise and tread of a staircase are 15 cm and 25 cm and weight of slab on slope is 5 $\frac{KN}{m^2}$. The dead weight of horizontal area will be ....

40. The creep coefficient for 7 days is .... than creep coefficient for 28 days.

41. The most commonly used admixture which prolongs the setting time and hardening time is .... ?

42. Time of transit .... affects the workability of concrete.

43. Modulus of elasticity of steel as per IS : 456-1978 shall be taken as ....

44. The higher modular ratio shows .... compressive strength of concrete.

45. A T-shaped retaining wall mainly consists of .... cantilevers.

46. The load factor for Dead load is .... than the load factor for live load.

47. If the compacting factor of a concrete is 0.90, then the workability according to Indian standards is termed as ....

48. Loss of stress with time at constant strain in steel is called ....

49. The radial moment at the boundary of a simply supported circular slab of radius R, subjected to UDL of w will be ....

50. The main reinforcement of a RC slab consists of 10 mm bars at 10 cm spacing. If 10 mm bars have to be replaced by 12 mm bars, the spacing should be increased to ....

51. In case of a 2-way slab, the limiting deflection of the slab is primarily a function of .... span.
52. The load carrying capacity of a column designed by working stress method is 500 KN. The collapse load of the column will be ....
53. In RC slab 75 mm thick the diameter of Reinforcement bars should not be more than .... mm.
54. The number of stirrups resisting shear force in RC-beam is given by formula ....
55. If L is lever arm in RC beam, S is the total perimeter of steel bars and F is the shear force, then bond stress around steel reinforcement is ....
56. The diameter of lateral ties, in no case has to be less than .... mm.
57. The partial factor of safety for concrete as per IS : 456 : 2000 is ....
58. Modulus of rupture of concrete is a measure of .... tensile strength.
59. Maximum quantity of water needed for 50 kg of cement for M 15 grade of concrete is ....
60. Most common method of prestressing used for factory product is .... Line Method.

ANSWERS

Multiple Choice Questions

1. d	**2.** a	**3.** d	**4.** a and c	**5.** b	**6.** c	**7.** a	**8.** b	**9.** b
10. d	**11.** d	**12.** a	**13.** a	**14.** d	**15.** a	**16.** c	**17.** c	**18.** d
19. a	**20.** C	**21.** a	**22.** d	**23.** b	**24.** b	**25.** d	**26.** a	**27.** a
28. c	**29.** c	**30.** b	**31.** c	**32.** a	**33.** b	**34.** c	**35.** b	**36.** a
37. c	**38.** b	**39.** b	**40.** c	**41.** D	**42.** a	**43.** c	**44.** c	**45.** d
46. b	**47.** b	**48.** c	**49.** a	**50.** a	**51.** b	**52.** b	**53.** a	**54.** c
55. d	**56.** a	**57.** c	**58.** c	**59.** b	**60.** a	**61.** d	**62.** d	**63.** b
64. c	**65.** c	**66.** d	**67.** c	**68.** b	**69.** b	**70.** b	**71.** b	**72.** a
73. c	**74.** d	**75.** d	**76.** b	**77.** c	**78.** a	**79.** c	**80.** b	**81.** d
82. c	**83.** C	**84.** c	**85.** c	**86.** b	**87.** b	**88.** c	**89.** a	**90.** b
91. b	**92.** b	**93.** b	**94.** a	**95.** a	**96.** b	**97.** a	**98.** c	**99.** a
100. a	**101.** c	**102.** a	**103.** d					

True & False

1. F Compressive Strength is More

2. T

3. F Only upto about half of Ultimate strength

4. T

5. T

6. F Above

7. T

8. T

9. F 2.5 $\frac{N}{mm^2}$

10. F Are caused

11. T

12. T

13. T

14. F In steel

15. T

16. F 1.5 Times

17. T

18. T

19. F 5000 $\sqrt{f_{ck}}$

20. T

21. F Vertical or Inclined

22. T

23. F Bond stresses

24. T

25. F 30 ϕ

26. T

27. F Slip is prevented

28. $\frac{1.0\ N}{mm^2}$

29. F 16 ϕ

30. T

31. T

32. F Total Compression

33. F Exceeds 12

34. T

35. F At 90°

36. T

37. T

38. T

39. F $\frac{w L^2}{10}$

40. F Placed above

41. T

42. T

43. F Deep

44. F 12.5 cm

45. T

46. T

47. T

48. T

49. F Small magnitude

50. F Increased by about 22%

51. T

52. F $\frac{F}{BZ}$

53. T

Fill in the Blanks

1. Critical
2. Twenty (20)
3. 2.5
4. Seven (7)
5. Two (2)
6. $\dfrac{3.5\text{ N}}{\text{mm}^2}$
7. 10,000 M Pa
8. $\dfrac{25000\text{ N}}{\text{mm}^2}$
9. $\dfrac{9\text{ N}}{\text{mm}^2}$
10. 158 KN – m
11. 4%
12. 15%
13. 0.1%
14. 75 mm
15. Slab
16. Over
17. Three
18. Downwards
19. 1375 mm
20. 25 : 16
21. Key
22. Friction
23. Thinner
24. 2.5
25. Thee (3)
26. 750 KN
27. Twelve (12)
28. Uniform
29. 15 cm
30. 50 mm
31. Compressive
32. Increase
33. Mass
34. Tensile
35. Increases
36. M 20
37. Allowable Stress
38. M 20
39. $\dfrac{5.83\text{ N}}{\text{mm}^2}$
40. More
41. Gypsum
42. Inversely
43. $\dfrac{200\text{ N}}{\text{mm}^2}$
44. Lower
45. Three (3)
46. Less
47. Medium
48. Relaxation
49. Zero
50. 14.4 cm
51. Short
52. 750 KN
53. 8 mm
54. $\dfrac{\text{Lever Arm}}{\text{Spacing of stirrups}}$
55. $\dfrac{F}{SL}$
56. 5 mm
57. 1.5
58. Flexural
59. 32 Litres
60. Long

Section 16

River Engineering

Definitions – 25

Multiple Choice Questions – 51

True & False – 25

Fill in the Blanks – 25

DEFINITIONS

Aggradation (or Alluviation): Aggradation occurs in areas in which they supply of sediment is greater than the system is able to transport. It happens when in a reach of stream inflow of sediment is more that the outflow.

Bed Forms: A bed form is a feature that develops at the interface of fluid and movable bed. Various bed forms have been identified– like Ripples, Dunes, Antidunes, Chute and Pools. These forms are according to increasing Froude Number of flow.

Braided Channels: It is one which flows in two or more channels around alluvial islands. In plan it contains– main channels, subsidiary channels, sand bars and islands. At high flows most bars are inundated.

Cohesive Sediment: Particles of cohesive soils like loam, clay and silt, bond to one another. Due to this bond they offer resistance to erosion.

Degradation: It refers to the lowering of a fluvial surface, such as a stream bed or flood plain, through erosion process. Degradation is characteristic of channel networks in which either bed material erosion is taking place in systems that are sediment starved lift more material than deposited.

Erosion Cycle: It is the natural process of demolition of the surface of the earth and removal of material by water or wind. The demolition of mountains and transportation of products to the sea is a one way movement. However, the earth is geologically active and thus forms new rocks by uplift.

Fall Velocity: It is an important parameter in the sediment transport study on suspension and deposition of sediment. It is defined by the equation giving equilibrium between gravity force, resistance to the fall and buoyant force. The expression for fall velocity (ω) of a particle diameter (d) is given by $\omega^2 = \frac{4}{3\,C_D}\,g\,d\,\Delta$, where $\Delta = \frac{\rho_f - \rho_w}{\rho_w}$, C_D is called Drag coefficient depends on Reynold's Number (R_e).

Gradually Varied Flow (GVF): It is the case of non-uniform flow in which only small change in depth occurs in long stretches of the channel. Therefore the curvature of streamlines is very small and distribution of pressure in vertical is assumed to be hydrostatic.

Incipient Movement: As the strength of flow increases a condition is reached when some particles from the bed start moving. This is called incipient motion condition. It is usually expressed in terms of Critical Shear Stress (τ_{oc}).

Mature River: The middle part of the river is called a mature river. It is of U-shape. It may be deep and run fast. It is much wider than young river.

Meandering Channels: The meandering river in plan consist of a series of bends of alternate curvatures connected by straight reaches called crossings. These are unstable channels as the banks cave in the downstream reaches of concave bends.

Non-Cohesive Sediment (material) like sand and gravel, the particles consist of discreet particles and their movement solely depends upon their size, shape and density.

Normal Depth (y_n or y_O) or Equilibrium Depth (y_e): Uniform and steady flow conditions actually reflect a condition of equilibrium. The main feature is that-the Total Energy Line Slope (S_f), The Water Surface Line Slope (S_w) and Channel Bed Slope (S_O), all are equal. $S_f = S_w = S_O$.

Old River: As river further goes down and flows over flat land, it becomes wider and slower. As it makes up its banks (levees), it also meanders (twists and turns).

Oxbows: An oxbow lake is formed when a wider meander of a river is cut off, creating a free-standing U-shaped body of water. Kanwar lake is the largest oxbow lake in India located in Begusarai district of Bihar.

Rating Curve: In order to avoid frequent discharge measurement, at a suitable river cross-section, where a unique relationship exists between discharge and water level a curve between Discharge (Q) and water elevation (h) is prepared. It can be drawn on linear scale or log-log scale.

Rough Boundary: When the ratio of equivalent sand roughness (k_s) and thickness of Laminar sub-layer (δ') is more than 6, the surface is hydraulically rough.

Shape Factor: In terms of three mutually perpendicular axes, shape factor is defined as, $SF = \frac{c}{\sqrt{a\,b}}$.

Sinuosity: An essential characteristic of meandering channel is sinuosity. It is quantified as the ratio of thalweg length and valley length. sinuosity greater than 1.5 represents a meandering channel.

Smooth Boundary: When laminar sub-layer (δ') is much larger than equivalent sand grain roughness (k_s), the boundary is called smooth boundary.

Spurs or Groynes: These are structures in the rivers to protect the river bank and embankments (marginal bunds) against heavy currents. They are constructed in lateral direction of the river, due to spurs width of river is reduced and depth increased.

Thalweg: It is the line as drawn on a map connecting the lowest points of a valley.

Tri-axial Size: In case of sediment particles if the shape is like an ellipsoid having a, b and c as its longest, intermediate and shortest mutually perpendicular axes respectively, then representative diameter is $(a\,b\,c)^{1/3}$ and Index diameter = d.

Velocity Distribution in Alluvial Streams: The velocity distribution formulae were developed by Prandtl-Karman as,

$$\frac{u}{u_*} = 5.75 \log \frac{y}{\nu} + 5.5 \qquad \text{for smooth surface.}$$

$$\frac{u}{u_*} = 5.75 \log \frac{y}{k_s} + 8.5 \qquad \text{for rough surface.}$$

where, u = Velocity at distance y from bed.

u_* = Shear velocity = $\sqrt{\frac{\tau_0}{\rho}}$, $\tau_0 = \gamma R S_0$.

Young or Youthful River: The part of the river that is near to its source is called young river. It is often of V-shaped and flows with high velocity over stones and rocks. These rivers have many waterfalls and rapids.

MULTIPLE CHOICE QUESTIONS

1. For stable shallow meandering alluvial Rivers width (w) is given as,
   (a) 4.83 Q (b) $4.83\ Q^2$ (c) $4.83\ Q\frac{1}{2}$ (d) $4.83\ Q\frac{1}{3}$
2. According to (IS 6966-1989) for fixing Maximum Flood Discharge the flood frequency taken as,
   (a) 100-150 years (b) 250 years (c) 50-100 years (d) 150-200 years
3. In a river the average velocity normally occurs at
   (a) Surface (b) Near Bottom (c) 0.6 depth (d) 0.8 depth
4. Repelling groynes are inclined
   (a) Upstream (b) Downstream (c) Normal to bank (d) Parallel to bank
5. Currentmeter is used to measure
   (a) River Discharge (b) River velocity (c) River Depth (d) Width of river
6. Which Indian river is said to bring sorrow to West-Bengal?
   (a) Damodar (b) Kosi (c) Ganga (d) Brahmputra
7. For estimation of flood in Dicken's Formula $Q = C\ A\frac{3}{4}$, C is
   (a) 11.5 (b) 10.5 (c) 9.5 (d) 11.0
8. Which countries have vast hydro-power potential?
   (a) India-China (b) Russia-Nepal
   (c) Japan-New Zealand (d) Hungary-France
9. Because water can dissolve many things, it is called
   (a) Great dissolver (b) The winner (c) Polluter (d) Universal solvent
10. The process of avulsion is caused by
    (a) Erosion of channel (b) Sediment deposition
    (c) Widening of channel (d) Levee failure
11. The change in elevation of a stream per unit length is called
    (a) Gradient (b) Base level (c) Surface elevation (d) Discharge intensity
12. Eco-sounder is used in rivers to measure
    (a) Depth (b) Width (c) Velocity (d) Slope
13. Which of the following is not for finding flood discharge
    (a) Dicken's (b) Lacey's
    (c) Nawab Jung Bahadur (d) Ryve, Formula
14. In meandering process on which side of the bank erosion takes place?
    (a) Convex side (b) Bottom of river
    (c) Concave side (d) Due to sediment load

15. In afflux, the rise in water level is due to
(a) Obstruction by the dam
(b) Obstruction by bridge piers
(c) Obstruction by waterway
(d) Obstruction by human activities

16. Guide banks are provided in rivers to
(a) Increase the depth of flow
(b) Decrease the velocity of flow
(c) Channelise the flow of river
(d) Reduce flood peak

17. The spacing between adjacent spurs in River Training is generally kept between .... times the spur length
(a) 1.5 to 2.5 (b) 1 to 2 (c) 2 to 2.5 (d) 2.5 to 3.5

18. A flood of certain magnitude has a return period of 20 years. The probability of exceedance is
(a) 4% (b) 5% (c) 20% (d) 10%

19. The cross-section of straight reach of the river is
(a) Rectangular (b) Trapezoidal (c) Circular (d) Trough

20. In a bend of a river shoals are formed on
(a) Bottom (b) Middle (c) Concave (d) Convex

21. Between two consecutive curves of a meander the straight portion of river is called
(a) Cut-off (b) Meander belt (c) Crossing (d) Bend

22. River originating from hill and flowing in all directions, the pattern developed is
(a) Radial (b) Dendritic (c) Trellis (d) Parallel

23. Which type of river tends to develop bends?
(a) Alluvial (b) Flashy (c) Virgin (d) Himalayan

24. Which stage of river is formed by erosion?
(a) Aggrading (b) Rocky (c) Tidal (d) Boulder

25. Particles that roll and slide along the river bottom are called
(a) Bed load (b) Suspended load (c) Dissolved load (d) Wash load

26. The part of river near the source is called
(a) Virgin river (b) Young river (c) Rapid (d) Hill river

27. Longest river in the world is
(a) Amazon US (b) Hwang Ho (c) Nile (d) Indus

28. Pioneer work on sediment transport is credited to
(a) While (b) Shield (c) Chezy (d) Strikler

29. In an aqueduct, natural drainage (River, Streams) is .... the canal
(a) Parallel to (b) Below (c) Above (d) At the level of

30. If RL of canal bed and HFL of river are 212.0 m and 210.0 m respectively, then the cross Drainage (CD) work will be
(a) Superpassage
(b) Syphon
(c) Siphon aqueduct
(d) Aqueduct

31. The uplift pressure on the roof of a siphon aqueduct is maximum when
    (a) Canal is running at F.S.L and drain is dry
    (b) Drain running at the HFL and canal dry
    (c) Canal dry
    (d) Drain dry

32. Velocity of drainage water in the barrels of siphon aqueduct is normally limited to
    (a) 1 to 2 $\frac{m}{s}$ (b) 2 to 3 $\frac{m}{s}$ (c) 3 to 4 $\frac{m}{s}$ (d) 4 to 5 $\frac{m}{s}$

33. Which of the following structures is commonly used to control river flooding
    (a) Barrage (b) Dam (c) Levee (d) Weir

34. By which technique natural path of any river can be altered
    (a) Channelization (b) Dredging (c) Revetment (d) Riprap

35. By which technique river bank erosion can be stopped
    (a) Sluice gate (b) Breakwater (c) Groynes (d) Riprap

36. Which term refers to the building up of sediment at the mouth of river, where river meets sea?
    (a) Delta (b) Estuary (c) Tributary (d) Flood plain

37. A backwater curve in river engineering is
    (a) Curve representing natural flow
    (b) Water surface profile of depths upstream of any obstruction
    (c) Graph of sediment transport rate
    (d) Energy gradient line.

38. For a typical stream Mass curve period is
    (a) 6 years (b) 5 years (c) 4 years (d) 2 years

39. Temporary spurs constructed every year post-flood period are also known as
    (a) Weirs (b) Canals (c) Bunds (d) Levees

40. In terms of Manning's 'N' Chezy's C is given by
    (a) $\frac{1}{N}\sqrt{R}$ (b) $\frac{1}{N}R^{2/3}$ (c) $\frac{1}{N}R^{1/3}$ (d) $\frac{1}{N}R^{1/6}$

41. The normal depth in a 3 m wide rectangular stream having a discharge of 1.5 cumec on slope 0.0004 will be, (given Mannings N = 0.013)
    (a) 0.39 m (b) 0.58 m (c) 1.0 m (d) 1.58 m

42. Critical flow in a stream occurs at Froude Number
    (a) = 1.0 (b) > 1.0
    (c) < 1.0 (d) Can occur at any Froude Number

43. A Gradually Varied Flow (GVF) is valid when
    (a) Channel slope is small (b) Channel is prismatic
    (c) Pressure distribution is hydrostatic (d) All these

44. Which one of the following generally decreases downstream along the length of a stream
(a) Channel width (b) Channel depth (c) Gradient (d) Velocity

45. What type of flow in rivers can carry gravel and cobbles?
(a) Laminar (b) Turbulent (c) Both a & b (d) Neither a, nor b

46. What type of material is most likely to be transported as suspended load
(a) Clay (b) Silt
(c) Sand (d) Depends upon energy of stream

47. What is the probability of a 50 years flood occurring next year?
(a) 2% (b) 10% (c) 50% (d) Cannot say

48. If sea level were to rise the slope of the longitudinal profile of many rivers would
(a) Increase (b) Decrease
(c) First increase then decrease (d) First decrease then increase

49. Which of the following processes is the most important erosional force
(a) Streams (b) Glaciers (c) Wind (d) Waves

50. In the figure here arrow indicates

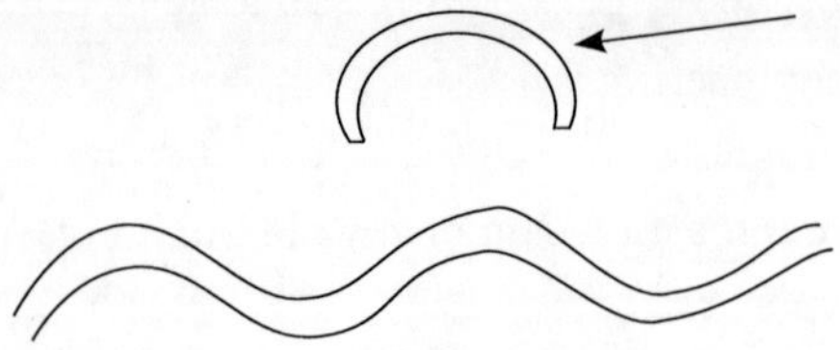

(a) Meander (b) Point bar (c) Natural Levee (d) Oxbow lake

51. Natural Levees are formed by
(a) Beavers (b) Humans (c) Floods (d) Erosion

TRUE OR FALSE QUESTIONS

1. Attracting groyens are built normal to the bank. (T or F)
2. Retrogression of the bed level downstream of weir occurs due to small percentage of silt. (T or F)
3. The sinuosity of meander is the ratio of Curved Length and straight distance. (T or F)
4. A deficit of sediment charge in river may cause meandering. (T or F)
5. The meandering of river generally occurs in boulder stage of the river. (T or F)
6. Wind speed is measured by current meters. (T or F)
7. According to CWC average river flow in India is 1880 km^3. (T or F)
8. Drainage density and stream density are same. (T or F)
9. The total length of streams in a catchment of area 8000 km^2 is 3200 km, the drainage density (Dd) is $0.4 \frac{m}{m^2}$. (T or F)

10. The total number of streams in Kosi basin is 6583 km and its basin area is 69,300 km^2, the stream density (Ds) is 10.5 $\frac{m^2}{m}$. (T or F)
11. If a river has catchment area of 70,000 km^2 and the Length is 700 km, its form factor is 0.142. (T or F)
12. A river before joining the sea or ocean gets divided into branches. (T or F)
13. A river which does not change its alignment or slope is called stable river. (T or F)
14. Meander ratio is ratio of Meander length (L) to meander belt (B). (T or F)
15. A braided river flows into two or more channels. (T or F)
16. Sinuosity is defined as the ratio of Valley length to Thalweg length. (T or F)
17. Pioneer work on sediment transport is done by Shield. (T or F)
18. River Engineering does not include sediment study. (T or F)
19. Dams are made only to generate electricity. (T or F)
20. River engineering projects often include the creation of artificial wetlands to improve water quality and provide wildlife habitat. (T or F)
21. To improve navigable waterways dredging is done. (T or F)
22. Check dams are small dams to slow down water flow and reduce soil erosion. (T or F)
23. Flood forecasting and early warning systems are not part of river engineering. (T or F)
24. River engineering projects do not require any Environmental Impact Assessment. (T or F)
25. Floodplain zoning is a non-structural river engineering approach that restricts certain types of development in flood-prone areas. (T or F)

FILL IN THE BLANKS

1. Sediment transported by the river is divided as Bed Load and ....
2. Percentage of inflowing sediment which is retained in the reservoir is termed as, ...
3. Pul's and Muskinghum method are used in ....
4. The tractive force per unit area (τ) on the channel is ....
5. In the simplest form the critical shear stress (τ_c) in a channel is given by Shield as ....
6. The sediment moving in rivers is termed as Suspended Load and ....
7. Primary cause of meandering in river is excessive ....
8. A repelling Groyne is inclined towards ....
9. In general the spacing of Groyne is .... to .... times the length of Groyne.
10. The soil in Indo-Gangetic plains is ....
11. The velocity in Rivers is measured by ....

12. Line joining the lowest points on the bed of river is called ....
13. Sinuosity is the ratio of Thalweg length and .... length.
14. Under a backwater curve sediment .... takes place.
15. The lines of equal velocity in a river cross-section are called ....
16. Maximum sediment load is carried by river ....
17. For Manning's N, strickler gave the formula ....
18. In terms of shear stress at river bed (τ_0) and density of water (ρ), shear velocity (u_*) is expressed as ....
19. For a drawdown curve $\frac{dy}{dx}$ is ....
20. Shield's curve is drawn between Shear Reynold's Number and .... function.
21. Antidunes are formed at Froude Number of around ....
22. For suspension of sediment in the flow .... is responsible.
23. The most popular method for discharge measurement is the .... method.
24. Synthetic Unit hydrograph is prepared for .... basins.
25. Under the National Perspective Plan (NPP), Ministry, National Water Development Agency has identified .... links under Himalayn Rivers component.

ANSWERS

Multiple Choice Questions

1. c	**2.** c	**3.** c	**4.** a	**5.** b	**6.** a	**7.** a	**8.** b	**9.** d
10. b	**11.** a	**12.** a	**13.** b	**14.** c	**15.** a & b	**16.** c	**17.** c	**18.** b
19. d	**20.** d	**21.** c	**22.** a	**23.** a	**24.** b	**25.** a	**26.** b	**27.** c
28. b	**29.** b	**30.** d	**31.** b	**32.** b	**33.** c	**34.** a	**35.** d	**36.** a
37. b	**38.** a	**39.** c	**40.** d	**41.** b	**42.** a	**43.** d	**44.** c	**45.** b
46. d	**47.** a	**48.** b	**49.** a	**50.** d	**51.** c			

True & False

1. F	**2.** T	**3.** T	**4.** F
5. F	**6.** F	**7.** T	**8.** F
9. T	**10.** F	**11.** T	**12.** T
13. T	**14.** F $\frac{B}{L}$	**15.** T	**16.** F $\frac{\text{Thalweg}}{\text{Valley}}$
17. T	**18.** F	**19.** F It is multipurpose	
20. T	**21.** T	**22.** T	**23.** F
24. F	**25.** T		

Fill in the Blanks

1. Suspended Load	**2.** Trap efficiency	**3.** Flood Routing	**4.** $\gamma R S_0$
5. 0.98 d	**6.** Bed Load	**7.** Sediment Load	**8.** Upstream
9. 2 to 2.5	**10.** Alluvial Soil	**11.** Current meters	**12.** Thalweg
13. Valley	**14.** Deposition	**15.** Isovels	**16.** Mississippi
17. $\frac{(d_{50})^{1/6}}{21}$	**18.** $\sqrt{\frac{\tau_0}{\rho}}$	**19.** Negative	**20.** Entrainment
21. 1 (One)	**22.** Turbulence	**23.** Area-velocity	**24.** Ungauged
25. 14			

SECTION 17

Soil Mechanics and Foundation Engineering

Definitions – 51

Multiple Choice Questions – 121

True & False – 102

Fill in the Blanks – 102

DEFINITIONS

Active and Passive Earth Pressures: Acting earth pressure is the earth pressure acted by the soil on the back of the retaining wall. Passive earth pressure is the pressure acted by the retaining wall on the soil. In other wards active pressure occurs when the wall tends to move away from the soil.

Angle of Repose (Critical Angle): For a granular material it is steepest angle of descent or dip relative to the horizontal plane on which the material can be piled without slumping. For dry sand it is about 35° and for wet sand 45°.

Bearing Capacity: The supporting power of a soil or rock is referred to as its bearing capacity. It depends upon the type of soil, its density and shear strength.

Black Cotton Soil: It is a unique soil type highly favourable for cultivation of cotton. Its high clay content and black colour due to presence of titaniferous magnetite make is suitable for growing cotton. It is highly plastic and has low bearing capacity.

Bulk Density of Soil (Moist Density): It is defined as the dry weight of the soil divided by its volume $\rho = \frac{M}{V}$. Typically expressed in terms of $\frac{g}{cm^3}$ or $\frac{kg}{m^3}$.

Capillary Rise: The interstices or the pores of soil mass may be looked upon as a series of capillary tubes, extending vertically above water table. The rise of water in capillary tubes or the fine pores of the soil, is due to the existence of surface tension. The height of rise depends upon the type of soil. For finer soils rise is more.

Coefficient of Absolute Permeability (K): While coefficient of permeability depends on properties of soil mass and also on properties of permeant (i.e. water), the coefficient of absolute permeability depends only on properties of soil. It is defined by expression $K = k\left(\frac{\mu}{\gamma_w}\right)$. (The units are m^2 or L^2 or Darcy) 1 Darcy = 0.987×10^{-8} cm^2.

Coefficient of Consolidation: It is defined as the rate at which saturated soil undergoes consolidation. Its unit is $\frac{cm^2}{s}$. Formula used is $C_v = 0.197\frac{d^2}{t_{90}}$.

Coefficient of Permeability (k): In Darcy's law (V = ki), k represents coefficient of permeability, it is defined as average velocity of flow that will occur through the total cross-section area of soil under unit hydraulic gradient. The units are same that of velocity ($\frac{m}{s}$ or LT^{-1}).

Contact Pressure: It is defined as the vertical pressure acting at the surface of contact between the base of a footing and the underlying soil mass. The contact pressure lies between the extreme conditions for uniform and non-uniform distribution for flexible and rigid footings.

Coulomb-Terzaghi Equation: It is written as, $S = C + \sigma \tan \phi$, where C is Cohesion and ϕ, angle of internal friction or shear resistance, σ is normal stress. S gives soil strength.

Darcy's Law: Darcy on the basis of his experiments for laminar flow condition found that in a saturated soil, the rate of flow (discharge per unit time) is proportional to **hydraulic gradient**, q = k i A.

Degree of Saturation: It is defined as the ratio of the volume of water present in a soil mass to the total volume of voids. $S = \frac{V_w}{V_v}$.

Discharge Velocity and Seepage Velocity: The velocity of flow (v) is the rate of discharge of water per unit of total cross-sectional area (A) of soil. Since the flow takes through the voids, the actual or the true velocity of flow will be more than the discharge velocity. This true velocity is called the seepage velocity. Seepage velocity > Discharge velocity. $v_s = \frac{v}{n}$.

Dry Density of Soil: It is the ratio of mass of solids per unit of total volume (prior to drying) of the soil mass. $\rho_d = \frac{M_d}{V}$.

Effective and Neutral Pressures: At any plane in a soil mass, the total stress (or unit pressure) σ is the total load per unit area. It has two distinct components-Intergranular Pressure (Effective Pressure) and Neutral Pressure (Pore Pressure). Effective pressure σ′ is the pressure transmitted from particles through their point of contact. Pore pressure is transmitted through the pore fluid (u). $\sigma = \sigma' + u$.

Flocculent Soil Structure: It is present in the clay particles which have large surface area. These are charged particles which have positive charge at the edge and negative charge on the face. The positive charge attracts the negative charge resulting in flocculent structure.

Grouting: The process of injecting a fluid material (Grout) into soil or rock to fill voids, improve stability, reduce permeability and fill cracks. An important grout is water-cement slurry.

Honeycomb Soil Structure: Such structure exists in grains of silt or rock flour smaller than 0.02 mm diameter and larger than 0.0002 mm. The grains coming in contact are held until miniature arches are formed, bringing over relatively large void spaces forming honeycomb structure. The structure has high void ratio.

Hydraulic Gradient: When a soil sample of length (L) and cross-sectional area (A) is subjected to differential head of water (h_1 and h_2) the hydraulic gradient $i = \frac{h_1 - h_2}{L}$, then velocity (v) will be, $v = k\,i$ and discharge, $q = k\,i\,A$.

Isobars: It is a curve or contour joining all points below the ground surface of equal vertical pressure. It is of the shape of a bulb. The zone in a loaded soil mass bounded by an isobar of given vertical pressure intensity is called Pressure Bulb.

Liquid Limit (LL): It is the water content where the soil starts to behave like a liquid. At this the soil has very low shear strength. At liquid limit soil changes from plastic stage to liquid stage.

Liquidity Index (LI): It is given by the following formula. $LI = \frac{(PL - \text{Natural Water Content})}{PI}$. Soils with LI = 1 or more will be closer to the liquid state. A LI of 0 or lower Indicate that soil is harder and brittle.

Mohr's Circle: It is an important tool for finding the stress or strain on any plane for a stressed body. It helps to visualise the normal stress and shear stress on any plane and helps to determine Principal stress and maximum shear stresses along with their planes.

Optimum Moisture Content (OMC): Proper compaction of soil ensures strong load bearing surface. For soil to compact properly it must be enough moist i.e. at which water content it attains maximum dry density. The moisture content at which soil achieved maximum dry density is called Optimum Moisture Content. The moisture content should neither be more nor less than OMC.– for best soil strength.

Pile Foundation: It is a kind of deep foundation, is actually a slender column or cylinder made of concrete or steel. It supports the structure's load by end bearing or skin friction.

Plastic Index (PI): It is given by Atterberg and is expressed as, PI = PL – LL. It indicates the size of the range between the two boundaries. Soils with high PI have higher clay content.

Plastic Limit (PL): It is the water content at the change from a plastic to semi-solid state. The test involves repeatedly rolling a soil sample into a thread until it reaches a point where it crumbles.

Porosity: For a given soil sample it is the ratio of the volume of voids to the total volume. $n = \frac{Vv}{V}$.

Principal Plane: In a loaded soil mass three mutually perpendicular planes can be identified where stress is wholly normal and shear stress is zero. These are known as Principal planes.

Pycnometer: It is a device used to measure the density of solids, liquids and soils. It is a glass flask with a ground glass stopper and a capillary tube. It uses Archimede's Principle to find density. It is also known as density meter.

Quick Sand: When flow is in upward direction, effective pressure is reduced. When seepage pressure is equal to submerged weight, effective pressure becomes zero. The soil loses all its shear strength and soil particles have a tendency to move up. The lifting of particles is called quick condition, boiling condition or quick sand.

Raft Foundation (or Mat Foundation): It is a combined footing that covers the entire area below a structure and supports all the walls and columns. It is basically a large concrete slab resting on a large area of soil reinforced with steel.

Range of Particle Size: The range of particles sizes is very large from Boulders with size of over 300 mm down to Clay particles that are below 0.002 mm. Some clay particles are even smaller 0.001 mm which behave as colloids and do not settle in water.

Relation between Porosity and Void Ratio: By the definition of both the relation can be easily derived as, $e = \frac{n}{1-n}$ and $n = \frac{e}{1+e}$.

Safe Bearing Capacity: It is the maximum load per unit area which the soil can withstand without any displacement or settlement or shear failure. It is given is (k Pa).

Saturated Density of Soil: When the soil is saturated, its bulk density is called saturated density (ρ_{sat}).

Shallow and Deep Foundation: According to Terzaghi, a foundation is shallow if its depth is equal to or less than its width. The examples of shallow foundation are- spread footing, strap footing and of deep foundation are– pier foundation, pile foundation or well foundation.

Shrinkage Limit (SL): It is the water content where the further loss of moisture does not cause a decrease in specimen volume.

Shrinkage of Soils: Due to decrease in water content in the soil shrinkage takes place. When a staurated soil is allowed to dry, it leads to a compression in the soil structure and consequent reduction in volume. Shrinkage is more prominent in clays. Presence of sand reduces the shrinkage.

Sieve Analysis: It is the method used to determine the particle size distribution of granular soils by passing the soil through a series of sieves.

Soil as a 3-Phase System: Soil is a three phase system consisting of solid particles (soil grains), Liquid (water and gas (air). The void space between soil grains is filled with water and air.

Soil Classification Based on Grain Size: In the Indian Standard Soil Classification System (ISSCS), soils are classified into 3 groups, Very coarse, Coarse and Fine Soils.

Soil Cohesion: It is the property of soil that describes how well the soil particles stick together. It is influenced by the factors such as water content, clay content, organic matter and cementation. Cohesive soils are fine grained soils.

Soil Consolidation: It refers to the process by which the volume of a partially or fully saturated soil decreases due to an applied stress. When a load is applied to a low permeability soil, it is initially carried by the pore water resulting in rapid increase in pore water pressure. This excess pore water pressure is dissipated as water drains away and the pressure is transferred to the soil skeleton, which is gradually compressed resulting in settlement or consolidation of soil mass.

Soil Division: Soils are broadly divided into three divisions– Coarse grained soils, Fine grained soils and Highly Organic Soils.

Soil Liquefaction: A phenomenon where saturated soil loses strength and stiffness due to applied stress, often occurring during seismic events, leading to potential foundation failure.

Subgrade: The layer of soil or other material beneath a pavement or foundation that provides support for the structure above.

Submerged Density (or Buoyant density): It is the submerged mass of soil solids per unit of the total volume of solid mass. $\rho' = \frac{M_d\,(sub)}{V}$, Also $\rho' = \rho_{sat} - \rho_w$.

Swelling of Soils : When water is added to a soil which has shrunk, there is a reduction in compressive stresses in solid grains. This results in elastic expansion of soil mass, causing swelling. It is a reversible process.

Void Ratio: For a given soil mass it is the ratio of the volume of voids to the volume of soil solids. $e = \frac{V_v}{V_s}$.

Water Content (w): Also called moisture content is defined as the ratio of weight of water (W_w) to the weight of solids (W_s or W_d) in a given mass of soil, $w = \frac{W_w}{W_s} \times 100$.

MULTIPLE CHOICE QUESTIONS

1. Soil is formed by
   (a) Flood (b) Earthquakes (c) Weathering (d) Landslide
2. Who is called 'Father of Soil Mechanics'?
   (a) Atterberg (b) Terzaghi (c) Hazen (d) Rankine
3. The volume of voids (Vv) is equal to the Sum of
   (a) Volume of air + volume of solids (b) Volume of air + Volume of water
   (c) Volume of solids + Volume of water (d) Volume of water + Weight of water

4. The Units of coefficient of absolute permeability (K) is

(a) $\frac{kg}{cm}$ (b) $\frac{m}{s^2}$ (c) m^2 (d) $\frac{cm}{s}$

5. The minimum water content at which the soil just begins to crumble when rolled into 3 mm diameter threads, is known as

(a) Liquid limit (b) Plastic limit (c) Shrinkage limit (d) Permeability limit

6. The quantity of seepage through a soil is proportional to

(a) Coefficient of permeability (b) Head loss
(c) Neither a nor b (d) Both a and b

7. The ratio of volume of Voids (V_v) to volume of Solids (V_f) is called

(a) Porosity (b) Specific gravity (c) Void Ratio (d) Water Content

8. To Coefficient of compressibility of soil is the Ratio of

(a) $\frac{\text{Stress}}{\text{Strain}}$ (b) $\frac{\text{Strain}}{\text{Stress}}$ (c) $\frac{\text{Stress}}{\text{Settlement}}$ (d) $\frac{\text{Loading}}{\text{Settlement}}$

9. The water content of soil is defined as the Ratio of

(a) $\frac{\text{Volume of water}}{\text{Volume of soil}}$ (b) $\frac{\text{Volume of water}}{\text{Volume of Void}}$

(c) $\frac{\text{Weight of water}}{\text{Weight of air in voids}}$ (d) $\frac{\text{Weight of water}}{\text{Weight of Solids}}$

10. Accurate determination of water content is made by

(a) Calcium carbide method (b) Sand bath method
(c) Alcohol method (d) Oven – drying method

11. The soil transported by wind is called?

(a) Loess (b) Talus (c) Drift (d) Dune sand

12. Void Ratio (e) of any soil can be

(a) $e < 1$ (b) $e > 1$ (c) $e \geq 1$ (d) All these

13. Which of the following is finest particles of sediment?

(a) Clay (b) Fine sand (c) Silt (d) Colloid

14. For a well graded soil uniformity coefficient (Cu) is

(a) Cu = 1 (b) Cu = 8 to 11 (c) Cu > 4 to 6 (d) Cu < 4

15. Uniformity coefficient is defined

(a) $\frac{d_{60}}{d_{10}}$ (b) $\frac{d_{50}}{d_{10}}$ (c) $\frac{d_{40}}{d_{10}}$ (d) $\frac{d_{25}}{d_{10}}$

16. Coarse grained soils are best compacted by a

(a) Rubber tyre roller (b) Sheep foot roller
(c) Drum roller (d) Vibratory roller

17. A cohesive soil specimen is subjected to a vertical compressive load, the inclination of the cracks to the horizontal is

(a) 90° (b) 45° (c) 22.5° (d) 0°

18. Total lateral earth pressure is proportional to

(a) Depth of soil (b) Square of depth of soil
(c) Angle of Internal friction (d) None of these

19. The ratio of the undisturbed shear strength to the remolded shear strength in cohesive soils undermined condition is

(a) Zero (b) One (c) > 1 (d) < 1

20. The passive earth pressure in a soil mass is proportional to

(a) $\tan^2\left(45° + \frac{\phi}{2}\right)$ (b) $\tan^2\left(45° - \frac{\phi}{2}\right)$ (c) $\tan^2(45° - \phi)$ (d) $\tan^2\phi$

21. According to IS specification, the minimum depth of foundation in sand and clay should be respectively

(a) 600 mm and 700 mm (b) 1 m to 800 mm
(c) 1 m to 1.2 m (d) 800 mm to 900 mm

22. For uniform spherical particles of sand, the ratio of void ratio in loose and dense states is

(a) 2.6 (b) 3.5 (c) 4.6 (d) 3.0

23. A soil having particles of nearly the same size is known as

(a) Well graded (b) Uniformity graded
(c) Poorly graded (d) Gap graded

24. Soils most susceptible to liquefaction are

(a) Saturated dense sands
(b) Saturated clays of uniform size
(c) Saturated gravel
(d) Saturated fine and medium sands of uniform particle size

25. The upstream slope of an earth dam under steady seepage condition is

(a) Equipopential line (b) Phreatic line
(c) Flow line (d) Seepage line

26. Westergaard's analysis for stress distribution beneath loaded areas is applicable to

(a) Sandy soils (b) Clayey soils (c) Stratified soils (d) Silty soils

27. The process by which a mass of saturated soil is caused by external forces to suddenly lose its shear strength and to behave as a fluid is called

(a) Piping (b) Slide (c) Quick condition (d) Liquefaction

28. When the degree of consolidation is 50%, the time factor is

(a) 0.2 (b) 0.5 (c) 1.0 (d) 2.0

29. Lime stabilization is very effective in treating

(a) Sandy soils (b) Silty soils (c) Non-plastic soils (d) Plastic clayey soils

30. The water content of soil, which represents the boundary between plastic state and liquid state, is known as

(a) Liquid limit (b) Plastic limit (c) Shrinkage limit (d) Plasticity index

31. The plasticity index is given by
(a) Plastic limit – Liquid limit
(b) Plastic limit – Shrinkage limit
(c) Shrinkage limit – Liquid limit
(d) Liquid limit – Plastic limit

32. A soil has a discharge velocity of $6 \times 10^{-7} \frac{m}{s}$ and void ratio of 0.5, what is its seepage velocity?
(a) $3 \times 10^{-7} \frac{m}{s}$
(b) $6 \times 10^{-7} \frac{m}{s}$
(c) $12 \times 10^{-7} \frac{m}{s}$
(d) $18 \times 10^{-7} \frac{m}{s}$

33. Contact pressure for flexible footing on any type of soil is
(a) Maximum of center
(b) Maximum at edged
(c) Uniform
(d) No trend of variation

34. Due to rise in water table, the effective stress in soil
(a) Increases
(b) Decreases
(c) May increase or decrease
(d) Does not change

35. Who is said to be father of soil mechanics?
(a) Hazen
(b) Darcy
(c) Rankine
(d) None of these

36. The Porosity (n) and Void Ration (e) are related as
(a) $n = \frac{1+e}{e}$
(b) $n = \frac{1-e}{e}$
(c) $n = \frac{e}{1+e}$
(d) $\frac{e}{1-e}$

37. In sieve analysis as per IS : 460 - 1962 the minimum aperture in mm is
(a) 1.0
(b) 0.5
(c) 0.075
(d) 0.045

38. The terminal velocity of sinking of a spherical particle is proportional to
(a) Diameter of particle D
(b) D^2
(c) $D^{0.64}$
(d) $D^{1/3}$

39. A penetrometer is used to determine
(a) Liquid Limit
(b) Plastic Limit
(c) Density
(d) Moisture Content

40. Bulk density of soil is
(a) $\frac{\text{Mass of soil solids}}{\text{Volume of solids}}$
(b) $\frac{\text{Mass of solids}}{\text{Total volume before drying}}$
(c) $\frac{\text{Total mass of soil}}{\text{Total volume}}$
(d) $\frac{\text{Submerged mass}}{\text{Total volume}}$

41. Capillary rise in clays is of the order
(a) 30 – 100 cm
(b) 10 to 30 m
(c) > 30 m
(d) < 10 m

42. Coefficient of permeability is maximum in which of the following soils?
(a) Silt
(b) Fine sand
(c) Coarse sand
(d) Silty sand

43. Darcy's Law is valid for Reynolds Number
(a) ≤ 1
(b) > 1
(c) Upto 2000
(d) Upto 5

44. Absolute permeability (K) is given by expression
(a) $K = k\left(\frac{\eta}{\gamma_w}\right)$
(b) $K = k\left(\frac{\gamma_w}{\eta}\right)$
(c) $K = \frac{1}{k}\left(\frac{\eta}{\gamma_w}\right)$
(d) $K = k\,(\eta\gamma_w)$

(where, K is permeability, η is viscosity)

45. Which of the following soils has more plasticity Index?
(a) Sand (b) Silt (c) Clay (d) Gravel

46. Effective stress on soil
(a) Increases void ratio and decreases permeability
(b) Increases both void ratio and permeability
(c) Decreases both void ratio and permeability
(d) Decreases void ratio and increases permeability

47. Which of the following soils is expected to have permeability of $0.8 \frac{mm}{s}$?
(a) Gravel (b) Silt (c) Clay (d) Sand

48. Total number of stress components at a point within a soil mass loaded at its boundary is?
(a) 3 (b) 9 (c) 6 (d) 16

49. Rate of consolidation
(a) Increases with decrease in Temperature (b) Increase with increase in Temperature
(c) Does not depend on Temperature (d) Is unaffected by permeability of soil

50. Rankine's theory of earth pressure assumed that the back of the wall is
(a) Plane & Smooth (b) Plane & Rough
(c) Vertical & Smooth (d) Vertical & Rough

51. An oven-dried sample of soil has a volume of 250 cm³ and mass 0.5 kg. Its dry density is
(a) $\frac{2g}{cm^3}$ (b) $1.25 \frac{g}{cm^2}$ (c) $5 \frac{g}{cm^3}$ (d) $12.5 \frac{g}{cm^3}$

52. If a soil sample has void ratio e = 0.54 and G = 2.7, the shrinkage limit will be
(a) 13.5% (b) 54% (c) 27% (d) 20%

53. If a soil sample has dry density $= 1.82 \frac{g}{cm^3}$ the shrinkage Ratio (SR) is
(a) 18.2 (b) 1.82 (c) 0.182 (d) 1.082

54. A soil deposit has a void ratio of 0.89. If the void ratio is reduced to 0.59 by compaction, the percentage volume loss is approximately
(a) 5.9% (b) 8.9% (c) 15.8% (d) 30%

55. In a standard Proctor Test volume of the mould is
(a) 945 ml (b) 459 ml (c) 549 ml (d) 1000 ml

56. Sheep foot rollers are suitable for compacting
(a) Coarse sand (b) Cohesive soil (c) Fine sand (d) Gravel

57. A soil has a discharge velocity of $5.9 \times 10^{-7} \frac{m}{s}$ and a void ratio of 0.49, its seepage velocity is
(a) $12 \times 10^{-7} \frac{m}{s}$ (b) $3 \times 10^{-7} \frac{m}{s}$ (c) $6 \times 10^{-7} \frac{m}{s}$ (d) $18 \times 10^{-7} \frac{m}{s}$

58. Due to rise in water table the pore pressure in soil
(a) Decreases (b) Does not change (c) Increases (d) Can change either way

59. A saturated soil sample has a dry unit weight of 1850 $\frac{kgf}{m^3}$ and specific gravity 2.64. The water content of soil will be

(a) 0.16 (b) 0.55 (c) 0.36 (d) 0.48

60. The angle of internal friction is least for

(a) Angular grained loose sand (b) Angular grained dense sand
(c) Round grained loose sand (d) Clays

61. The Neutral Stress in a soil mass is

(a) Force per Neutral area (b) Force per effective area
(c) Stress taken by pore water (d) Stress taken by grains

62. The property of soil which allows it to deform rapidly without rupture, elastic rebound and volume change is called

(a) Porosity (b) Plasticity (c) Permeability (d) Ductility

63. Flow net in a soil mass is utilized to find

(a) Exit gradient (b) Seepage
(c) Hydrostatic Pressure (d) All these

64. If the coefficient of active pressure k_a in a soil is $\frac{1}{3}$, the passive pressure will be

(a) 3 (b) $\frac{1}{3}$ (c) $\frac{2}{3}$ (d) 1

65. For a loose sand sample and a dense sand sample consolidated to same Effective Stress

(a) Ultimate strength is same and also peak strength is same
(b) Ultimate strength is not same but peak strength is same
(c) Ultimate strength is same but peak strength of dense sand is greater than loose sand
(d) Ultimate strength is same but peak strength is not same

66. For determination of permeability of clayey soil which method is best suited

(a) Constant head (b) Horizontal permeability test
(c) Falling head (d) Capillary test

67. Physical properties of a permeant which influance permeability are

(a) Viscosity only (b) Viscosity and unit weight
(c) Unit weight only (d) None of these

68. If the volume of voids is equal to volume of solids in a soil mass, then the value of porosity and void ratio respectively are

(a) 1.0 and 0.0 (b) 0.0 and 1.0 (c) 0.5 and 1.0 (d) 1.0 and 0.5

69. If permeability of a soil is 0.8 $\frac{m}{s}$, the type of soil is

(a) Gravel (b) Sand (c) Silt (d) Clay

70. If voids of a soil mass are full of air only, it is called

(a) Dry soil (b) Dehydrated soil (c) Air entrained soil (d) Saturated soil

71. If the degree of saturation of partially saturated soil is 60%, then air content of soil is
(a) 60% (b) 100% (c) 80% (d) 40%

72. For accurate determination of the water content of soil the most accurate method is
(a) Pycnometer Method (b) Sand bath Method
(c) Oven drying Method (d) Calcium-Carbide Method

73. For which of the following Plastic Index is zero
(a) Clay (b) Sand (c) Silt (d) Clayey silt

74. Which of the following soil has maximum Plastic Index?
(a) Clay (b) Sand (c) Silt (d) Clayey silt

75. According to IS-classification, the range of silt size particle is
(a) 4.75 mm to 2.00 mm (b) 2 mm to 0.425 mm
(c) 0.425 mm to 0.075 mm (d) 0.075 mm to 0.002 mm

76. Cohesive grained soils are best compacted by roller,
(a) Drum (b) Rubber type (c) Vibratory (d) Sheep foot

77. Which of the following grain size is finest?
(a) D_{15} (b) D_{10} (c) D_{85} (d) D_{50}

78. The angle that Coulomb's failure envelope makes with horizontal is called
(a) Cohesion (b) Angle of Internal friction
(c) Angle of Repose (d) Capillary potential

79. Soil deposits formed by wind transport are called
(a) Alluvial (b) Glaciofluvial (c) Gravity (d) Aeolian

80. The plasticity index of a highly plastic soil is about
(a) 10 – 20 (b) Greater than 40 (c) 20 – 40 (d) Less than 10

81. In which soil, the capillary rise will be least
(a) Fine gravel (b) Coarse sand (c) Medium sand (d) clay

82. The standard penetration test is useful to measure
(a) Shear strength of soft clays (b) Shear strength of sands
(c) Consistency of clays (d) All these

83. At shrinkage limit, the soil is
(a) Dry (b) Saturated (c) Partially saturated (d) None of these

84. A shallow foundation is usually defined which has depth
(a) Less than 60 cm (b) Less than its width
(c) Less than 1 m (d) Less than 30 cm

85. The weakest bond in soils is
(a) Ionic (b) Covalent (c) Hydrogen (d) Metallic bond

86. Cohesionless soils are formed due to
(a) Oxidation (b) Hydration
(c) Physical disintegration (d) Chemical decomposition

87. According to IS-classification, the soils can be classified into number of groups

(a) 18 (b) 3 (c) 7 (d) 15

88. Permeability of soil varies

(a) Inversely as square of grain size (b) As square of grains size

(c) As grain size (d) Inversely as void ratio

89. Valid range for n, the percentage voids is

(a) $0 < n < 100$ (b) $0 < n \leq 100$ (c) $n > 0$ (d) $n \leq 0$

90. If Plasticity Index of a soil mass is zero, soil is

(a) Sand (b) Silt (c) Clay (d) Clayey silt

91. The critical hydraulic gradient i_c of a soil mass of specific gravity G and void ratio e is given by

(a) $\frac{G+1}{1-e}$ (b) $\frac{G-1}{1+e}$ (c) $\frac{G+1}{1+e}$ (d) $\frac{G-1}{1-e}$

92. The consistency of a saturated cohesive soil is affected by

(a) Water content (b) Particle size distribution

(c) Density Index (d) Coefficient of Permeability

93. If porosity of a soil sample is 20%, the void ratio will be

(a) 0.20 (b) 0.80 (c) 1.00 (d) 0.25

94. If soil is dried beyond its shrinkage limit, the volume change will be

(a) Large (b) Moderate (c) low (d) None

95. Relationship between submerged unit weight and saturated weight involves which principle?

(a) Equilibrium of floating bodies (b) Archimedes Principle

(c) Stokes Law (d) Darcy's Law

96. A soil sample has a void ratio 0.5, its porosity will be nearly

(a) 50% (b) 66% (c) 100% (d) 33%

97. The Liquid Limit (LL), Plastic Limit (PL) and Shrinkage Limit (SL) of cohesive soil satisfy the relation

(a) LL > PL > SL (b) LL > PL < SL

(c) LL < PL < SL (d) LL < PL > SL

98. The ratio of saturated unit weight to dry unit weight of a soil is 1.25. If the specific gravity of Solids (Gs) is 2.65. The void ratio of the soil is

(a) 0.625 (b) 0.944 (c) 0.663 (d) 1.325

99. The void ratio of the densest, loosest and the natural states of a sand deposit are 0.2, 0.6 and 0.4 respectively. The relative density will be

(a) 50% (b) 75% (c) 100% (d) 25%

100. From a liquid test on a soil sample following data were obtained,

Number of blows	17	22	25	28	34
Water content %	63.8	63.1	61.9	60.6	60.5

The Liquid Limit of the soil is

(a) 63.1 (b) 62.3 (c) 61.9 (d) 60.6

101. A soil having particles of nearly the same size is known as

(a) Well graded (b) Uniformly graded (c) Poorly graded (d) Gap graded

102. The shape of clay particle is usually

(a) Angular (b) Tubular (c) Rounded (d) Flaky

103. If the shearing stress is zero on two planes, then the angle between the two is

(a) 90° (b) 45° (c) 225° (d) 135°

104. The shear strength of a soil

(a) Is directly proportional to the angle of internal friction

(b) Is inversely proportional to the angle of internal friction

(c) Decrease with increase in normal stress

(d) Decrease with decrease in normal stress

105. A combined footing is generally used when

(a) Number of columns is more than two and spaced apart

(b) Number of columns is two and are spaced close

(c) Number of columns is two spaced far apart

(d) There is only one column

106. Negative skin friction on a pile

(a) Acts downward and increases load carrying capacity of pile

(b) Acts upward and increases the load carrying capacity of pile

(c) Acts downwards and reduces load carrying capacity of pile

(d) Acts upwards and reduces load carrying capacity of pile

107. Greater skin friction

(a) Retrards sinking of well (b) Accelerates sinking

(c) Does not effect sinking (d) First increases then slows

108. Terzagi's equation of ultimate bearing capacity for a strip footing may be used for square footing resting on pure clay soil with the correction factor of

(a) 0.4 (b) 0.6 (c) 1.2 (d) 1.3

109. A cohesionless soil having an angle of shearing resistance of (ϕ) and standing on a slope (i). The factor of safety will be ?

(a) $\frac{\tan i}{\tan \phi}$ (b) $\tan i - \tan \phi$ (c) $\frac{\tan \phi}{\tan i}$ (d) $\tan \phi - \tan i$

110. In a saturated clay layer undergoing consolidation with single drainage at its top, the pore water pressure would be the maximum at its

(a) Top (b) Middle (c) Bottom (d) Both top and bottom

111. The void Ratio and specific gravity of a soil are 0.65 and 2.72 respectively. The degree of saturation corresponding to water content of 20% is (approximately)

(a) 65% (b) 21% (c) 84% (d) 54.5%

112. A 4 m deep river flowing over sand bed with saturated unit weight of 20,000 $\frac{N}{m^3}$ and unit weight of water equal to 9810 $\frac{N}{m^3}$. The effective vertical stress at 4 m from the top of sand bed will be

(a) 41 $\frac{kN}{m^2}$ (b) 119 $\frac{kN}{m^2}$ (c) 83 $\frac{kN}{m^2}$ (d) 51 $\frac{kN}{m^2}$

113. Valid range for Degree of Saturation (S) of soil in percentage is

(a) $S > 0$ (b) $0 \le S \le 100$ (c) $0 < S < 100$ (d) 50

114. Degree of consolidation is (d = drainage path and t = time) is proportional to

(a) $\frac{t}{d}$ (b) $\frac{t}{d^2}$ (c) $\frac{d}{t}$ (d) $\frac{d^2}{t}$

115. If the water content of a fully saturated soil mass is 100% then the void ratio (e) will in terms of specific Gravity (G)

(a) $e < G$ (b) $e = G$ (c) $e > G$ (d) Independent of G

116. If the volume of voids (V_v) is equal to volume of solids (V_S) in a soil mass, then the value of porosity (n) and void ratio (e) will be

(a) 1.0 and 0.0 (b) 0.0 and 1.0 (c) 0.5 and 1.0 (d) 1.0 and 0.5

117. The value of compression index (C_c) for a remoulded sample whose liquid limit is 50% is

(a) 0.28 (b) 0.028 (c) 0.36 (d) 0.036

118. For determination of permeability of a coarse-gravel soil the best method is

(a) Constant Head (b) Falling Head
(c) Both (a) and (b) (d) None these

119. Uniformity coefficient of a soil is

(a) Always < 1 (b) Always = 1 (c) ≤ 1 (d) ≥ 1

120. Which of the following soil has more Plasticity Index (PI)?

(a) Sand (b) Silt (c) Clay (d) Gravel

121. For a mass of 3.192 g of soil with void ratio 0.532, the saturated unit weight is

(a) 20.44 $\frac{kN}{m^3}$ (b) 76.22 $\frac{kN}{m^3}$ (c) 23.55 $\frac{kN}{m^3}$ (d) 50.43 $\frac{kN}{m^3}$

TRUE OR FALSE QUESTIONS

1. Buoyant Unit weight equals the saturated density minus the weight of water. (T or F)
2. The maximum shear stress occurs on the filament which makes an angle of 45° with the horizontal plane. (T or F)
3. A partially saturated soil is classified as four phase soil. (T or F)
4. The active earth pressure of a soil is proportional to $\tan^2\left(45° - \frac{\phi}{2}\right)$, ϕ is angle of friction. (T or F)

5. Sand is more porous than clays. (T or F)
6. The lateral earth pressure on a retaining wall is proportional to the internal friction of the soil. (T or F)
7. The cohesion or inter molecular attraction in a soil is more in well compacted soil. (T or F)
8. The quantity of water flowing through the soil depends only on coefficient of Permeability of soil. (T or F)
9. In a flow net if N_f = Number of flow channels, N_d = Number of Potential drops and H = Total hydraulic head difference, the discharge $q = kH\frac{N_d}{N_f}$, k is constant. (T or F)
10. Under-reamed piles are generally pre-cast piles. (T or F)
11. Ratio of volume of voids to the volumes of soil solids in a given soil mass is known as Void Ratio. (T or F)
12. Kaolinite is most stable clay. (T or F)
13. If in a soil sample specific Gravity = G and voids = e the hydraulic gradient $i = \frac{G+1}{1+e}$. (T or F)
14. The Westergaard Analysis is used for Sandy Soils. (T or F)
15. The effective size of particle of soil is denoted by D_{10}. (T or F)
16. The coefficient of compressibility of soil is the Ratio of stress to strain. (T or F)
17. Rankine's theory of Active Earth Pressure assumes that soil mass is homogeneous, dry and cohesionless. (T or F)
18. The water content of soil is defined as the ratio of weight of water to weight of solids of given mass of soil. (T or F)
19. Accurate water content in soil is determined by Alcohol method. (T or F)
20. Stoke's Law states that fall velocity of any particle depends only on its weight. (T or F)
21. Liquid limit minus plastic limit is known as Plasticity Index of the soil. (T or F)
22. The liquid limit and plastic limit exist in clayey soils. (T or F)
23. The ultimate settlement of a soil is directly proportional to depth of soil strata and compressive index. (T or F)
24. The seepage force in a soil, is tangential to the equipotential lines of flow net. (T or F)
25. Failure plane carries maximum shear stress. (T or F)
26. Geological Cycle for the formation of soil, is Weathering → Transportation → Deposition → Upheaval. (T or F)
27. A phreatic line is defined as the line within a dam section below which there is positive hydrostatic pressure. (T or F)
28. A decrease in water content results in a reduction of the volume of a soil. (T or F)
29. For a shallow foundation the depth width ratio should be more than 2. (T or F)
30. The specific gravity of Quartz is about 2.9. (T or F)

31. Plasticity Index is the numerical difference between Plastic Limit and shrinkage limit. (T or F)
32. The most desirable soil structure is Granular structure. (T or F)
33. In permeable soils the chances of waterlogging are more. (T or F)
34. Water content of soil can be greater than hundred percent. (T or F)
35. A fully saturated soil is a three phase system. (T or F)
36. If the voids of a soil mass are full of air the soil is termed as dehydrated soil. (T or F)
37. Sand particle are made of rock minerals. (T or F)
38. Coefficient of permeability of soil increases with increase in temperature. (T or F)
39. Compressibility of sandy soil is more than that of clayey soil. (T or F)
40. Coarse grained soils are best compacted by rubber tyred roller. (T or F)
41. The shear strength of a soil decreases with decrease in normal stress. (T or F)
42. If a Cohesive soil specimen is subjected to a Vertical compressive load, the inclination of Cracks to the horizonal will be 90°. (T or F)
43. If the shearing stress is zero on two planes, then the angle between these planes will be 90°. (T or F)
44. The angle of internal friction is maximum for angular grained dense sand. (T or F)
45. A soil mass is said to be in plastic equilibrium if it is on the verge of failure. (T or F)
46. The angle of Internal friction of clays is nearly 30° to 45°. (T or F)
47. The water content in a soil sample when it continues to loose weight without losing volume is called shrinkage limit. (T or F)
48. Density of soil in SI units is expressed as $\frac{N}{m^3}$. (T or F)
49. The soil often found in areas of poor drainage with high evaporation is Alkaline. (T or F)
50. In quick sand or sand boiling, net effective pressure is zero. (T or F)
51. Soil containing organic matter are of spongy nature. (T or F)
52. Water content of soil cannot be greater than 100%. (T or F)
53. Quick sand is a type of sand. (T or F)
54. At OMC, soil attaints maximum dry density (MDD). (T or F)
55. OMC is used to gauge the amount of soil moisture required to achieve full soil compaction. (T or F)
56. Air content (A_c) is the Ratio of volume of Voids (V_v) and Volume of air, (V_a) i.e. $\left(\frac{V_v}{V_a}\right)$. (T or F)
57. Degree of Saturation (S) is defined as the Ratio of the volume of Pore space (V_p) and Volume of water (V_w) i.e. $\left(\frac{V_p}{V_w}\right)$. (T or F)
58. Degree of saturation (S) is 1 for dry soil. (T or F)

59. Black cotton soils are deficient in Nitrogen (N). (T or F)
60. The porosity of compacted sand is about 5 to 15%. (T or F)
61. According to IS classification system soil can be classified into 18 Groups. (T or F)
62. Finer the particle more is the total surface area. (T or F)
63. Sand bath method is the most accurate for determining water content of soil. (T or F)
64. At liquid limit shear strength is small. (T or F)
65. The maximum permissible eccentricity of a retaining wall of width (B) to avoid failure in tension is $\frac{B}{3}$. (T or F)
66. Upstream slope of an earth dam is a seepage line. (T or F)
67. If the permeability of a soil is about $10^{-3}\,\frac{m}{s}$, the type of soil is clean sand. (T or F)
68. Contact pressure beneath a rigid footing resting on a cohesive soil is uniform throughout. (T or F)
69. The effective angle of shearing resistance increases as the size of particles increases. (T or F)
70. The average coefficient of permeability parallel to strata is more than in vertical. (T or F)
71. For heavy loaded isolated columns the best foundation is Grillage foundation. (T or F)
72. A soil is said to be well-graded if it has most of the particles of about the same size. (T or F)
73. Shear strength of a soil is a unique function of Effective stress only. (T or F)
74. If degree of saturation of a partially saturated soil is 60%, the air content of the soil is 20%. (T or F)
75. The water content of soil can be more than 100%. (T or F)
76. A fully saturated soil is a 2-phase system. (T or F)
77. Bearing capacity of a soil is independent of rate of loading. (T or F)
78. The greater the unit weight greater will be permeability. (T or F)
79. Soil is defined as disintegrated rock. (T or F)
80. The water content of a soil is the ratio of weight of solids to weight of water. (T or F)
81. The units of coefficient of permeability are same as velocity. (T or F)
82. The coefficient of permeability of soil varies as $\sqrt{D_{10}}$. (T or F)
83. Volume of soil does not affect the permeability of soil. (T or F)
84. The seepage pressure depends on permeability. (T or F)
85. Peat and muck are Cumulose soils. (T or F)
86. The Ratio of Plasticity index to clay fraction is known as activity of clay. (T or F)
87. The measure of soil compaction is its wet density. (T or F)
88. The compression of the clay layer in mm due to lowering of water table will be 100. (T or F)
89. 50% of consolidation at a site subjected to a stress level of 200 kPa occurred in 2 months. If the site was laded with 400 kPa, 50% consolidation would occur in 4 months. (T or F)
90. Value shear Test is not normally used for determining in situ shear strength of soft clays. (T or F)

91. For a sample of dry cohesionless soil with friction angle ϕ, the failure plane will be inclined to the major Principal plane by an angle equal to $45° + \frac{\phi}{2}$. (T or F)

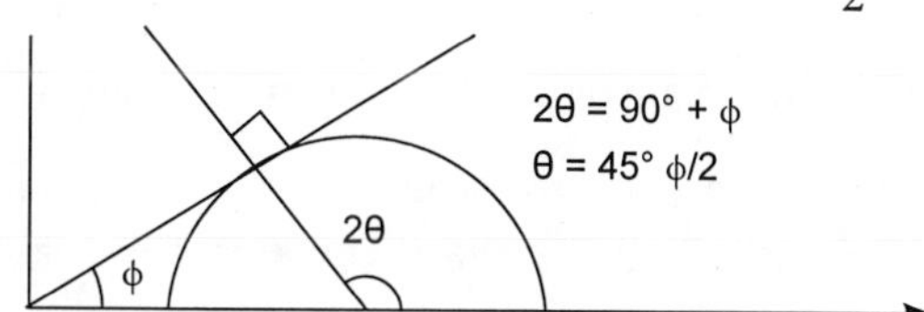

92. The problem of slope stability analysis assuming circular slip surfaces and considering equilibrium of the free body as a whole is statistically indeterminate. (T or F)

93. A foundation is said to be deep if $\frac{\text{Depth}}{\text{Width}}$ Ratio is more than 2. (T or F)

94. The value of cohesion C for clays is about 27 kPa. When the confined compressive strength of saturated sample clay is 54 kPa. (T or F)

$$C = \frac{q_u}{2} = \frac{54}{2} = 27 \text{ kPa.}$$

95. A strip footing is resting on the surface of a purely clay deposit. If the width is doubled the ultimate bearing capacity will not change. (T or F)

96. The coefficient of permeability decreases with increase in temperature. (T or F)

97. The pressure exerted by water on soil through which it percolates is called seepage pressure. (T or F)

98. The seepage pressure always acts in the direction of flow. (T or F)

99. The direction of seepage is always parallel to the equipotential lines. (T or F)

100. The rate of expulsion of pore water is directly dependent on permeability of the soil. (T or F)

101. In a clay layer if pressure increases slightly from p to p + δp and void ratio decreases from e_0 to e, the coefficient of permeability will be $\frac{e_0 - e}{\delta p}$. (T or F)

102. Honeycomb like pattern can be seen in silt. (T or F)

FILL IN THE BLANKS

1. The downward water entry into the soil is called ....
2. The father of soil Mechanics is ....
3. The soil particles of sizes 0.002 mm to 0.075 mm are categorised as ....
4. Mont morillionite is present in .... soil.
5. Most permeable soil is ....
6. For normally consolidated Soil Over Consolidated Ratio (OCR) is equal to ....

7. In Quick sand condition Net Effective Pressure is ....
8. The minimum water content at which the soil just begins to crumble when rolled into threads of 3 mm diameter is known as ....
9. The ratio of volume of voids to the total volume of a soil mass is called ....
10. Kaolinite is also called .... clay.
11. The ratio of settlement at any time 't' to the final settlement is known as degree of ....
12. In a liquid limit test, the moisture content at 10 blows was 70% and at 100 blows was 20%, the liquid limit of the soil is .... %.
13. While defining the shape of particle the coefficient of curvature is given by the formula .... ?
14. The ratio of the weight of given volume of soil solids to the weight of an equal volume of distilled water at a given temperature, is known ....
15. A soil has bulk density 2.3 $\frac{g}{cm^3}$ and water content 15%, the dry density of the sample, is ....
16. The weight of a pycnometer containing 400 g sand and water full to top is 2150 g. The weight of pycnometer full of clean water is 1950 g. If the specific gravity of soil is 2.5, the water content is .... %.
17. The soil moisture driven off by heat is called .... water.
18. The equation for shear strength of soil $\tau = C + \sigma \tan \phi$ is given by ....
19. According to Indian Standards the specific Gravity is the Ratio of Unit weight of soil solids to that of water at a temperature of .... °C.
20. Water formed transported soil is called ....
21. Geological formation which stores water and is able to transport is called ....
22. Porous geological formation but of impermeable structure like clay are called ....
23. According to Darcy's Law V = k i, where i has the Dimensions of ....
24. Seepage velocity, $V_s = \frac{V}{n}$ where n is ....
25. According to I.S. classification system, the soil is classified into how many groups ....
26. A soil has a discharge velocity of 6×10^{-7} $\frac{m}{s}$ and void ratio of 0.5, the seepage velocity will be ....
27. If volume of voids is equal to the volume of solids in a soil mass the void ratio is equal to ....
28. If the degree of saturation of a partially saturated soil is 60%, then the air content of the soil is ....
29. The Plasticity Index (PI) of sand is ....
30. If the permeability of a soil mass is about 0.8 $\frac{mm}{s}$, the type of soil is ....
31. Loess is silty clay formed by the action of ....
32. Mechanical stabilization of soil is done with the help of ....

33. In Quick sand condition or sand boiling seepage pressure is equal to .... weight.
34. The rate at which soil mass loses its shear strength with increase in water content is known as ....
35. The Ratio of plasticity index (I_P) to flow Index (I_F) is called .... Index.
36. The soils which fall above A-line in a plastic chart are ....
37. The 'air content' is defined as volume of air voids to the volume of ....
38. For a dry soil degree of saturation is equal to ....
39. For completely saturated soil degree of saturation is equal to ....
40. The degree of saturation (S) of soil lies between ....
41. If the voids of a soil mass are full of air only the soil is termed as .... soil.
42. The measure of particle size range is .... coefficient.
43. If Plastic limit > Liquid limit, then plasticity Index is ....
44. If Plasticity Index is zero, the soil is ....
45. Degree of consolidation is directly proportional to ....
46. Degree of consolidation is inversely proportional to .... of drain path.
47. Compressibility of sandy soil is much .... than clayey soil.
48. In OMC, O stands for ....
49. The quantity of water flowing through the soil depends upon total head loss through the soil and coefficient of ....
50. The effective size of particles of soil is denoted by ....
51. Boussinesq Equation assumes that soil mass is elastic medium and Modulus of Elasticity (E) is ....
52. The zone in a loaded soil mass bounded by an isobar of given vertical pressure intensity is called a pressure ....
53. Decrease in Water content of a saturated soil results in ....
54. Due to load increment on the soil the pore pressure ....
55. Among two parameters of shear strength, one is (ϕ) angle of Internal friction and the other is (C) called ....
56. Shear strength test is either field test or .... test.
57. For a sand having Internal friction $\phi = 30°$, the ratio of passive to active lateral earth pressure will be ....
58. Water content of mixed soil of 1 kg with water content 100% and 1 kg with water content 50%, will be ....
59. A normally consolidated clay settled. 10 mm when effective stress was increased from 100 to 200 $\frac{KN}{m^2}$. If the effective stress is further increased from 200 to 400 $\frac{KN}{m^2}$, then settlement will be ....

60. A soil has bulk density of 22 $\frac{KN}{m^3}$ and water content 10%, the dry density will be ....
61. A vane is 10 cm long (H) and 6 cm diameter (D) was pressed in soft clay the remoulded soil was sheared at torque (T) of 18 N – m. The cohesion (S) of clay is ....
62. Specific Gravity of water is ....
63. The maximum permissible exit gradient is the critical gradient divided by ....
64. Shearing strength of soil formula $\tau = c + \sigma \tan \phi$, σ is ....
65. The foundation whose depth is not more than its width is called .... foundation.
66. According to Indian standard the smallest sieve size is ....
67. Of a soil mass Liquid Limit is 20% and Plastic Limit is 25%, then the plasticity index of the soil is ....
68. The measure of soil compaction is its .... density.
69. Compaction by vibratory roller is the best method of compaction for well graded dry ....
70. OMC is Optimum .... content.
71. Consolidation of soil is a function of .... stress.
72. A saturated undisturbed sample from a clay strata has moisture content of 22.22% and specific weight 2.7. Taking $\gamma_w = 10 \frac{KN}{m^3}$, the void ratio is ....
73. For medium plasticity, Plasticity Index (PI) is ....
74. Of a soil sample Bulk density is 1.68 g/cc and water content is 30%. It's dry density is ....
75. The notation for well graded sand is ....
76. For low Bearing capacity soil the most suitable foundation is ....
77. Core test usually performed in the .... type of soil.
78. In direct shear test of a soil sample, if the compressive stress at failure is P, the shear strength of the soil is ....
79. A fully saturated soil is said to be a .... phase system.
80. For a partially saturated soil the degree of saturation is 60%, the air content of the soil is ....
81. The most likely Relative density of compacted sand is approximately equal to ....
82. Units of the coefficient of consolidation are ....
83. For an anisotropic soil, permeability in x-direction is Kx and in y-direction Ky in a 2-D flow, the effective permeability will be ....
84. Under a given load a clay layer attains 30% degree of consolidation in 100 days. The time required for the same to attain 60% of consolidation will be ....
85. A normally consolidated clay layer settles 1 cm when the pressure increases from 1 $\frac{kg}{cm^2}$ to 2 $\frac{kg}{cm^2}$. Additional settlement will be, if pressure is further increased from 2 to 4 $\frac{kg}{cm^2}$ will be ....

86. For a local soil Plastic Index (PI) = 15, PI of sand = 0, for a desired PI of 6, the percentage of sand in the mix should be ....

87. For fully saturated soil degree of saturation is equal to ....

88. The time factor corresponding to 25% degree of consolidation is ....

89. The exit gradient should not be more than ....

90. The void ratio of a soil sample is 1.0, the corresponding porosity is ....

91. The porosity (n) = 0.7 and degree of saturation = 40%. The volume of air in a 100 m^3 soil sample is ....

92. The notation 'SC' as per IS, soil classification system refers to ....

93. In Darcy's Law V = Ki, the units of (k) are ....

94. On the phreatic line the pressure is ....

95. A sampling tube has an inner diameter of 72 mm and outer diameter 75 mm, the area ratio is ....

96. Phreatic line in earthen dams has a shape of a ....

97. The coefficient of Earth Pressure for a loose sand (angle of internal friction of 30°) is ....

98. In a flownet diagram, the length of the flow line in the last square is 2 m, the total head loss is 18 m and the number of potential drops is 12. The value of exit gradient is ....

99. Sum of consistency index and .... index is always equal to unity.

100. An oven dried soil of 100 cm^3 weighs 175 g, the shrinkage ratio will be ....

101. Undisturbed soil samples are required for .... test.

102. Undisturbed soil samples are obtained by .... samplers.

Explanation: $\frac{k_p}{k_a} = \left(\frac{1+\sin\phi}{1-\sin\phi}\right)^2$.

ANSWERS

Multiple Choice Questions

1. c **2.** b **3.** b **4.** c **5.** b **6.** d **7.** c **8.** b **9.** d
10. d **11.** a **12.** d **13.** d **14.** c **15.** a **16.** d **17.** b **18.** b
19. c **20.** a **21.** d **22.** a **23.** b **24.** d **25.** a **26.** c **27.** d
28. a **29.** d **30.** a **31.** d **32.** d **33.** c **34.** b **35.** d **36.** c
37. d **38.** b **39.** a **40.** c **41.** b **42.** c **43.** a **44.** a **45.** c
46. c **47.** d **48.** b **49.** b **50.** c **51.** a **52.** d **53.** b **54.** c
55. a **56.** b **57.** d **58.** c **59.** a **60.** d **61.** c **62.** b **63.** d
64. a **65.** c **66.** c **67.** b **68.** c **69.** a **70.** a **71.** d **72.** c
73. b **74.** a **75.** d **76.** d **77.** b **78.** a **79.** d **80.** c **81.** a
82. b **83.** b **84.** b **85.** c **86.** c **87.** a **88.** b **89.** a **90.** a
91. b **92.** a **93.** d **94.** d **95.** b **96.** d **97.** a **98.** c **99.** a
100. c **101.** b **102.** d **103.** a **104.** d **105.** b **106.** c **107.** a **108.** d
109. c **110.** c **111.** c **112.** a **113.** b **114.** b **115.** b **116.** c **117.** a
118. a **119.** d **120.** c **121.** a

True & False

1. T **2.** T **3.** F 3-Phase **4.** T
5. F Sand .25 - .50 Clay .4 – .7 **6.** T **7.** T
8. F And gradient (Head Difference) **9.** F $q = KH\frac{N_f}{N_d}$ **10.** F Bored piles
11. T **12.** T **13.** F $i = \frac{G-1}{1+e}$ **14.** F Stratified
15. T **16.** F $\frac{\text{Strain}}{\text{Stress}}$ **17.** T **18.** T
19. F Oven drying **20.** F Also on shape and size **21.** T
22. T **23.** T **24.** F Perpendicular **25.** FDoes not carry
26. T **27.** T **28.** T **29.** F $\frac{D}{B} \leq 1$
30. F 2.65 **31.** F Liquid Limit-Plastic Limit **32.** T
33. F Less **34.** T **35.** F Two Phase **36.** F Dry soil
37. T **38.** T **39.** F Less than **40.** F By vibrators
41. T **42.** F 45° **43.** T **44.** T
45. T **46.** F 5° to 20° **47.** T **48.** F $\frac{kg}{m^3}$
49. T **50.** T **51.** T

52. F Can be greater than 100% **53.** F Liquefied soil **54.** T

55. T **56.** F $A_C - \frac{V_a}{V_v}$ **57.** F $S = \frac{V_w}{V_p}$ **58.** F S = 0

59. T **60.** F 30% to 40% **61.** T **62.** T

63. F Oven Drying **64.** T **65.** F $e = \frac{b}{6}$ **66.** F Equipotential Line

67. T **68.** F Pressure is more at Edges **69.** T

70. T **71.** T **72.** F Wide Range of Particles

73. T **74.** F 40% **75.** T **76.** T

77. T **78.** T **79.** T **80.** F $WC = \frac{W_w}{W_s}$

81. T **82.** F as D_{10}^2 **83.** T **84.** F Hydraulic Gradient

85. T Organic Matter **86.** T **87.** F Dry unit weight **88.** T

89. F **90.** T **91.** T **92.** F

93. F 4 to 5 **94.** T **95.** T **96.** F Decrease in temperature

97. T **98.** T **99.** F Perpendicular **100.** T

101. T **102.** T

Fill in the Blanks

1. Infiltration **2.** Karl Terzaghi **3.** Silt **4.** Black cotton

5. Coarse gravel **6.** 1 (One) **7.** Zero **8.** Plastic Limit

9. Porosity **10.** China **11.** Consolidation **12.** 65%

13. $\frac{D_{30}^2}{D_{60} \times D_{10}}$ **14.** Specific Gravity **15.** $\frac{2g}{cm^3}$ **16.** 20%

17. Hygroscopic **18.** Coulomb **19.** 27° C **20.** Alluvium

21. Aquifer **22.** Aquiclude **23.** $M^oL^oT^o$ (Dimensionless)

24. Porosity **25.** 18 (Eighteen) **26.** $18 \times 10^{-7} \frac{m}{s}$ **27.** 1.0 (One)

28. 40% **29.** Zero **30.** Sand **31.** Wind

32. Bitumen **33.** Submerged **34.** Flow Index **35.** Toughness

36. Clays **37.** Voids **38.** Zero **39.** 1.0 (One)

40. 0 and 100 **41.** Dry soil **42.** Uniformity **43.** Zero

44. Sand **45.** Time **46.** Square **47.** Less

48. Optimum **49.** Permeability **50.** D_{10} **51.** Constant

52. Bulb **53.** Consolidation **54.** Increases **55.** Cohesion

56. Penetration **57.** 9 (Nine) **58.** 71% **59.** 10 mm

60. $\frac{20\ KN}{m^3}$	**61.** $\frac{2.65\ N}{cm^2}$	**62.** 1 (One)	**63.** Factor of Safety (FS)
64. Normal Stress	**65.** Shallow	**66.** 0.045 mm	**67.** (– Five)$^{-5}$
68. Dry	**69.** Sand	**70.** Moisture	**71.** Effective Stress
72. 0.6	**73.** 7 (Seven)	**74.** $1.3\ \frac{g}{cc}$	**75.** SW
76. Raft foundation	**77.** Clay	**78.** $\frac{P}{2}$	**79.** Two
80. 40%	**81.** 0.95	**82.** $\frac{cm^2}{S}$	**83.** $\sqrt{K_x\ Ky}$
84. 400 days	**85.** 2 cm	**86.** 60 %	**87.** 1 (One)
88. $\frac{\pi}{64}$	**89.** 1 (One)	**90.** 0.5	**91.** 42 m^3
92. Clayey Sand	**93.** $\frac{m}{s}$	**94.** Atmospheric	**95.** 8.5%
96. Parabola	**97.** $\frac{1}{3}$	**98.** 0.75	**99.** Liquidity
100. 1.75	**101.** Consolidation Test		**102.** Thin Walled

SECTION 18

Steel Structures

Definitions – 42

Multiple Choice Questions – 100

True & False – 51

Fill in the Blanks – 52

DEFINITIONS

Annealing Process: It is a heat treatment process that changes the physical and sometimes also the chemical properties of a material to increase ductility and reduce the hardness to make it more workable. The annealing process requires the material above its recrystallizaiton temperature for a set amount of time before cooling.

Arc Welding: It is welding using the heat of an arc as a heat source. It is used to join metal to metal by using electricity to create high temperature enough to melt the metal and the melted metals, when cooled result in binding of the metals. Electric arc is generated between electrodes and base metal.

Bearing Plates: It is a plate placed under one end of a truss beam, girder, or column to distribute the load on larger area.

Bolt: It is an externally helical threaded fastener capable of being tightened or released by a twisting force (Torque) to a matching nut with a pre-formed female thread.

Boom: It refers to a long cantilevered structural element, typically designed to withstand significant bending loads, commonly found in cranes, excavators, certain bridges, extending as a large, extended beam that extends outwards to lift or support heavy objects at its tip.

Bottom Lateral Bracing or Bottom Chord Bracing: It is used to maintain the truss spacing, reduce vibration, and to provide lateral support for the bottom chord to resist buckling in the event of stress reversal and unbalanced loading of trusses with intermediate supports.

Brittleness of Steel: Steel can become brittle depending upon the heat-treating process it goes through, the carbon content, and other factors. Steel with a higher carbon content is harder and stronger, but also more brittle.

Butt Weld: It is one of the simplest and versatile type of weld joint designs. The joint is formed simply by placing two pieces of metal end-to-end and then welding along the join. This joint is strong and can be applied to wide range of metals. It is prone to cracking.

Cold-Rolled Steel: Essentially it is hot rolled steel that has been further processed by being allowed to cool at room temperature and annealed or temper rolled. Cold rolling produces steel with closer dimensional tolerances and a wider range of surface finishes than hot rolled through the use of strength hardening.

Combined Footing: It is a type of foundation that supports two or more columns or walls by distributing the load to wider area of soil. These are used when columns are very close to each other and soil's bearing capacity is limited.

Critical Buckling Load: The load at which the column just buckles, is known as buckling load or critical load or crippling load. The buckling load is less than the crushing load for long columns. Therefore, long columns should be designed for buckling load. On the other hand short columns are designed with respect to crushing load.

Dead Load: Dead loads are structural loads of a constant magnitude over time. They include the self-weight of structural members, such as walls, plasters, ceiling, floors, beams, columns, and roofs. Dead load also include the loads of fixtures that are permanently attached to the structure.

Effective Length of Colum: It is the distance between its zero moment or the distance between the inflection points. It is very important parameter in the design of columns because it determines the critical buckling load of the column.

Fatigue: Material fatigue is the progressive damages and eventual failure of a material due to repeated or cyclic loading. Failure occurs when a material is subjected to repeated stress, even when the stress is below the material's static strength.

Fillet Welds: It is a triangular weld that joins two pieces of metal plates that are perpendicular or inclined. These are commonly used in T-joints, lap joints, and corner joints. Examples of fillet welds include-shear tabs, cover plates, bracing connections, and column bases.

Grillage Footing: It is a type of footing that is used to distribute heavy load from columns to soil with low bearing capacity. It consists of one or more layers of beams usually steel, placed on the type of a concrete layer. The beams are arranged perpendicular to each other and the layers are encased in concrete. They are economical and can be used in variety of settings.

Gusset Plate: It is a thin plate used to connect beams and girders to columns. A Gusset plate can be fastened to a permanent member either by bolts, rivets or welding or a combination of the three. They are used in bridges and buildings.

Hot Rolled Steel: Hot rolling is the process to make hot rolled steel at high temperatures (typically over 926° C). These high temperatures are above steel's recrystallization temperature allowing the steel to be easily shaped, formed and made in larger sizes. It is used where precise shapes and tolerances are not required such as rail tracks, I-beams, metal buildings, sheet metals etc. This is cheap compared to cold rolled steel.

Lacing: Lacing in steel structure is a method of joining two or more components of a column, shaft or a composite beam to function as single member. Lacing is done with a network of flat surfaces. Lacing panels are usually 55 mm to 70 mm side and 6 mm to 10 mm thick.

Leg Length: In a steel structure **leg length** refers to the length of one-side of an angled steel member, typically used in components like angles or channels where the cross-section has two distinct angled legs. Essentially, it is the measurement of the length of one side.

Lintel Beam: It is horizontal beam that spans on opening, effectively distributing the weight of the structure above it to the surrounding vertical supports. Its significance lies in its ability to bear the load, prevent sagging, and preserve the overall strength of the structure.

Live Load: Live loads are also called imposed loads that do not have any **impact acceleration**. All these loads are part of what an occupant brings in the building. These items are normally furniture and movable partitions. Moving vehicles on bridges also among major live loads.

Load Factor: It is defined as the ratio of the collapse load to the working load. Load Factor, $F = \frac{W_u}{W_f}$. It can be shown that, F = Factor safety × S, where S is called shape factor.

Pitch of Rivets: It is the distance between the centre of two adjacent rows of rivets that are parallel to the direction of force application. The minimum pitch of the rivet should be atleast 2.5 times the gross diameter. The maximum pitch should not exceed 32 times the thinner outside plate or 300 mm whichever is less.

Plastic Hinge: It is a deformation in a beam that occurs when plastic bending happens. This phenomenon occurs when a beam's cross-section is subjected to yield stresses throughout its depth, preventing it from taking any more loads. When additional load is applied, the beam rotates at that section, which is known as a plastic moment.

Plate Girder: It is a structural element made of welded steel plates that is used to support heavy loads over long spans. These are commonly used in bridge construction, industrial buildings, and other large structures.

Purlins: A metal purlin is a lightweight, horizontal beam or bar that is used for structural support in roofing. In addition to providing support metal purlins also increase building's resistance to heavy winds. Purlins are a key component to the secondary frame to the commercial, residential and industrial metal buildings.

Rivet: A rivet is a permanent mechanical fastener that joins two pieces of material together. It has a smooth cylindrical shaft with a head on one end and tail on the other. When installed, the end opposite the head is deformed and called the **shop-head** or **buck head**. They are made of stainless steel, aluminium, brass or copper.

Secondary Stresses: In steel structures secondary stresses refer to additional stresses developed often arising from factors like joint rigidity, eccentricity of loads, temperature variations or lack of fit, which are not considered in a simple analysis and can induce bending or twisting moments in members primarily designed for axial loading.

Shape Factor: Shape factor of any section is the ratio of plastic moment capacity to its yield moment capacity and it is a function of cross sectional geometry. For example– shape factor (S) for triangular section is 2.34, for rectangular 1.5, for solid circular 1.7 and for T-section 1.90 to 1.95, irrespective of dimensions.

Shear Joints: These are bolted joints where the applied loading is at right angle to the fastener axis. In shear joints, the joint strength depends solely on the shear strength of the bolts. Failure in shear joints occur when joint numbers slip sideways past each other and cut the fastener.

Spandrel Beam: It is a horizontal beams that runs along the outside of a building's floor or roof, providing support and distributing loads. They support weight of walls and roofs. Spandrel beams are added between columns and run along the outside edge of the floor or roof.

Splice: A splice in a steel structure is a joint that connects two or more steel members to create a continuous structural element. Splices are often used in beams, columns, trusses and girders, where longer structural members are required splices provide flexibility to the structure and also accommodate temperature changes.

Steel Stack: It is a structure made of steel that releases combustion gases into the atmosphere. They are used to avoid damaging air pollution control devices and other equipments. They are made of both carbon or stainless steel and in heights from 6 to 150 m and with diameter upto 6.5m. The minimum thickness of plates in steel stack should be 6 mm.

Stiffeners: These are secondary plates or sections which are attached to beam webs on flanges to stiffen them against out of plane deformations. Almost all bridge beams are provided with stiffeners.

Stringer Beam: These are structural components that provide longitudinal support for a floor deck. They are typically inclined secondary beams. It is a crucial part of building of a staircase.

Strut in Steel Structure: It is a structural member designed to resist compressive forces, essentially acting as brace or support that prevents buckling or lateral movement within the structure. It is often used in roof trusses, bracing systems and excavation shoring.

Sway Bracing: Sway braces are used for controlling vibrations, absorbing shock, leading/guiding/restraining the movement of pipe resulting from thermal expansion and bracing a pipeline against sway. They contain springs that provide a restrain force in both a tension and compression mode.

Tie: It refers to a structural component design to resist tension forces. It is opposite of a strut which resists compressive forces. Ties are commonly used in trusses and other steel frameworks to connect different elements and distribute loads effectively.

Top Lateral Bracing: These are provided to increase the torsional stiffness of U-shaped steel trapezoidal box girders during construction. It is typically mounted at the top flange level to form a quasi closed section.

Tubular Steel: Tubular steel includes hollow and solid bars that are circular, rectangular or other geometric shape. These can be composed of multiple alloying elements, with varying carbon contents and wide choice to suit particular application. Tubular steel is pervasively used in the construction industry. They have excellent mechanical stability and good impact resistance.

Web Crippling: It is a local failure that occurs when the web of beam buckles due to compressive forces. Web crippling is caused by concentrated loads or reactions applied to a short length of a beam. It often occurs at the support of a beam, where bottom flange is resting on support, and the top flange is holding up the load, such as on a bridge abutment.

MULTIPLE CHOICE QUESTIONS

1. The difference between gross diameter and nominal diameter for the rivets upto 25 mm diameter is
   (a) 1.0 mm (b) 1.5 mm (c) 2.0 mm (d) 2.5 mm
2. In fillet weld material which of the following is the mode of failure?
   (a) Tension (b) Shear (c) Bearing (d) Crusting
3. Load on connection is not eccentric for
   (a) Lap joint (b) Single cover butt joint
   (c) Double cover butt joint (d) All above joints
4. The centre to centre maximum distance between bolts in tension member of thickness 10 mm is
   (a) 200 mm (b) 160 mm (c) 120 mm (d) 100 mm
5. The thickness of plate to be connected by a rivet is 16 mm, then the suitable size of the rivet as per Unwin's formula will be
   (a) 16 mm (b) 20 mm (c) 24 mm (d) 27 mm
6. Which of the following does not represent a weld type?
   (a) Butt (b) Plug (c) Zig-zag (d) Lap

7. The type of stress induced in the foundation bolts fixing a column to its footing is
   (a) Pure tension (b) Bearing
   (c) Pure compression (d) Bending
8. The dead load includes
   (a) Self weight of structure (b) All superimposed load
   (c) Weight of stationary equipments (d) Weight of furniture
9. The value of Factor of Safety (FS) is decided keeping in view
   (a) Strength of material (b) Design loads
   (c) Internal forces (d) All these
10. A strut is a compression member which is?
   (a) Small in length (b) Loaded lightly
   (c) Vertical or inclined (d) All of these
11. The effective area of a butt weld is taken as the .... of the effective throat thickness and the effective length of butt weld?
   (a) Sum (b) Difference (c) Product (d) Ratio
12. Effective length of a column is the length between the points of?
   (a) Maximum moments (b) Zero shear
   (c) Max shear (d) None of these
13. Factor of safety is the number by which the Yield stress of material is dived to?
   (a) Bearing stress (b) Bending stress (c) Shear stress (d) Working stress
14. When one is placed above the other and they are connected by means of rivets, the joint is known as?
   (a) Lap joint (b) Single cover butt joint
   (c) Double cover butt joint (d) None of these
15. Number of rivets required in a riveted joint is equal to
   (a) $\frac{\text{Force}}{\text{Rivet value}}$ (b) $\frac{\text{Force}}{\text{Shearing strength}}$
   (c) $\frac{\text{Force}}{\text{Rivets strength in bearing}}$ (d) $\frac{\text{Force}}{\text{Plate strength in shearing}}$
16. The strength at which steel fails under repeated applications of load is known as?
   (a) Impact (b) Tensile (c) Yield (d) Fatigue
17. Rolled steel angles are classified as series
   (a) Two (b) Three (c) Four (d) Five
18. When load line coincides with Centre of Gravity (CG) of the rivet group, then rivets are subjected to?
   (a) Shear only (b) Tension only (c) Bending only (d) Both shear and Tension
19. When load is exerted or transferred from one surface to another in contact, the stress is known as
   (a) Direct stress (b) Bending stress (c) Bearing stress (d) Shear stress

20. The assumption that all rivets shear equally a non-eccentric load is valid at a load
    (a) Below working load (b) Equal to working load
    (c) Above working load (d) Equal to failure load
21. When the same number of rivets are used in joints, then which pattern is most efficient?
    (a) Chain riveting (b) Zig-zag riveting (c) Diamond riveting (d) Any of these
22. The maximum allowable stress for rolled steel beams and channels is, (in MPa)
    (a) 128.5 (b) 156.5 (c) 181.5 (d) 288.5
23. A channel section consists of?
    (a) Two webs (b) Two flanges
    (c) Two web and two flanges (d) One web and two flanges
24. As per Indian standards rolled steel I-sections are classified into how many series
    (a) Four (b) Five (c) Six (d) Seven
25. The Ratio of Transverse strain to the longitudinal strain under axial load is known as
    (a) Tangent modulus of Elasticity (b) Modulus of Rigidity
    (c) Bulk Modulus of Elasticity (d) Poissons Ratio
26. For the same material, height, weight and conditions for vertical columns, the buckling load will be largest for the column cross-section?
    (a) Solid square (b) Thin hollow circle (c) Solid circle (d) I-section
27. Junction between the flange and web of a beam is known as
    (a) Fillet (b) Lap-joint (c) Butt-joint (d) Shear-joint
28. For rivets upto 24 mm diameter, the diameter of the rivet hole is larger than that of rivet by
    (a) 1 mm (b) 1.5 mm (c) 2 mm (d) 2.5 mm
29. Which of the following section is a best compression member section?
    (a) Single angle (b) Double angle (c) Tubular (d) I-section
30. The efficiency of a riveted joint having minimum pitch as per IS : 800 is
    (a) 40% (b) 50% (c) 60% (d) 70%
31. In comparison to field rivets, the shop rivets are
    (a) Stronger (b) Weaker (c) Equally strong (d) Any of these
32. As per IS : 800 the maximum deflection in a beam of length (L) should not exceed
    (a) $\frac{L}{180}$ (b) $\frac{L}{250}$ (c) $\frac{L}{325}$ (d) $\frac{L}{360}$
33. Minimum pitch of the rivets should not be less than, (diameter of rivets = d)
    (a) 1.5 d (b) 2.0 d (c) 2.5 d (d) 3.0 d
34. Bolts are most suitable to carry
    (a) Shear (b) Axial tension (c) Bending (d) Shear and bending
35. For standard 45° fillet, the ratio of size of fillet to throat thickness is
    (a) 1 : 1 (b) $1:\sqrt{2}$ (c) 2 : 1 (d) $\sqrt{2}:1$

36. The slenderness ratio of a column supported throughout its length by a masonry wall is
(a) Zero (b) 10 (c) 100 (d) Infinity

37. The effective length of a battened column is increases by
(a) 5% (b) 10% (c) 15% (d) 20%

38. The maximum slenderness ration of a compression member carrying both dead plus superimposed load is
(a) 180 (b) 200 (c) 250 (d) 350

39. The best arrangement to provide unified behaviour in built-in steel columns is by
(a) Lacing (b) Battening (c) Tie plates (d) Perforated cover plates

40. The use of tie plates in laced columns is
(a) Prohibited (b) Not prohibited
(c) Permitted between two parts of Lacing (d) Permitted at start and end of lacing system only

41. Rankine-Gordon Formula for buckling load is adopted for columns having slenderness ratio?
(a) More than 120 (b) More than 120 but less than 200
(c) Less than 120 (d) For any value

42. Allowable working stress for struts may be assumed as
(a) $60\ \frac{N}{mm^2}$ (b) $80\ \frac{N}{mm^2}$ (c) $100\ \frac{N}{mm^2}$ (d) $120\ \frac{N}{mm^2}$

43. Intermediate butt welds are used to resist?
(a) Shear stresses (b) Dynamic stresses (c) Alternate stresses (d) All of these

44. An eccentrically loaded column is subjected to
(a) Bending stress design of an eccentrically loaded (b) Direct stress
(c) Shear stress (d) Both (a)and (b)

45. In roof trusses, the horizontal beams spanning between the two adjacent trusses are known as
(a) Spandrel beam (b) Rafters (c) Purlins (d) All of these

46. The weakest plane in a fillet weld is
(a) Side parallel to force (b) Side normal to force
(c) Along the throat (d) Normal to throat

47. Large deflections in beam
(a) May result in cracking of ceiling plaster (b) Indicate the lack of rigidity
(c) May cause distortions in connections (d) All these

48. The beams resting in purlins are known as
(a) Spandrel beams (b) Rafters (c) Trimmers (d) Stringers

49. The buckling load in steel column is
(a) Related to length
(b) Directly proportional to slenderness ratio
(c) Inversely proportional to slenderness ratio
(d) Non-linearly related to slenderness ratio

50. Which among the following is not a tension member?
(a) Cable (b) Bar (c) Tie (d) Boom

51. If 'd' is the distance between flange angles, the maximum spacing of vertical stiffeners is
(a) 1.33 d (b) 1.25 d (c) 1.5 d (d) 1.75 d

52. The range of economical spacing of trusses varies from
(a) $\frac{L}{3}$ to $\frac{L}{5}$ (b) $\frac{L}{4}$ to $\frac{2L}{5}$ (c) $\frac{L}{3}$ to $\frac{L}{2}$ (d) $\frac{2L}{3}$ to $\frac{3L}{5}$

53. To minimise the total cost of roof truss, the ratio of cost of truss to the cost of purlins shall be
(a) 1 (b) 2 (c) 3 (d) 4

54. The relation between intensity of wind pressure (p) and velocity of wind (v) is taken as
(a) $p \alpha v$ (b) $p \alpha v^2$ (c) $p \alpha \frac{1}{v}$ (d) $p \alpha \sqrt{v}$

55. The live load for a sloping roof with 15° slope, where access is not provided to roof is taken as
(a) $0.65 \frac{KN}{m^2}$ (b) $0.75 \frac{KN}{m^2}$ (c) $1.35 \frac{KN}{m^2}$ (d) $1.5 \frac{KN}{m^2}$

56. Shear buckling of web in a plate girder is prevented by using
(a) Vertical intermediate stiffener (b) Bearing stiffener
(c) Horizontal stiffener at neutral axis (d) None of these

57. Gantry girders are designed to resist
(a) Lateral loads (b) Longitudinal and vertical loads
(c) Lateral, Longitudinal and Vertical loads (d) Lateral and Longitudinal loads

58. Generally purlins are placed at the panel points so as to avoid
(a) Axial force in rafter (b) Shear force in rafter
(c) Deflection of rafter (d) B.M. in rafter

59. The percentage of carbon in steel decreases its
(a) Hardness (b) Ductility (c) Strength (d) Brittleness

60. Which of these have least carbon?
(a) Wrought iron (b) Cast iron (c) Mild steel (d) Pig steel

61. As per Indian standards specifications, which of the following physical properties of structural steel is taken irrespective of its grade?
(a) Unit mass of steel (b) Modulus of elasticity
(c) Poisson's ratio (d) All of these

62. The strength at which steel fails under repeated application of load is known as
(a) Impact (b) Tensile (c) Yield (d) Fatigue

63. The size of rivet is expressed by the
(a) Length of shank (b) Diameter of shank
(c) Type of head (d) All these

64. The rivet line is also known as
(a) Scrieve line (b) Back line (c) Gauge line (d) All these

65. The hot driven rivets are

(a) Power driven shop rivets (b) Hard driven rivets
(c) Power driven field rivets (d) All these

66. For a rivet of 36 mm diameters, the hole diameter should be

(a) 36 mm (b) 37.5 mm (c) 38 mm (d) 38.5 mm

67. The distance between two consecutive rivets of adjacent chains and measured at right angles to the direction of the force in structural member is known as

(a) Pitch of rivet (b) Staggered pitch of rivet
(c) Gauge distance of rivet (d) Any of these

68. For a rivet diameter upto 24 mm, the diameter of rivet hole is larger than the rivet by

(a) 1 mm (b) 1.5 mm (c) 2 mm (d) 2.5 mm

69. The strength of Field rivets as compared to shop rivets is

(a) Same (b) 75% (c) 89% (d) 90%

70. The efficiency of a riveted joint is equal to

(a) $\dfrac{\text{Least strength of riveled joint}}{\text{Strength of solid plate}}$ (b) $\dfrac{\text{Greatest strength of riveled joint}}{\text{Strength of solid plate}}$

(c) $\dfrac{\text{Least strength of solid plate}}{\text{Least strenght of Riveted joint}}$ (d) $\dfrac{\text{Least strength of solid plate}}{\text{Greatest strenght of Riveted joint}}$

71. The working shear stress in gross area of power driven field rivets as per the IS : 800-2007 in $\dfrac{\text{N}}{\text{mm}^2}$ is

(a) 78.5 (b) 94.5 (c) 102.5 (d) 157.5

72. Minimum pitch in a riveted joint should not be

(a) Less than 200 mm
(b) More than 200 mm
(c) Less than 2.5 times the gross diameter of rivet
(d) Less than 2.5 times the gross diameter of rivet

73. If same number of rivets are used in the joints, then which pattern will Yield highest efficiency?

(a) Chain riveting (b) Zig-zag riveting
(c) Diamond riveting (d) All these

74. The rivets which are driven at atmospheric temperature are known as

(a) Power driven shop rivets (b) Hand driven rivets
(c) Power driven field rivets (d) Cold driven rivets

75. Arc welding is a

(a) Fusion welding (b) Pressure welding (c) Thermit welding (d) All these

76. The effective area of butt weld is taken as the .... of effective throat thickness and the affective length of butt weld

(a) Sum (b) Difference (c) Product (d) Ratio

77. A butt weld is specified by
    (a) Leg length (b) Plate thickness
    (c) Effective throat thickness (d) Penetration thickness
78. The web crippling due to excessive bearing stress can be avoided by
    (a) Increasing the web-thickness
    (b) Providing suitable stiffeners
    (c) Increasing the length of the bearing plates
    (d) None of these
79. As per IS : 800, for compression flange, the outstand of flange plates should not, in terms thickness (t) of thinnest flange plate, exceed
    (a) 12 t (b) 16 t (c) 20 t (d) 25 t
80. As per IS : 800, the maximum B.M. for design of Purlins can be taken as
    (a) $\frac{wL}{6}$ (b) $\frac{wL}{8}$ (c) $\frac{wL}{10}$ (d) $\frac{wL}{12}$
    (where w is the total distributed load and L centre-to-centre distance of supports)
81. Steel tanks are mainly designed for
    (a) Weight of tank (b) Wind pressure (c) Water pressure (d) Earthquake force
82. Which among the following sections should preferably be used at place where torsion may occur?
    (a) angle section (b) Channel section (c) Box type section (d) Any of these
83. The capacity of smallest pressed steel tank is
    (a) 1000 litres (b) 1650 litres (c) 1950 litres (d) 2450 litres
84. The minimum thickness of plates in a steel stack should be
    (a) 4 mm (b) 6 mm (c) 8 mm (d) 10 mm
85. The least dimension in case of a circular column of diameter D is taken as
    (a) 0.5 D (b) D (c) 0.68 D (d) 0.88 D
86. If the floor is supported at or near the bottom but top chords of a bridge are not braced, then the bridge is called
    (a) Deck type (b) Through type (c) Half through type (d) Double deck type
87. The bracing provided in the plane of end post is called
    (a) Sway bracing (b) Portal bracing
    (c) Top lateral bracing (d) Bottom lateral bracing
88. Load factor is
    (a) $\frac{\text{Ultimate Load}}{\text{Yield Load}}$ (b) $\frac{\text{Yield Load}}{\text{Working Load}}$ (c) $\frac{\text{Ultimate Load}}{\text{Working Load}}$ (d) None of these
89. Which of the following section has larger shape Factor?
    (a) Rectangle (b) I-section (c) Solid circular (d) Diamond
90. The Ratio of Plastic section modulus to Elastic section modulus is
    (a) Always more than one (b) Equal to one
    (c) Always less than one (d) Can be less or more

91. The Ratio of allowable stress in bending compression to that of in bending tension in steel beams is

(a) 1 (b) ≤ 1 (c) ≥ 1 (d) More than 1

92. The effective length of the column AB is

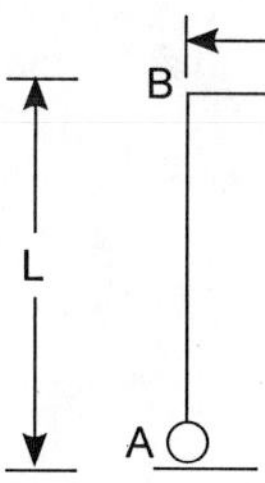

(a) 0.7 L (b) Less than 0.7 L
(c) More than 0.7 L but less than1 (d) Equal to 1

93. Upper Yield point in stress-strain curve in structural steel can be avoided by

(a) Cold working (b) Hot working (c) Quenching (d) Galvanizing

94. A simply supported beam of span L supports a concentrated load W at its mid-point. If cross-section of beam is an I-section then the length of Elastic-Plastic zone of the plastic hinge will be

(a) $\frac{L}{8}$ (b) $\frac{L}{4}$ (c) $\frac{L}{2}$ (d) $\frac{3}{4}L$

95. For an I-beam the shape Factor is 1.12. The Factor of safety in bending is 1.5. If allowable stress is increased by 20% for wind and Earth quake. Load Factor becomes

(a) 1.10 (b) 1.25 (c) 1.35 (d) 1.40

96. Factor of safety adopted by IS:800-1984 while arriving at the permissible stress in axial compression is

(a) 2.0 (b) 1.0 (c) 1.67 (d) 1.50

97. A steel beam supporting loads from floor slab as well as from wall is termed as

(a) Stringer beam (b) Lintel beam (c) Spandrel beam (d) Header beam

98. Which of the following is not a compression member?

(a) Strut (b) Tie (c) Rafter (d) Boom

99. What is the effective length of the column shown here?

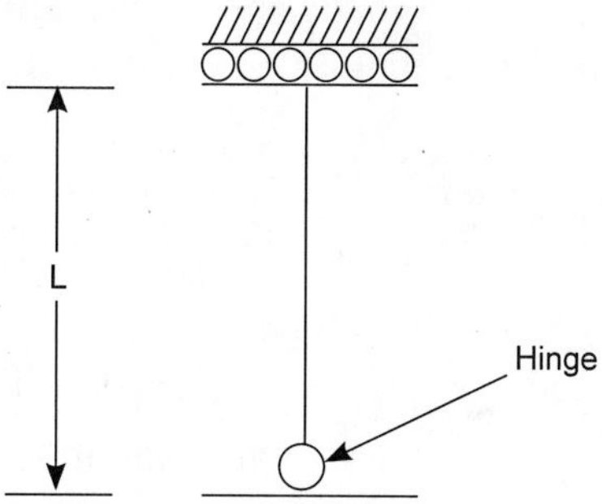

(a) 0.8 L (b) L (c) 1.2 L (d) 2 L

100. Arrange the following cross-sections in the increasing of their shape factors.

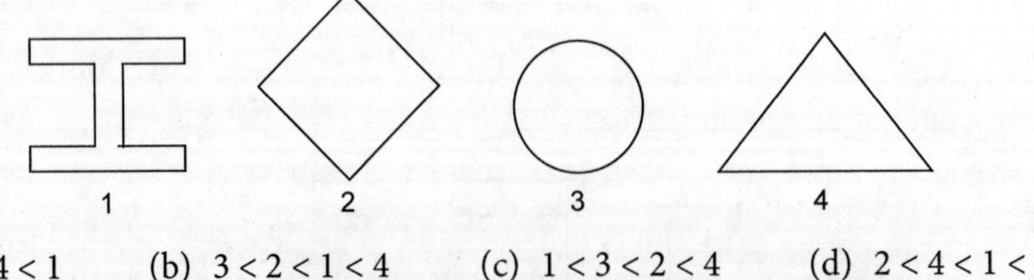

(a) 2 < 3 < 4 < 1 (b) 3 < 2 < 1 < 4 (c) 1 < 3 < 2 < 4 (d) 2 < 4 < 1 < 3

TRUE OR FALSE QUESTIONS

1. The basic wind speed is specified at a height of 10 m above mean ground level in an open terrain. (T or F)
2. In order to analyse the statically indeterminate structures, the equations of statical equilibrium are not enough to determine all the forces on the structure. (T or F)
3. The rivet line is an imaginary along which rivets are placed. (T or F)
4. The power driven shop rivets are stronger compared to power driven field rivets. (T or F)
5. The impact strength of steel is measure of its ability to absorb energy at high rate of loading. (T or F)
6. A space structure is three dimensional. (T or F)
7. A structure large in two dimensions but small in third dimension is called a space structure. (T or F)
8. The modulus of elasticity of concrete is more than steel. (T or F)
9. Cast iron is manufactured by remelting pig iron with coke and lime stone. (T or F)
10. As per Indian standards, rolled steel I-sections are classified into five categories. (T or F)
11. The channels are subjected to torsion because of the absence of symmetry of sections with regard to the axis perpendicular to the web. (T or F)
12. The diameter of cold rivet measured before driving is known as Gross diameter. (T or F)
13. Bearing failure of rivet occurs when the rivet is crushed by the plate. (T or F)
14. In the design of a riveted joint the bearing stress is assumed to be uniform between the contact surfaces of plate and rivet. (T or F)
15. If plate thickness is t mm and the nominal diameter of rivet is d mm, then according to Unwin formula 6 = 6t. (T or F)
16. A fillet weld is known as special fillet weld when its cross-section is 45° isosceles triangle. (T or F)
17. A fillet weld stressed in longitudinal shear is known as side fillet weld. (T or F)
18. The failure of column depends upon length of the column. (T or F)
19. If the cross-sectional area of a column is kept constant the load required to cause failure due to direct compression decreases as the length of the column is increased. (T or F)

20. Buckling is defined as the sudden bending of elements or members under compressive stresses. (T or F)
21. The tension members in a roof truss are struts. (T or F)
22. The 'Tie' is not a compression member. (T or F)
23. The effective length of a column is its supported length. (T or F)
24. The columns having large slenderness ratios fail in elastic buckling. (T or F)
25. The tubular section of a column has Low local buckling strength. (T or F)
26. A plate connecting two or more members intersecting each other is called Gusset plate. (T or F)
27. IS 800 : 2007 is based on Limit state method. (T or F)
28. Economic depth of a plate girder corresponds to maximum weight. (T or F)
29. The shape factor for a solid circular section of diameter D is equal to $\frac{16}{3\pi}$. (T or F)
30. The square root of the ratio of moment of inertia of the cross-section to its cross-sectional area is called Radius of gyration. (T or F)
31. When the bolts are subjected to reversal of stresses, the best bolt is, black bolt. (T or F)
32. If a bolt is of 18 mm diameter then gross diameter for standard clearance hole as per IS : 800 : 2007 is 19.0 mm. (T or F)
33. In case of connection of two plates Double cover Butt Joint is most efficient. (T or F)
34. Steel of Yield strength 400 MPa has been used in a structure, the maximum allowable tensile strength will be 200 MPa. (T or F)
35. Lateral buckling of beam starts from top flange due to compression in the top flange. (T or F)
36. In steel bridges for span less than 9 m and class AA loading for wheeled vehicles, the impact factor is taken as 25% (T or F)
37. A building with gabled roof will experience pressure on its leeward slope as always positive. (T or F)
38. As per IS : 875 for the purpose of specifying basic wind velocity, India has been divided into 6 zones. (T or F)
39. The number of seismic zones in which India is divided is 6 zones. (T or F)
40. Shape factor is a property of section depends only on its geometry. (T or F)
41. Load Factor is always less than Factor of safety. (T or F)
42. Apart from gravity loads only lateral loads are considered in the design of gantry girder located within an industrial building. (T or F)
43. Intermediate vertical stiffeners are provided in plate girders to eliminate web buckling. (T or F)
44. The Thickness of web for unstiffened plate girder with clear distance d between the flanges shall not be less than $\frac{d}{100}$. (T or F)
45. The stress at which a material fractures under large number of reversal of stress is called Endurance Limit. (T or F)

46. The ratio of collapse load of a propped cantilever of span (l) carrying a UDL throughout the span to that of simply supported beam carrying same load is 1.5 (T or F)
47. The maximum slenderness ratio of a compression member which carry dead and superimpsed load is 180. (T or F)
48. The best compression member section is Tubular section. (T or F)
49. A discontinuous compression member extends between two adjacent joints only. (T or F)
50. The minimum radius of gyration of the single angle section is much more than other sections of same cross-sectional area. (T or F)
51. The number of batten plates should be such that it divides the column longitudinally in at least four parts. (T or F)
52. The width of the bearing plate is equal to the width of the flange column. (T or F)
53. When the bearing capacity of soil is poor, then grillage footing is recommended for column footing. (T or F)
54. A beam is a structural member subjected to transverse loads. (T or F)

FILL IN THE BLANKS

1. The value of Poisson's Ratio for steel within Elastic region ranges from .... to ....
2. Shear modulus of elasticity is also known as modulus of ....
3. The slenderness ratio of a single angle strut should be less than ....
4. A piece of round steel forged in place to connect two or more steel members is known as a ....
5. A skeleton structure is one in which member is represented by a .... ?
6. Butt joint with double cover plates are free from .... stresses.
7. The Euler's formula for columns is valid for large .... ratio.
8. A butt weld is specified by effective .... thickness.
9. The stability of structure is ensured if restoring moment is .... than the maximum overturning moment.
10. Bending stress may develop in case of .... joint in steel structure.
11. The average axial stress in column at its failure is .... than the yield strength of the material.
12. When secondary stresses are taken into account alongwith primary stresses than allowable stress is increased by .... %.
13. The overlap of batten plates with the main members in welded connections should be more than .... times the thickness of batten plate (t).
14. The effective length of a fillet weld of Length L and Size S is given by ....

15. The actual thickness of butt weld as compared to the plate thickness is usually ....
16. When two plates are placed end to end and are joined by two cover plates the joint is known as double cover .... joint.
17. Bolts are most suitable to carry axial ....
18. As per IS-800 efficiency of riveted joint having minimum pitch, is .... %.
19. The order of elongation which a speciment of mild steel undergoes before fracture is .... %.
20. The Ratio of the Plastic moment capacity of a beam section to its yield moment capacity is termed as .... factor.
21. When 20 mm rivets are used in lacing bars, then the minimum width of lacing bar should be .... mm.
22. For a compression member having the same effective length about any cross-sectional axis, the most preferred section from strength point of view is a .... section.
23. The permissible bending stress in slab base $\left(f_y = 250\frac{N}{mm^2}\right)$ is .... $\frac{N}{mm^2}$.
24. Lacings are subjected to .... loading only.
25. By what percentage is permissible stress in a grillage beam be increased if it is encased in concrete .... %.
26. A channel section has one web and two ....
27. Rolled steel flats are used as .... members.
28. With increase of carbon the brittleness of steel ....
29. The ductility of steel .... with increase in carbon in steel.
30. For wind velocity (V) in $\frac{Km}{Hour}$. The intensity of wind pressure is directly proportional to ....
31. Shear modulus of elasticity is also known as modulus of ....
32. As per IS : 800-2007, the factor of safety adopted with respect to yield stress of steel is ....
33. Poisson's Ratio for steel within elastic region ranges between ....
34. The cold driven rivets need large pressure to form the head and complete the ....
35. The strength of field rivets as compared to shop rivets is ....
36. The axial load sufficient to keep the column in a slight deflected shape is called critical load, crippling load or .... load.
37. A column splice is used to .... the length of a column.
38. In a grillage footing the maximum shear force occurs at the .... of base plate.
39. In combined footing, the shape for the base of footing should be .... to support two equal column loads.
40. In a steel beam web carries ....
41. The gross-section of the web of a beam is equal to Depth of beam × ....
42. In a web crippling, local buckling of web occurs immediately adjacent to a .... of stress.

43. For a simply supported beam the maximum permitted deflection as per IS 800 : 2007, is .... of the span.

44. The shape factor of the given shape is ....

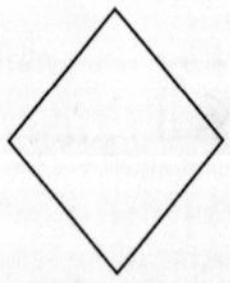

45. Among wrought iron, steel and cast iron the minimum coal content is in ....

46. The most common type of steel used in general construction is .... steel.

47. Working stress or permissible stress is obtained by dividing Yield stress by the ....

48. The intermediate support for the roof is provided by ....

49. As per IS 800 : 1984 the minimum pitch of rivets in a row is recommended as the diameter of the rivet times ....

50. The effective length of a circular electric pole of Length (L) and constant diameter erected on ground is ....

51. The depth of plate girder for long span is usually .... of span.

52. At the location of a plastic hinge curvature is ....

ANSWERS

Multiple Choice Questions

1. b	2. b	3. c	4. b	5. c	6. c	7. a	8. a	9. d
10. d	11. c	12. b	13. d	14. a	15. a	16. d	17. a	18. a
19. c	20. d	21. c	22. d	23. d	24. b	25. d	26. b	27. a
28. b	29. c	30. c	31. a	32. c	33. c	34. b	35. d	36. a
37. b	38. a	39. a	40. d	41. b	42. a	43. a	44. d	45. c
46. c	47. d	48. b	49. a	50. d	51. c	52. a	53. b	54. b
55. a	56. a	57. c	58. d	59. b	60. a	61. d	62. d	63. b
64. d	65. d	66. c	67. c	68. b	69. d	70. a	71. b	72. d
73. c	74. d	75. a	76. c	77. d	78. c	79. b	80. c	81. c
82. c	83. c	84. b	85. d	86. c	87. b	88. c	89. d	90. a
91. b	92. c	93. a	94. a	95. d	96. c	97. c	98. b	99. d
100. c								

True & False

1. T
2. T
3. T
4. F Are weaker
5. T
6. T
7. F Surface Structure
8. F Of steel is more
9. T
10. T
11. F Parallel to the web
12. F Nominal diameter
13. T
14. T
15. F $d = 6\sqrt{t}$
16. F Standard Fillet weld
17. T
18. F Slenderness Ratio
19. T
20. T
21. F Compression Members
22. T
23. F Pin-end Length
24. T
25. F Has High
26. T
27. T
28. F Minimum weight
29. T
30. T
31. F High Strength Bolt
32. F 20 mm
33. T
34. F 240 MPa
35. T
36. T
37. F Negative
38. T
39. F 5 Zones
40. T
41. F More than
42. F Lateral and Longitudinal Loads
43. T
44. F $\frac{d}{85}$
45. T
46. F 1.457
47. T
48. T
49. T
50. F Less than
51. F Three Parts

Fill in the Blanks

1. 0.25 to 0.33
2. Rigidity
3. 180
4. Rivet
5. Line
6. Bending
7. Slenderness
8. Throat
9. Greater or More
10. Lap
11. Less
12. $16\frac{2}{3}\%$
13. 5 t
14. L–25
15. More
16. Butt
17. Tension
18. 60%
19. 10%
20. Shape
21. 60 mm (About three times)
22. Box
23. $185 \frac{N}{mm^2}$
24. Axial
25. 33%
26. Flanges
27. Tension
28. Increases
29. Decreases
30. Proportional to V^2
31. Elasticity
32. 1.67
33. 0.25 and 0.33
34. Driving
35. Less (About 90%)
36. Buckling
37. Increase
38. Edge
39. Rectangular
40. Shear
41. Web Thickness
42. Concentration
43. $\frac{1}{300}$
44. 2 (Two)
45. Wrought Iron (< 0.15%)
46. Mild
47. Factor of Safety
48. Purlins
49. 2.5
50. 2 L
51. $\frac{1}{8}$
52. Infinite

SECTION 19

Strength of Materials

Definitions – 51

Multiple Choice Questions – 107

True & False – 57

Fill in the Blanks – 53

DEFINITIONS

Arch: An arch distributes loads by transferring weight along their curve to the supports. The load path helps reduce tensile stress, making arch highly efficient supporting structure. Its shape allows it to handle compressive forces predominantly, which are more easily managed by most construction materials.

Arching of Beam: It reduces the bending moment throughout the length of the beam.

Bending Stress: It is the stress experienced by a material when subjected to bending. As a horizontal beam is loaded, it generates bending moment stresses in top and bottom fibres of the section which can result in over stressing of the section and causing it to fail.

Brittle Material: A brittle material exhibits little deformation before fracture, the strain normally being less than 5%. These materials fail suddenly without visible warning. Examples are– Cast iron, ceramics, concrete, high strength steel etc.

Buckling Load: The maximum axial load that a column can withstand just before buckling occurs is referred to as the **Critical Buckling Load** or **Crippling Load**. Here the stress within the column is below Yield stress and column has not reached its ultimate strength.

Bulk Modulus: When a body is subjected to three mutually perpendicular stresses of equal intensity (σ), the ratio of direct stress to the corresponding volume strain is known as **bulk modulus**. Usually denoted by K. It is a measure of the ability of a substance to withstand changes in volume when pressed from all sides $K = \dfrac{\Delta P}{\left(\dfrac{\Delta_v}{V}\right)}$.

Composite Section: Composite sections are formed by combining two or more bars of equal length, but of different materials rigidly fixed so as to act as one unit. Examples are– bimetallic beams, RCC, plastic coated pipes, wooden beams reinforced with steel plate etc.

Ductile Material: It has the ability to undergo significant plastic deformation under tensile stress before rupture is called the ductile property of the material. Examples are nickel, copper, steel etc.

Eccentric Loading: It means when point of action of load is away from the neutral axis of columns. In practical sense when loads are acting such that they are not balancing each other, it is called eccentric loading and it may cause failure of structure.

Elastic Constants: Elastic constants of a solid include– Young's modulus, Bulk modulus and Rigidity modulus.

- Relation between Modulus of Elasticity (E), Modulus of Rigidity (G) and Poisson's Ratio (μ) is, $E = 2G(1 + \mu)$ Pascal.
- Relation between Young's modules (Y) and Bulk Modulus (K), Y or $E = 3K(1 - 2\mu)$.

Elasticity: It is the ability of a material to return to its original shape after the removal of an applied load. This behaviour is described by Hooke's Law, which states that stress is proportional to strain in the elastic region.

Fatigue: It refers to the weakening of a material caused by repeated loading and unloading cycles leading to development of cracks.

Flexural Rigidity: It is defined as the **force couple** required to bend a fixed non-rigid structure by one unit of curvature, or as the resistance offered by a structure while undergoing bending. It is expressed as (E I), I is Moment of Inertia. Unit of flexural rigidity are N. m².

Hooke's Law: It states that the strain of a material is proportional to applied stress within elastic limits of that material. When an elastic material is stretched, the atoms and molecules deform until stress is applied and as the stress is removed, they return to their original position.

Hoop Stress: It is a type of stress experienced by the walls of a circular or cylindrical object. It acts in the circumferential direction perpendicular to the axis of the structure. The analysis for thin and thick wall cylinder is different. For thin wall cylinder $\left(\frac{\text{Inner radius}}{\text{Wall thickness}} \geq 10\right)$ hoop stress is given by, $\sigma_h = \frac{pr}{t}$, p = internal gauge pressure, r = inner radius and t = with thickness.

Hoop Stress and Longitudinal Stress: Hoop stress = 2 × Longitudinal stress. The above relation is based on the assumption that the wall-thickness (t) is small compared to the radius of the cylinder. In such cases above relation is valid.

Hoop Stress for Thick Walled Vessels: For thick walled cylindrical vessels, hoop stress is given by Lame's Equation. $\sigma_n = \frac{r_i^2 p_i - r_0^2 p_0 - r_i^2 p_i (p_0 - p_i) r^2}{r_0^2 - r_i^2}$ where, r_i = inner radius, r_0 = outer radius p_i = inner pressure, p_0 = external pressure, r = radial variable where stress is to be known equation as $\sigma_r = A - \frac{B}{d^2}$, Radial stress.

Impact Test: An impact test is a technique used to determine a material's ability to resist deformation when subjected to a sudden shock or impulse load. Actually, it determines the amount of energy absorbed by a material during fracture. This absorbed energy is a measure of a given material's toughness. IZOD impact strength test is very popular.

Isotropic Materials: These are materials whose properties remain the same when tested in different directions. Isotropic materials differ from **anisotropic** material, which display varying properties when tested in different sides. Glass, plastic and metals are normally isotropic. Wood and composite materials are good examples of anisotropic.

Lame's Theorem: It is associated with the thick cylinder problems. The theorem is based on the theory that the material of cylinder is isotropic. The longitudinal strain is same at all points and is independent of radius.

Leaf Spring: It is a suspension system of cars and other wheeled vehicles. It consists of number of layer of leaves with a gradation in their size, the longer layer be on the top, with each layer joined to the other.

Linear Arch: It is a compression member in which shear force and bending moment is zero thought the span and it is subjected to normal thrust only.

Modulus of Rigidity: It is the elastic coefficient where a shear force is applied resulting in lateral deformation. It is equal to the ratio of shear stress with respect to shear strain. It is a physical property of material.

Mohr's Circle: It is a powerful graphical method used to visualise and analyse the stress state at a single point within a body. It allows to determine the normal and shear stress components for different orientation of the stress element graphically, instead of using the stress transformation equations. Lot of information can be found from Mohr's Circle, like maximum shear stress τ_{max} which is equal to the radius of the circle.

Moment of Inertia: It is known as angular mass or rotational inertia. Moment of Inertia is defined as, "the quantity expressed by the body resisting angular acceleration, which is the sum of the product of the mass of every particle with its distance square from the axis of rotation." Actually it is the amount of torque needed for a specific angular acceleration in a rotation axis. The SI unit is kgm^2.

Neutral Axis: In the cross-section of a beam (a member resisting bending) or a shaft along which there are no longitudinal stresses or strains. All fibres on one side of the neutral axis are in the state of tension, while those on the other side are in compression.

Perfect Frame: It may be defined as that one which is made up of members just sufficient to keep the frame in equilibrium, when loaded, without any change in its shape.

Plasticity: It is the ability of a solid material to undergo permanent deformation without fracture. Elasticity in comparison enables a solid to return to its original shape after the load is removed.

Poisson's Ratio: It is the ratio of transverse contraction strain to longitudinal extension strain in the direction of the stretching force. Here, compressive deformation is considered as negative and tensile deformation as positive.

$$\text{Poisson's Ratio} = \frac{\text{Transverse strain}}{\text{Longitudinal strain}}$$

$$\nu = \frac{\varepsilon_t}{\varepsilon_l}$$

Principal Plane: These are planes within the material such that the resultant stresses across them are wholly normal stresses or planes across which no shearing stresses occur. The normal stresses acting on Principal plane are called principal stresses.

Proof Resilience: It is defined as the maximum strain energy stored in a body-when the body is strained upto the elastic limit. It means that it is the ability of the body to store or absorb energy without permanent deformation. **Modulus of Resilience** is defined as proof resilience per unit volume.

Pure Bending Moment: When a beam is loaded such that there is no shear force on any cross-section normal to the longitudinal axis of the beam and hence it is subjected to constant bending moment, then the beam is said to be in pure bending and Bending Moment is called Pure Bending Moment.

Radius of Gyration (Kv): It is a geometric property which is frequently used for beams of various cross-sections. Actually, it is a measure of the elastic stability of a cross-section against buckling. This is the distance from the axis at which the area may be imagined to be concentrated to give the same Second Moment of Area (I).

$$\therefore \qquad I = AK \quad \text{or} \quad K = \sqrt{\left(\frac{I}{A}\right)}.$$

Resilience: In **material science** resilience is the ability of a material to absorb energy when it is deformed elastically, and release that energy on unloading.

Section Modulus (z): It is the ratio of Moment of Inertia (I) of the beam cross-section about the neutral axis to the distance (y_{max}) of extreme fibre from the neutral axis. The section modulus (z) of the cross-sectional shape is significant in designing beams. It is a direct measure of strength of the beam.

Shear Centre: It is a point through which if the external load passes, then there will not be any twisting of the section. The section will only be subjected to bending. It will not be subjected to Torsion. Also called "Centre of Twist".

Slenderness Ratio: It is the ratio of the length of the column and the least radius of gyration of its cross-section $\lambda = \frac{l_e}{k_{min}}$. For, short columns $\lambda < 32$, Medium columns $\lambda = 32$ tp 120, Long columns $\lambda > 120$.

Strain: When a system of forces act on a body, it undergoes some deformation. This deformation per unit of length is known as **Unit strain** or simply a **Stain,** $\varepsilon = \frac{\delta l}{l}$. It is a dimensionless quantity.

Strain Energy: It is defined as the energy stored in a body due to deformation. It is usually denoted by U. It is given by formula $U = F\frac{\delta}{2}$, F = Force applied, δ = Compression. In terms of Young's Modulus (E), strain energy is $U = \frac{\sigma^2}{2E \times V}$, σ = stress and V = Volume of body.

Stress: The internal force per unit area within materials, often expressed in Pascal (Pa). It can be tensile (pulling), Compressive (pressing) or Shear (sliding) $\sigma = \frac{F}{A}$.

Tensile Strength: It is the maximum amount of tensile (pulling) stress that a material can withstand without failure.

Thermal Stress: It is defined as the stress produced by any change in the temperature of the material. Thermal stress is induced in a body when the temperature of the body is raised or lowered and the body is not allowed to expand or contract freely. Thermal stress = α t E, where, α = Coefficient of Linear Expansion, t = Change in temperature, E = Modulus of Elasticity.

Three Hinged Arch: A 3-hinged arch has **hinge** at each support plus one more internally. The third hinge provides an additional static equation as moment there is zero. This makes the system statically determinate.

Torsional Rigidity: It is the object's torsional resistance to twisting as a torque is applied to the component. It is denoted by GJ, where G stands for shear Modulus of the material and J polar moment of Inertia of the cross-section.

Torsional Stress: When a material is subjected to twisting. It is particularly applied to shafts and rods. It can be defined as "Shear stress which acts on a transverse cross-section that is caused by the application of a twist. It can be expressed as, Torsional stress, $\tau = \frac{T.n}{J}$ where, T = Torque applied to circular section, r = radius of cross-section and J = Polar Moment of Inertia.

Two Hinged Arch: The two-hinged arch has only two hinges, one at each support, which allow for horizontal movement at the supports. Two-hinged arch is the statically indeterminate structure to degree one since there are four reaction components to be determined while the number of equations available from static equilibrium is only three.

Ultimate Compressive Strength: It is the stress required to rupture a specimen. It is force per unit area that a material can withstand in compression.

Ultimate Strength: It is the maximum stress a material can withstand before failure. This is critical for determining safe load limits.

Yield Locus: Yield criteria determine which combination of multi-axial stresses will cause Yielding in a material. The Tresca Yield Surface is an irregular hexagon and Von Mises Yield Surface is an ellipse. In simple terms Yield locus is the condition representing the transition from elastic to plastic deformation.

Yield Strength: It is the stress at which a material becomes to deform plastically, meaning it will not regain its original shape after the load is withdrawn.

Young's Modulus: It is a measure of the stiffness of a material, defined as the ratio of tensile stress to tensile strain $E = \frac{\sigma}{\varepsilon}$.

MULTIPLE CHOICE QUESTIONS

1. In the stress-strain diagram curve A represents

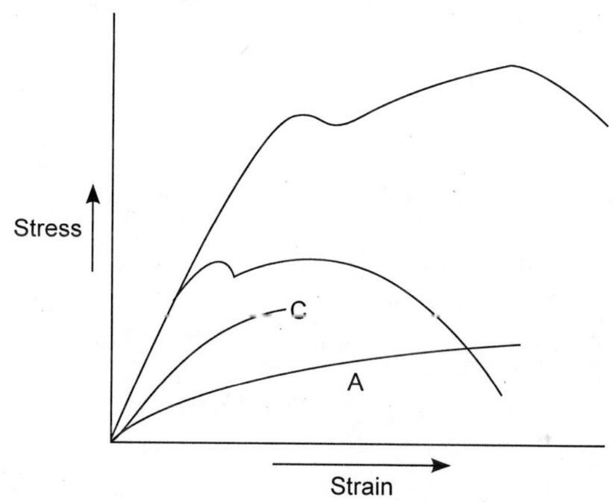

(a) Mild steel (b) Soft brass (c) Low carbon steel (d) Cold rolled steel

2. The S.I. unit of stress is

(a) $\frac{KN}{mm^2}$ (b) $\frac{N}{mm^2}$ (c) $\frac{N}{m^2}$ (d) $\frac{kg}{m^2}$

3. The dimensions of strain are

(a) $M L T^{-2}$ (b) $M^0 L T^0$ (c) $M^0 L^0 T^0$ (d) $M^0 L^{-1} T^0$

4. Hook's Law is valid upto

(a) Yield point (b) Breaking point (c) Plastic limit (d) Elastic limit

5. The ratio of linear stress to linear strain is called
   (a) Poisson's ratio (b) Modulus of Elasticity
   (c) Modulus of Rigidity (d) Bulk Modulus
6. For mild steel the modulus of Elasticity is about
   (a) $10 \dfrac{KN}{mm^2}$ (b) $80 \dfrac{KN}{mm^2}$ (c) $100 \dfrac{KN}{mm^2}$ (d) $210 \dfrac{KN}{mm^2}$
7. A mild steel bar of cross-sectional area (A) is loaded with an axial Tensile load (P). The load is now doubled and area reduced to half. The ratio of initial to final stress would be
   (a) 1 (b) 2 (c) 4 (d) 8
8. The ultimate tensile stress for mild steel is .... the ultimate compressive stress.
   (a) Equal to (b) Less than (c) More than (d) Dependent on loading
9. A bar of length 1.0 m extends 1.0 mm under a tensile force of (F) Newton. The strain produced is
   (a) 1 (b) 0.1 (c) 0.01 (d) 0.001
10. A steel bar is heated from 15° C to 55° C and is free to expand, the bar will induce.
   (a) No stress (b) Shear stress (c) Tensile stress (d) Compressive stress
11. When a composite bar of copper and steel is cooled, then copper bar will experience stress
   (a) Tensile (b) Compressive (c) Shear (d) NO
12. Shear stress on a Principal plane is
   (a) Zero (b) Maximum (c) Minimum (d) None of these
13. Among several prismatic beams of equal lengths, the strongest in flexure is the one having maximum
   (a) Moment of inertia (b) Section modulus
   (c) Tensile strength (d) Area of cross section
14. A prismatic bar when subjected to pure bending assumes the shape of
   (a) Catenary (b) Cubic parabola
   (c) Quadratic parabola (d) Arc of a circle
15. Deflection a leaf spring is more if it's
   (a) Strength is more (b) Strength is less
   (c) Stiffness is less (d) Stiffness is more
16. The maximum stress produced in a bar of tapering section is at
   (a) Smaller end (b) Larger end (c) Mid-section (d) Anywhere
17. The ratio of lateral to linear strain is called
   (a) Modulus of elasticity (b) Bulk modulas
   (c) Bulk modulus of Rigidity (d) Poisson's ratio
18. If Modulus of elasticity of a given material is twice the Modulus of rigidity then, bulk modulus will be
   (a) 2 C (b) 3 C (c) $\frac{2}{3}$C (d) $\frac{3}{2}$C

19. The ratio of Bulk modulus to Young's modulus for a Poisson's ratio of 0.25 will be

(a) $\frac{1}{3}$ (b) $\frac{2}{3}$ (c) 1 (d) $\frac{3}{2}$

20. For a material of Bulk modulus k, the Modulus of rigidity C, the Poisson's ratio will be

(a) $\frac{3k-4C}{6k+4C}$ (b) $\frac{3k+4C}{6k-4C}$ (c) $\frac{3k-2C}{6k+2C}$ (d) $\frac{3k+2C}{6k-2C}$

21. A localised compressive stress at the area of contact between two members is known as

(a) Tensile stress (b) Bending stress (c) Crushing stress (d) Shear stress

22. A body is subjected to two normal stresses $20\frac{KN}{m^2}$ (Tensile) and $10\frac{KN}{m^2}$ (Compressive) normal to each other. The maximum shear stress will be

(a) $5\frac{KN}{m^2}$ (b) $10\frac{KN}{m^2}$ (c) $15\frac{KN}{m^2}$ (d) $20\frac{KN}{m^2}$

23. The maximum shear stress is .... the algebraic difference of maximum and minimum normal stresses

(a) Equal to (b) One fourth (c) One-half (d) Twice

24. Strain energy is the

(a) Energy stored in a body when strained within Elastic limit

(b) Energy stored when strained upto breaking of specimen

(c) Maximum energy which can be stored in a body

(d) Proof resilience per unit volume of a material

25. The strain energy stored in a body when suddenly loaded is .... the strain energy stored when load is applied gradually

(a) Equal to (b) $\frac{1}{2}$ of (c) Twice (d) Four times

26. Any material having identical properties in all direction is called

(a) Homogenous (b) Isotropic (c) Orthotopic (d) Elastic

27. Any material with Modulus of elasticity zero is said to be

(a) Rigid (b) Elastic (c) Flexible (d) Plastic

28. A free bar of length (L) is uniformly heated from 0° C to t° C, α is the coefficient of linear expansion and E modulus of elasticity, the stress in the bar is

(a) $\alpha L E$ (b) $\alpha E \frac{t}{2}$ (c) Zero (d) L E

29. The deformation of a bar under its own weight as compared to that when subjected to direct axial load equal to its own weight will be

(a) The same (b) One-fourth (c) Half (d) Double

30. The stress below which a material has high probability of not failing under reversal of stress is known as

(a) Tolerance limit (b) Elastic limit

(c) Proportional limit (d) Endurance limit

31. A bar is subjected to axial tensile stress. If the volumetric strain in the bar is 0.44 times the axial strain, the Poisson's ratio of the material is

(a) 0.44 (b) 0.30 (c) 0.28 (d) 0.11

32. A bi-metallic strip is made of two metals having equal cross-sections. Due to temperature change, the stress developed in one strip is $-40\,\frac{N}{mm^2}$. The stress in the other component will be

(a) $-40\,\frac{N}{mm^2}$ (b) $20\,\frac{N}{mm^2}$ (c) $40\,\frac{N}{mm^2}$ (d) $-20\,\frac{N}{mm^2}$

33. In case of an I-section, the web resists mainly

(a) Axis Force (b) Shear Force (c) BM (d) Both SF & BM

34. A beam of square cross-section of side (B) is used as a beam with one diagonal horizontal. The location of maximum shear stress from Neutral axis will be at a distance of

(a) Zero (b) $\frac{B}{4}$ (c) $\frac{B}{4\sqrt{2}}$ (d) $\frac{B}{8}$

35. The radius of Mohr's circle of stress of a strained element is $20\,\frac{N}{mm^2}$ and minor principal tensile stress is $10\,\frac{N}{mm^2}$. The major principal stress is

(a) $30\,\frac{N}{mm^2}$ (b) $50\,\frac{N}{mm^2}$ (c) $60\,\frac{N}{mm^2}$ (d) $100\,\frac{N}{mm^2}$

36. In a stress-strain curve, the area upto elastic limit stress indicates the mechanical property

(a) Ductility (b) Strength (c) Resilience (d) None of these

37. Maximum shear stress in a circular cross-section is

(a) $\frac{9\,q_{av}}{8}$ (b) $\frac{4\,q_{av}}{3}$ (c) $\frac{3\,q_{av}}{2}$ (d) $\frac{8\,q_{av}}{3}$

38. The principal stresses σ_1, σ_2 and σ_3 at a point respectively are 80 M Pa, 30 M Pa and – 40 M Pa the maximum shear stress will be

(a) 25 M Pa (b) 35 M Pa (c) 55 M Pa (d) 60 M Pa

39. Principal strains at a point are 100×10^{-6} and -200×10^{-6}. The maximum shear strain at the point will be

(a) 300×10^{-6} (b) 200×10^{-6} (c) 150×10^{-6} (d) 100×10^{-6}

40. Which of the following Mohr's circles represents the state of pure shear?

(a) 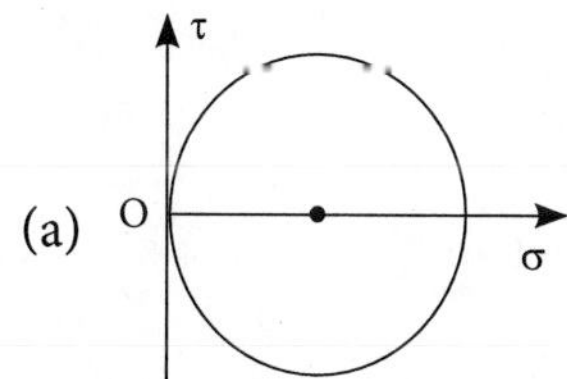

(b) 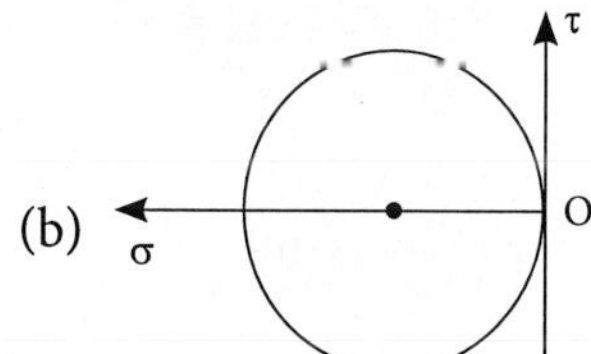

(c) 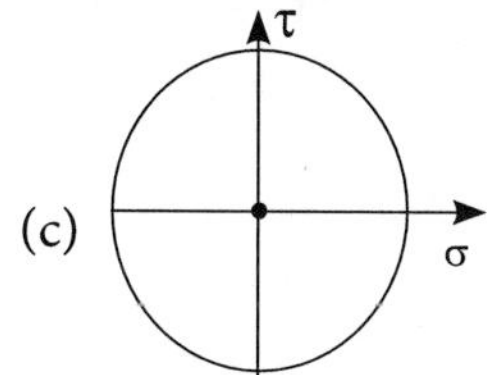

(d) 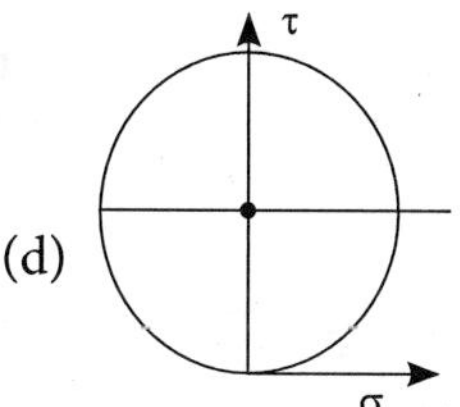

41. For ductile materials the suitable theory of failure is

(a) Maximum Principal stress theory (b) Maximum shear stress theory
(c) Both a and b (d) None of these

42. The compressibility of any material is proportional to

(a) Poisson's Ratio μ (b) Modulus of Elasticity E
(c) $\frac{1}{E}$ (d) $\frac{1}{\mu}$

43. Notched bar tests are frequently used for testing the

(a) Impact strength of material (b) Hardness of material
(c) Machinability of metal (d) Corrosion resistance

44. Major principal stress at a point is 220 M Pa. The radius of Mohr's stress circle is 70 M Pa. The minor principal stress at the point is

(a) 150 M Pa (b) 100 M Pa (c) 80 M Pa (d) 50 M Pa

45. The major and minor principal stresses at a point are 120 M Pa and 40 M Pa, what is the normal stress on a plane of τ_{max} stress

(a) 120 M Pa (b) 80 M Pa (c) 40 M Pa (d) 60 M Pa

46. The principal stresses developed at a point are + 60, – 60 and 0 M Pa. Using shear strain energy theory, factor of safety obtained is $\sqrt{3}$. The Yield stress of the material

(a) 60 M Pa (b) 120 M Pa (c) 60 $\sqrt{3}$ M Pa (d) 180 M Pa

47. The slope deflection of beam of varying flexural rigidity may be computed by which method

(a) Macaulay (b) Mohr (c) Conjugate beam (d) Moment Distribution

48. The commonly used technique of strengthening thin pressure vessel is

(a) Wire winding (b) Shrink fitting
(c) Autofretcage (d) Multi-layered construction

49. If a shaft is simultaneously subjected to torque (T) and BM (M), the ratio of maximum bending stress and maximum shear stress will be

(a) $\frac{2M}{T}$ (b) $\frac{M}{T}$ (c) $\frac{2T}{M}$ (d) $\frac{T}{M}$

50. Two shafts of same material and same length are connected in a series and subjected to a torque T. If $\frac{d_2}{d_1} = 2$, then $\frac{\tau_1}{\tau_2}$ is

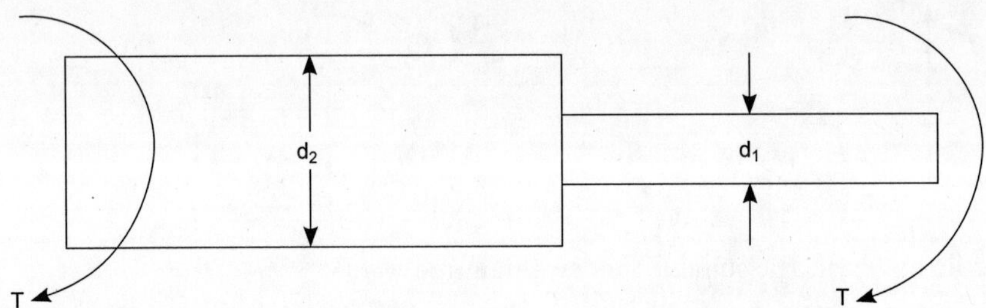

(where τ_1, τ_2 are extreme shear stresses in either of two segments)

(a) 4 (b) 2 (c) 8 (d) 16

51. For linear elastic systems, the type of displacement function for strain energy is

(a) Linear (b) Quadratic (c) Cubic (d) Quartic

52. The number of elastic constants for a linear elastic isotropic and homogenous material is

(a) 4 (b) 3 (c) 2 (d) 1

53. The dimensions for the flexural rigidity of a beam element in Mass (M), Length (L) and Time (T) are

(a) $M T^{-2}$ (b) $ML^3 T^{-2}$ (c) $ML^{-1} T^{-2}$ (d) $ML^{-1} T^2$

54. U_1 and U_2 are the strain energies stored in a prismatic bar due to axial tensile forces P_1 and P_2. The strain energy U stored in the same bar due to combined action of P_1 and P_2 will be

(a) $U = U_1 + U_2$ (b) $U = U_1 \times U_2$

(c) $U < U_1 + U_2$ (d) $U > U_1 + U_2$

55. For the state of stresses (in M Pa) shown here the maximum shear stress (in M Pa) is

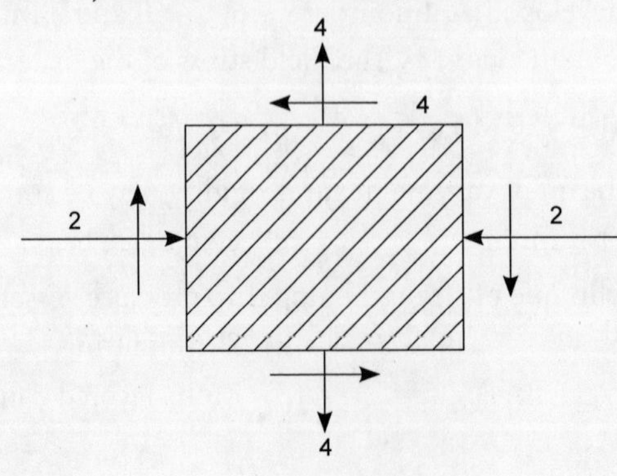

(a) 5.0 (b) 6.0 (c) 12.0 (d) 8.0

56. The Poisson's Ratio is defined as

(a) $\left|\dfrac{\text{axis stress}}{\text{lateral stress}}\right|$ (b) $\left|\dfrac{\text{lateral strain}}{\text{axial strain}}\right|$ (c) $\left|\dfrac{\text{lateral stress}}{\text{axial stress}}\right|$ (d) $\left|\dfrac{\text{axial strain}}{\text{lateral strain}}\right|$

57. A metal bar length 100 mm is inserted between two rigid supports and its temperature is raised by 10° C. If coefficient of thermal expansion is 12×10^{-6} per °C and the Young's modulus is 2×10^{5} M Pa, stress in the bar is

(a) Zero (b) 12 M Pa (c) 24 M Pa (d) 240 M Pa

58. An axially loaded bar is subjected to normal stress of 173 M Pa, shear stress in the bar is

(a) 75 M Pa (b) 86.5 M Pa

(c) 100 M Pa (d) 122.5 M Pa

59. An element is subjected to bi-axial normal tensile strains of 0.003 and 0.002. The normal strain in the plane of maximum shear strain is

(a) Zero (b) 0.005 (c) 0.001 (d) 0.0025

60. The state of stress represented by the Mohr's circle shown here is

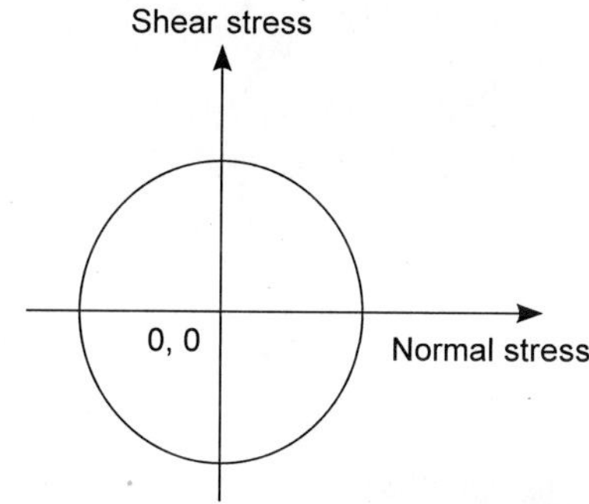

(a) Isotropic stress (b) Uniaxial tension

(c) Pure shear (d) bi-axial tension of equal magnitudes

61. An elastic isotropic body is in hydrostatic state of stress as shown here. For no change in the volume to occur, what should be its Poisson's ratio?

(a) 0.00 (b) 0.25 (c) 0.50 (d) 1.00

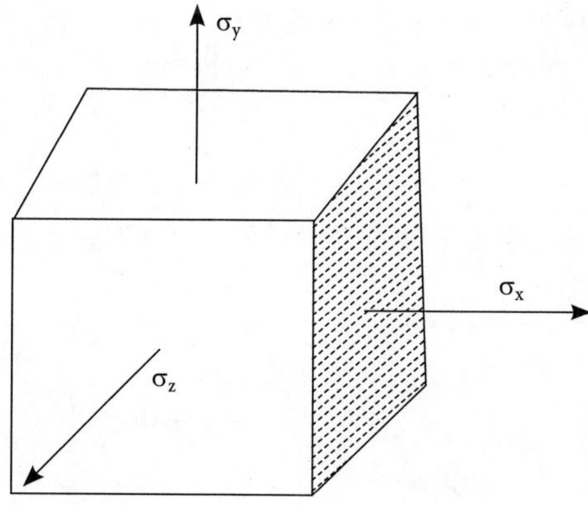

62. A plate in equilibrium is subjected to uniform stresses along edges with magnitude as shown. The Young's Modulus of material is $2 \times 10^{11} \frac{N}{m^2}$ and Poisson's ratio 0.3. If σ_{zz} is assumed to be zero then the strain ε_{zz} is

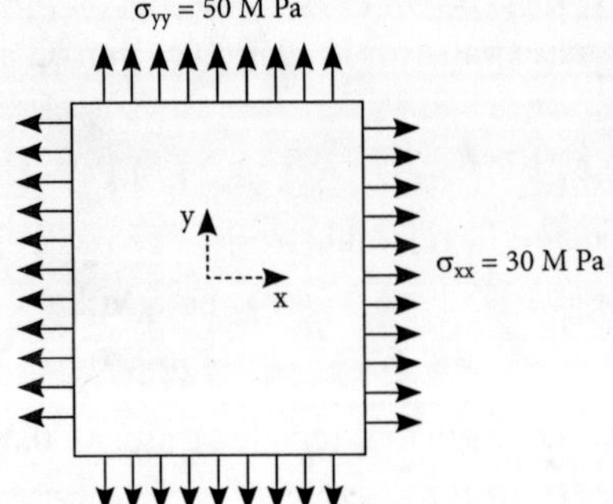

(a) -120×10^{-6} (b) -60×10^{-6} (c) 0.0 (d) 120×10^{-6}

63. The maximum value of Poisson's ratio for an elastic material is

(a) 0.25 (b) 0.5 (c) 0.75 (d) 0.1

64. If the Young's modulus of elasticity of a material is twice its modulus of rigidity, then the Poisson's ratio of the material is

(a) −1 (b) −05 (c) 0.5 (d) Zero

65. Limit of proportionality depends upon

(a) Area of cross-section (b) Type of loading
(c) Type of material (d) All these

66. The relationship between Young's modulus of Elasticity E, Bulk modulus K and Poisson's ratio μ is given by

(a) $E = 2K(1 - 2\mu)$ (b) $E = 3K(1 + \mu)$ (c) $E = 3K(1 - 2\mu)$ (d) $E = 2K(1 + \mu)$

67. Of the two prismatic beams of same material, length and flexural strength, one is circular and other is square in Cross-section. The ratio of weights of circular and square beams is

(a) 1.118 (b) 3.342 (c) 1.00 (d) 0.793

68. A beam of uniform strength has at every cross-section same

(a) Bending moment (b) Bending stress (c) Deflection (d) Stiffness

69. The shear centre in the channel section shown will be at

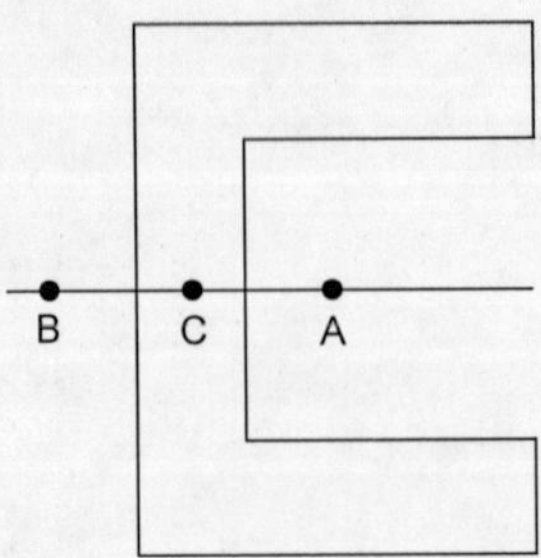

(a) A (b) B (c) C (d) None of these

70. The effect of arching a beam is
   (a) To reduce Bending Moment throughout (b) To increase Bending Moment throughout
   (c) To increase Shear Force (d) To decrease Shear Force
71. If the depth of a beam of rectangular section is reduced to half the strain energy stored in the beam becomes
   (a) $\frac{1}{4}$ times (b) $\frac{1}{8}$ times (c) 4 times (d) 8 times
72. A mild steel bar in three parts each 20 cm long having diameters 2 cm, 1 cm and 3 cm as shown. The bar is subjected to an axial pull of 4t. If $E = 2 \times 10^6 \frac{kg}{cm^2}$ and elongations in three parts are Δ_1, Δ_2 and Δ_3,

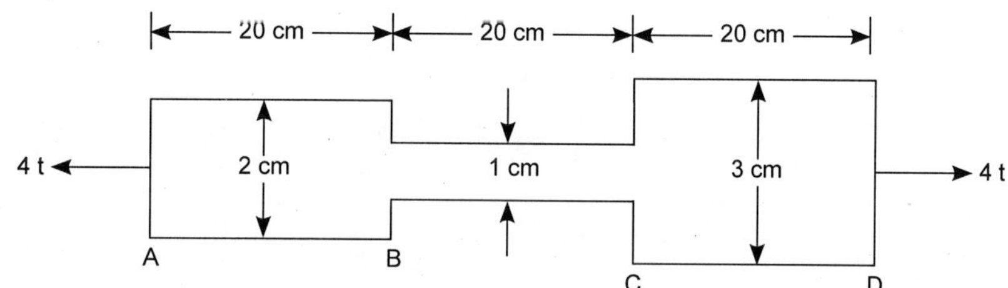

then the ratio of the greatest to the least of these elongations will bc
   (a) 9 (b) 4 (c) 3 (d) 2
73. Impact test enables to estimate the property of
   (a) Hardness (b) Toughness (c) Strength (d) Creep
74. The phenomenon of decreased resistance of a material to reversal of stress is called
   (a) Creep (b) Fatigue (c) Resilience (d) Plasticity
75. The property of metal which allows it to deform continuously at slow rate without any further increase in stress is known as
   (a) Fatigue (b) Resilience (c) Plasticity (d) Creep
76. For design of a cast iron member the most appropriate theory of failure is
   (a) Mohr's (b) Rankine's
   (c) Maximum strain (d) Maximum shear energy theory
77. A steel rod of cross-sectional area 1000 mm^2 is 5 m long. If a pull of 100 KN is suddenly applied to it, then the maximum stress intensity will be
   (a) $50 \frac{N}{mm^2}$ (b) 100 c (c) $200 \frac{N}{mm^2}$ (d) $400 \frac{N}{mm^2}$
78. A Mohr's circle reduces to a point when the body is subjected to
   (a) Pure shear
   (b) Uniaxial stress only
   (c) Equal and opposite axial stresses on two mutually perpendicular planes, the planes being free of shear
   (d) Equal axial stresses on two mutually perpendicular planes, the planes being free of shear

79. The Bending moment diagram for an overhanging beam is shown here.

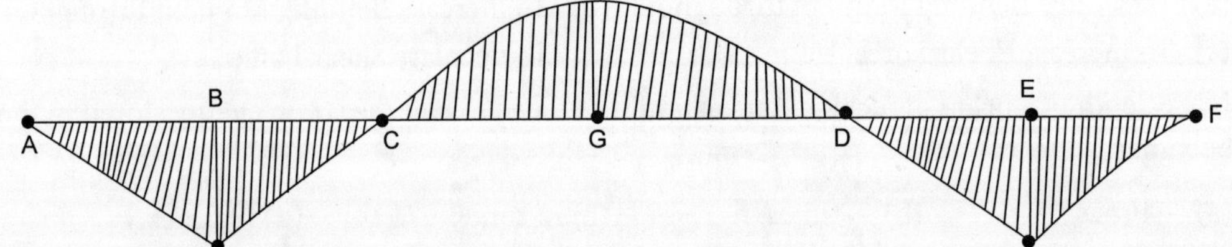

The point of contraflexure would include

(a) A nd F (b) B and E (c) C and D (d) A and D

80. An overhanging beam of Length (L) and uniform flexural rigidity EI is loaded by two concentrated loads at two ends. The beam is supported with overhangs (a) on each side as shown here,

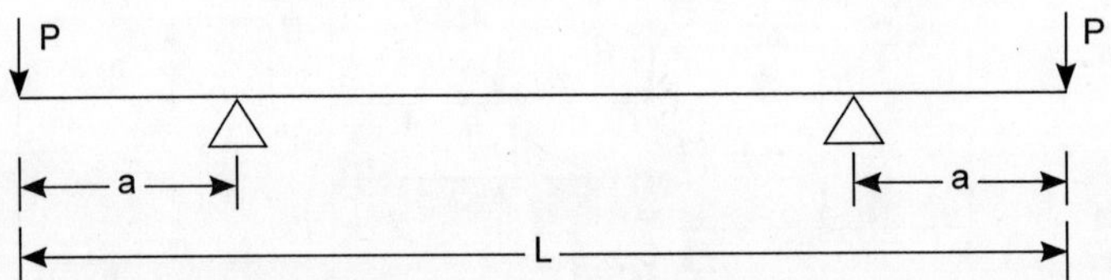

The vertical deflection upwards at the centre of the beam is given by

(a) $\frac{Pa\,(L-2a)^2}{8\,EI}$ (b) $\frac{Pa^2\,(L-2a)}{8\,EI}$ (c) $\frac{Pa\,L^2}{8\,EI}$ (d) $Pa^2\,(L-a).8EI$

81. In four vertical columns of the same material height and weight with same end conditions, the buckling load will be the largest for a column having the cross-section

(a) A solid square (b) A thin hollow circle

(c) A solid circle (d) An I-section

82. According to Tresca, yield locus is

(a) A rectangle (b) A circle (c) A hgexagon (d) Ellipse

83. If two springs of stiffness k_1 and k_2 are connected in series, the stiffness of the combined spring is

(a) $\frac{k_1\,k_2}{k_1+k_2}$ (b) $\frac{k_1+k_2}{k_1\,k_2}$ (c) k_1+k_2 (d) $k_1\,k_2$

84. A steel cube of volume 8000 cc is subjected to an all round stress of $\frac{1330\text{ kg}}{\text{cm}^2}$. The bulk modulus of the material is $\frac{1.33\times10^6\text{ kg}}{\text{cm}^2}$. The volumetric change is

(a) 8 cc (b) 6 cc (c) 0.8 cc (d) 10^{-3} cc

85. In terms of Bulk Modulus (k) and modulus of rigidity (c), the Poisson's ratio can be expressed as

(a) $\frac{2k-4c}{6k+4c}$ (b) $\frac{3k+4c}{6k-4c}$ (c) $\frac{3k-2c}{6k+4c}$ (d) $\frac{3k+2c}{6k-2c}$

86. For a circular column of length L and diameter D, having both ends hinged, the slenderness ratio is 160 the $\frac{L}{D}$ ratio of column is

(a) 80 (b) 57 (c) 40 (d) 20

87. Two closed coil springs of stiffness S and 2S are arranged in series in one case and in parallel in the other case. The ratio of stiffness of springs connected in series to parallel is

(a) $\frac{1}{3}$ (b) $\frac{1}{9}$ (c) $\frac{2}{3}$ (d) $\frac{2}{9}$

88. If the diameter of a shaft subjected to Torque alone is doubled, then Horse Power (P) can be increased to

(a) 16 P (b) 8 P (c) 4 P (d) 2 P

89. If the strain energy absorbed in a cantilever beam in bending under its own weight is (k) times greater than the strain energy absorbed in the identical simply supported beam in the bending under its own weight, then (k) is equal to

(a) 2 (b) 4 (c) 6 (d) 8

90. The material of rubber balloon has Poisson's ratio of 0.5. If uniform pressure is applied to blow the balloon, the volumetric strain of material will be

(a) 0.05 (b) 0.25 (c) 0.20 (d) Zero

91. For points in the neutral surface of the beam the directions of Principal stress across the x-axis is

(a) 90° (b) $\pm$ 45° (c) 45° (d) –45°

92. What is constant in Pure bending?

(a) Moment (b) Force (c) Deflection (d) Curvature

93. An element is subjected to two mutually perpendicular unlike equal stresses (p). The radius of Mohr's circle will be

(a) Zero (b) $\frac{p}{2}$ (c) p (d) 2 p

94. Effective length of a column with one end fixed and other end free having length (L) is

(a) L (b) $\frac{L}{2}$ (c) $\frac{3}{4}$L (d) 2 L

95. The effective length of a column of length (L) and both ends fixed will be

(a) L (b) 0.5 L (c) 0.75 L (d) 2 L

96. The Euler-load for a column is 1000 KN and crushing load is 1500 KN. The Rankine-load is equal to

(a) 600 KN (b) 1000 KN (c) 1500 KN (d) 2500 KN

97. A timber beam of rectangular section is simply supported at ends having a steel strip securely fixed at the top surface as shown here. The centroid of the 'equivalent' timber beam in this case from top of the surface is

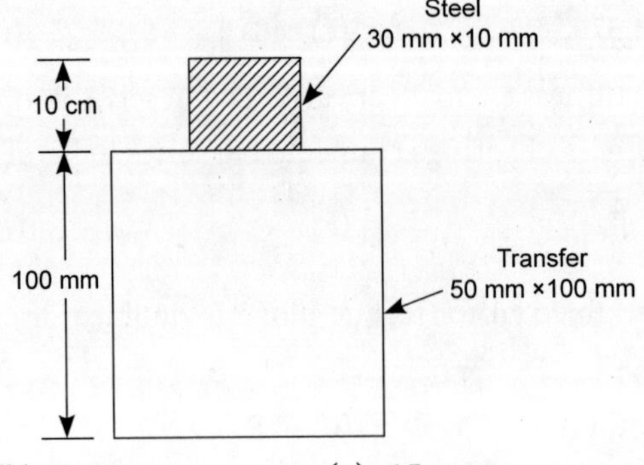

(a) Is 5 mm (b) Is 30 mm (c) 15 mm (d) 25 mm

98. In a particular material if the modulus of rigidity is equal to bulk modulus, then the Poisson's Ratio will be

(a) $\frac{1}{8}$ (b) $\frac{1}{4}$ (c) $\frac{1}{2}$ (d) 1

99. A solid circular beam radius 0.25 m and length 2 m is subjected to a twisting moment of 20 KN – m about z-axis at free end. The stress components τ_{xy} at point M in the cross-sectionn of the beam at a distance of 1 m from the fixed end is

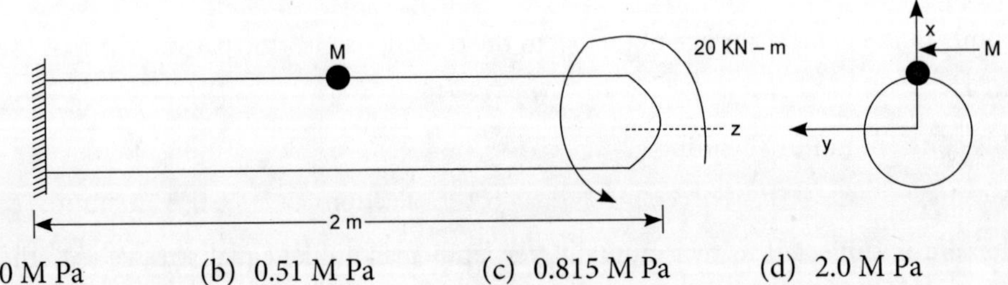

(a) 0.0 M Pa (b) 0.51 M Pa (c) 0.815 M Pa (d) 2.0 M Pa

100. Cross-section of a column consisting of two steel strips each of thickness (t) and width (b) is shown here. The critical load of the column with perfect bond and without bond between the strips are P and P_0. The ratio $\frac{P}{P_0}$ is

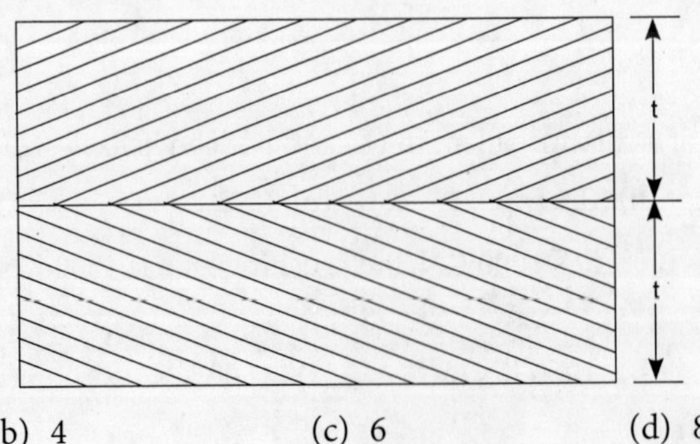

(a) 2 (b) 4 (c) 6 (d) 8

101. The point within the cross-sectional plane of a beam through which the resultant of the external loading on the beam has to pass through to ensure pure bending without twisting of the cross-section of the beam is called

(a) Moment centre (b) Centroid (c) Shear centre (d) Elastic centre

102. Which of the following materials is most elastic?

(a) Rubber (b) Plastic (c) Brass (d) Steel

103. The strain energy stored in a body, when the load is gradually applied, is (where σ = stress in the material of the body and E = Modulus of Elasticity and V = volume of the body)

(a) $\frac{\sigma E}{V}$ (b) $\frac{\sigma V}{E}$ (c) $\frac{\sigma^2 E}{2V}$ (d) $\frac{\sigma^2 V}{2E}$

104. The section modulus of a rectangular section about an axis through its CG is

(a) $\frac{b}{2}$ (b) $\frac{d}{2}$ (c) $\frac{bd^2}{2}$ (d) $\frac{bd^2}{6}$

105. If the depth is kept constant for a beam of Uniform strength, then its width will vary in proportional to

(a) M (b) $\sqrt{M}$ (c) M^2 (d) M^3

(M = Bending Moment)

106. Lame's theory is associated with

(a) Thin cylindrical shells (b) Thick cylindrical shells
(c) Direct and Bending stresses (d) None of these

107. When a column is subjected to an eccentric load, the stress induced is

(a) Direct stress only (b) Bending stress only
(c) Shear stress only (d) Both direct and bending stress

TRUE OR FALSE QUESTIONS

1. A body is subjected to two equal and opposite pushes, as a result of which the body tends to reduce its length, the stress and strain induced is compressive. (T or F)
2. When a material is loaded within elastic limit stress is inversely proportional to strain. (T or F)
3. The units of Young's Modulus are same as those of stress. (T or F)
4. A shear stress across a plane, is always accompanied by a balancing shear stress across the plane and normal to it. (T or F)
5. The modulus of elasticity of a material is twice its modulus of rigidity, then the Poisson's ratio of the material is zero. (T or F)
6. The planes which carry no shear stress are known as Principal Planes. (T or F)
7. The type of hoop stress in thin cylindrical shell is radial stress. (T or F)

8. The shear stress in a thick cylinder shell is less than the internal pressure acting on the shell. (T or F)
9. The hoop stress in a thick-cylindrical shell is maximum at the outer radius. (T or F)
10. In a thin cylindrical shell, the ratio of Longitudinal and hoop stress is 2. (T or F)
11. The slenderness ratio of a 5 m long column hinged at both ends and having a circular cross-section with diameter 16 cm is 125. (T or F)
12. Rankines formula is valid when slenderness ratio is less than 80. (T or F)
13. Three hinged arch is statically determinate. (T or F)
14. If central rise of a symmetrical parabolic arch is (h), then the rise of arch at quarter points is $\frac{h}{3}$. (T or F)
15. A linear arch has normal thrust only. (T or F)
16. In the pin-jointed plane frame shown here, the Force in BD is zero. (T or F)

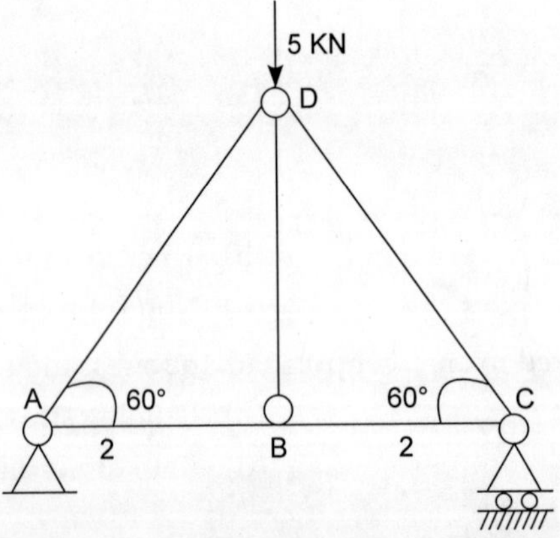

17. The rigid-jointed plane frame will sway to the left. (T or F)

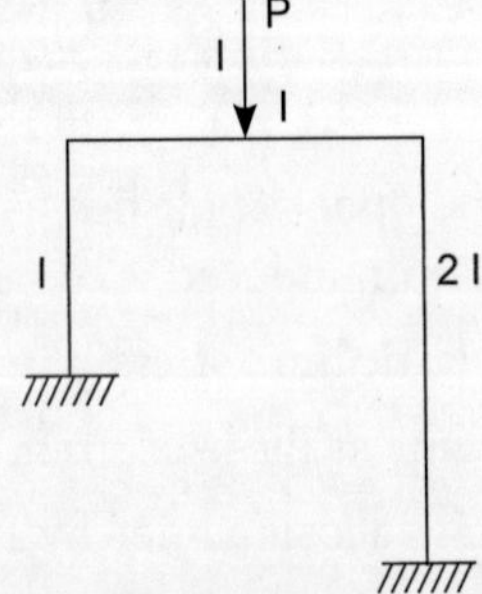

18. The stiffness of a helical spring is expressed as load per unit deflection. (T or F)
19. Strain energy stored in a member is given by 0.5 × stress × strain. (T or F)
20. If an element is subjected to pure shearing stress τ_{xy}, the maximum principal stress is equal to $2\tau_{xy}$. (T or F)
21. If Bulk Modulus of a material is equal to shear modulus, the Poisson's ratio is 0.125. (T or F)

22. The resultant cuts the base of a circular column of diameter (d) with eccentricity equal to c . The ratio between maximum compressive and maximum tensile stress is 5 (five). (T or F)

23. Two triangular wedges are glued together as shown here. The stress acting normal to the interface (∂_n) is zero. (T or F)

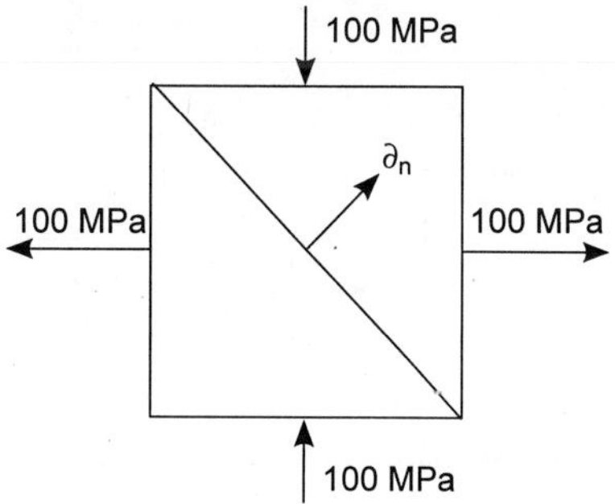

24. The strain energy stored in a body due to shear stress is $\frac{\tau V}{2c}$. (where τ = Shear stress, C is Shear Modulus and V = Volume of body) (T or F)

25. If in a beam, a single Load (W) is applied gradually and the deflection under the load point is δl, then the strain energy in the beam will be $W\delta l$. (T or F)

26. The shear force of a simply supported beam carrying a central point load changes sign at mid point. (T or F)

27. The maximum BM of a simply supported beam with middle point load lies near the support. (T or F)

28. The point of contraflexure occurs in overhanging beams. (T or F)

29. When a bar is subjected to change of temperature and its deformation is prevented, the stress induced in the bar is called Thermal Stress. (T or F)

30. The thermal stress in a bar indirectly proportional to change in temperature. (T or F)

31. The thermal stress depends upon the area of cross-section of the bar. (T or F)

32. Poisson's ratio is the ratio of linear strain to the volumetric strain. (T or F)

33. A simply supported beam is loaded as shown here. (T or F)

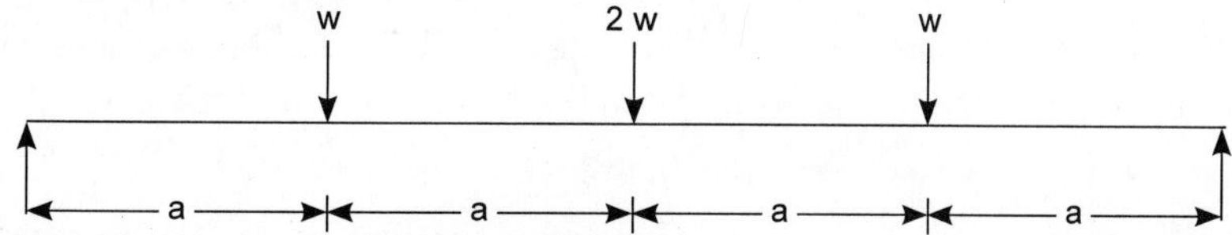

The maximum shear force in the beam will be 2W.

34. The bending moment at the end of a simply supported beam will be zero. (T or F)

35. The maximum bending moment of a cantilever beam of length (L) and carrying a UDL of (w) per unit length lies t the middle of its length. (T or F)

36. When the SF diagram between any two points is an inclined straight line, it indicates that there is a uniformly varying load between two points. (T or F)

37. In the simply supported beam shown here the Maximum deflection will be at D. (T or F)

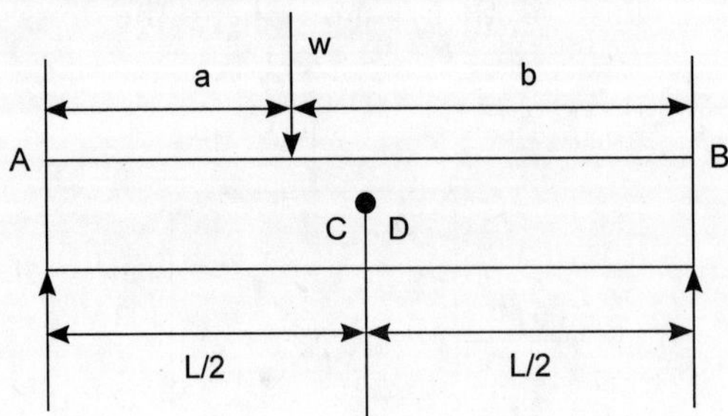

38. When there is no increase or decrease in Shear Force between two points, it indicates that there is no change in the bending moment between these points. (T or F)

39. The section modulus of a circular section about an axis through its CG is $\frac{\pi d^3}{16}$. (T or F)

40. The strength of any beam mainly depends on its weight. (T or F)

41. The Bending Stress in a beam is directly proportional to bending moment. (T or F)

42. The breaking stress is equal to ultimate stress. (T or F)

43. Short columns cause deflection in the structure. (T or F)

44. Fatigue failure occurs after repeated cyclic loading even if the maximum stress is below the Yield stress. (T or F)

45. A brittle material has low ability to absorb energy before fracture. (T or F)

46. The modulus of Elasticity and Young's modulus are the same. (T or F)

47. The Factor of safety is the ratio of working stress to ultimate stress. (T or F)

48. The bending stress in a beam is inversely proportional to section modulus. (T or F)

49. A beam of triangular section is placed with its base horizontal. The maximum shear stress will occur at centre of Gravity of the section. (T or F)

50. Two solid shafts A and B are made of same material. The diameter of A is 50 mm and of B is 100 mm. The strength of B is double as that of shaft A. (T or F)

51. Longitudinal stress in a thin cylinder is half of the hoop stress. (T or F)

52. Two closely coiled helical springs A and B are equal in all respects but the number of turns of spring A is half that of spring B, the Ratio of deflections in spring A and B is 2. (T or F)

53. The shear stress at the centre of a circular shaft under torsion is zero. (T or F)

54. For two shafts connected in parallel and subjected to twisting moments, the angle of twist of each shaft will be same. (T or F)

55. Two shafts (identical) of diameters (d_1) and (d_2) are subjected to torque (T). The ratio of strain energies stored will be $\left(\frac{d_1}{d_2}\right)^4$. (T or F)

56. The stiffness of a closely-coiled helical spring is inversely proportional to number of turns. (T or F)

57. The hoop stress in a thick cylindrical shell is maximum at the inner radius. (T or F)

FILL IN THE BLANKS

1. A simply supported RC beam of 10 m sags on shrinkage. For uniform curvature of 0.004 m^{-1} along the span, maximum deflection will be at mid span is .... m.
2. A symmetrical I-section having width of each flange = 50 mm, thickness of web = 10 mm of steel is subjected to a Shear Force = 100 KN. The magnitude of the shear stress $\frac{N}{mm^2}$ in the web at its junction with top flange will be .... M Pa.
3. A hollow circular shaft has an outer diameter of 100 mm and inner diameter 50 mm. If allowable shear stress is 125 M Pa, the maximum Torque in KN – m that the shaft can resist is ....
4. Buckling load for a given column depends upon least lateral dimension and .... of column.
5. Euler's formula for mild steel long column hinged at both ends is not valid for slenderness ratio less than ....
6. Effective length of a chimney of 20 m height is taken as .... m.
7. The radius of Mohr's circle for two equal unlike principal stresses of magnitude (p) is ....
8. Maximum Bending Moment in a beam occurs where shear force .... sign.
9. Rate of change of Bending Moment is equal to .... force.
10. A prismatic beam fixed at both ends carries a UDL. The ratio of Bending Moment at support to the Bending Moment at the mid span will be ....
11. The variation of BM in the portion of a beam carrying linearly varying load is a .... curve.
12. The maximum BM due to a moving load on a fixed ended beam occurs at a ....
13. If qav is the average shear stress, maximum shear stress in a circular cross-section will be ....
14. For no Torsion, the plane of bending should pass through .... centre of the section.
15. If the deflection at the free end of a uniformly loaded cantilever beam is 15 mm and the slope of the deflection curve at the free end is 0.02 radian, then the length of the beam is .... m.
16. A hollow shaft will transmit .... power than a solid shaft of same weight and material.
17. A basic perfect frame is of .... shape.
18. Method of joints is applicable only when the number of unknown forces under consideration is not more than ....

19. Force in the member AD of the pin-jointed frame shown here is ....

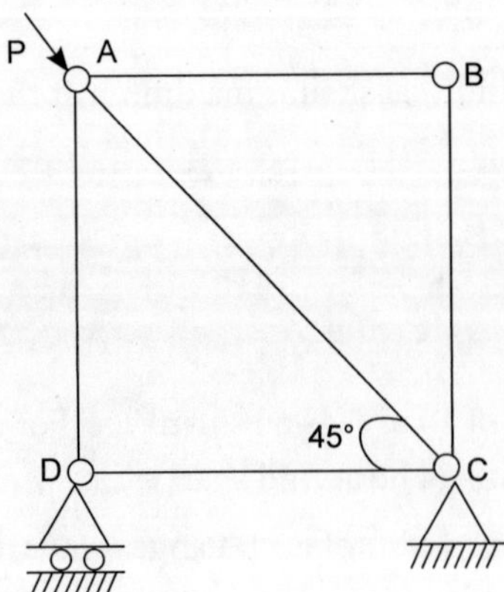

20. In a closed coil helical spring subjected to an axial load, other quantities remaining the same, if the diameter of the wire is doubled the stiffness of the spring compared to original one will become, ... times.

21. If all the dimensions of a prismatic bar elongating under its own weight are increased in proportion m : 1, then total elongation will increase in the ratio ....

22. If the crushing stress in the material of a mild steel column is $\frac{3300 \text{ kg}}{\text{cm}^2}$, Euler's formula for crippling load is applicable for slenderness ratio equal or greater than ....

23. The capacity of a strained body for doing work on the removal of the straining Force is called ....

24. The strain energy stored in a spring, when subjected to maximum load, without suffering permanent distortion is called .... resilience.

25. The Proof resilience per units volume of a material is known as .... resilience.

26. The Bending Moment diagram for a cantilever beam loaded with uniformly distributed load will be a .... curve.

27. The Shear Force of a cantilever beam of length (l) and carrying a gradually varying load from zero at the free end and w per unit length at fixed end will be at the fixed end equal to ....

28. The maximum Bending Moment for the beam shown here is ....

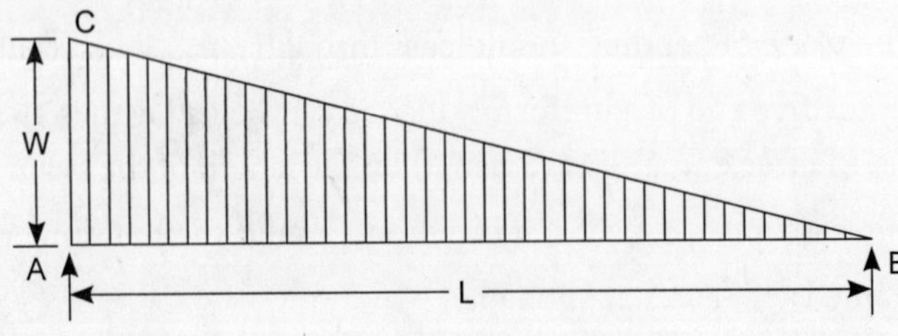

29. The maximum BM for the beam shown is previous problem lies at a distance from B is equal to ....

30. When a cantilever beam is loaded at its free end, the maximum compressive stress shall develop at .... fiber.
31. A beam of uniform strength has same .... stress at every section.
32. The line of intersection of the neutral layer with any normal cross section of the beam is called ....
33. The neutral axis of the cross-section of a beam is that axis at which the bending stress is ....
34. The compressive strength of any brittle material is ....
35. The compression test is carried on .... materials.
36. The tensile test is carried on .... materials.
37. Long columns fail due to ....
38. In short columns the slenderness ratio is less than ....
39. Rankin's constant for cast iron is ....
40. When a shaft is subjected to pure twisting the type of stress developed is .... stress.
41. The structural element which transfer load from beams and slabs to footing are mainly ....
42. Approximate percentage of reinforcement provided in a beam varies from .... %.
43. According to IS-456 : 2000, the column should be designed for minimum eccentricity of.................... mm.
44. According to IS – 456 : 2000, the minimum number of longitudinal bars to be provided in rectangular columns is ....
45. The point at which a material begins to deform plastically is called the ....
46. The measure of a material's ability to absorb energy and plastically deform without fracturing is known as ....
47. The stress at which a material fractures is known as its .... strength.
48. When force is applied an elastic member will ....
49. The extremities of any diameter on Mohr's circle represent normal stresses on planes at angle of ....
50. A coil is cut into two halves, the stiffness of cut coil will be ....
51. The shear stress at the outermost fibres of a circular shaft under torsion is ....
52. The Torsional Rigidity of a solid circular shaft of diameter (d) is proportional to ....
53. Two beams A and B carrying central point load (w) as shown. The deflection of A will be .... as compound to beam B.

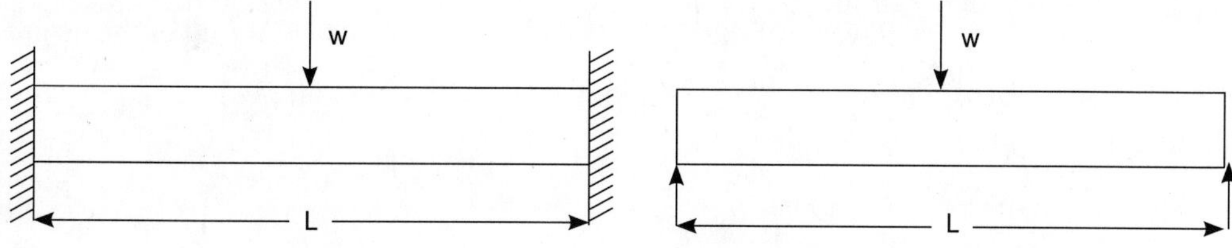

ANSWERS

Multiple Choice Questions

1. b 2. c 3. c 4. d 5. b 6. d 7. c 8. c 9. d
10. a 11. a 12. a 13. b 14. d 15. c 16. a 17. d 18. c
19. b 20. c 21. c 22. c 23. c 24. a 25. d 26. b 27. d
28. c 29. c 30. d 31. c 32. c 33. b 34. c 35. b 36. c
37. b 38. d 39. a 40. c 41. b 42. d 43. a 44. c 45. b
46. d 47. c 48. d 49. a 50. c 51. b 52. c 53. b 54. d
55. a 56. b 57. c 58. b 59. d 60. c 61. c 62. a 63. b
64. d 65. c 66. c 67. a 68. b 69. b 70. a 71. d 72. a
73. b 74. b 75. d 76. b 77. c 78. b 79. c 80. a 81. b
82. b 83. a 84. a 85. c 86. c 87. d 88. b 89. c 90. d
91. b 92. a 93. c 94. d 95. b 96. a 97. b 98. a 99. a
100. b 101. c 102. d 103. d 104. d 105. a 106. a 107. d

True & False

1. T 2. F Directly proportional 3. T
4. F 5. T 6. T 7. F Circumfrancial
8. F More 9. F 10. T 11. T
12. F Does not depend 13. T 14. F $\frac{3h}{4}$
15. T 16. T 17. F To the right 18. T
19. F 0.5 × Stress strain × volume 20. F = τ_{xy} 21. T
22. T 23. T 24. F $\frac{\tau^2 V}{2C}$ 25. F $\frac{w\delta l}{2}$
26. T 27. F At paws of loading 28. T
29. T 30. F Directly proportional 31. F Does not depend
32. F Lateral to Linear strain 33. T 34. T
35. F 36. T 37. F At C 38. T
39. F $\frac{\pi d^3}{32}$ 40. F Its section modulus 41. T
42. F Less than 43. F 44. T 45. T
46. T 47. F Reverse of this 48. T 49. F At mid of the height
50. F Eight times 51. T 52. F Is $\frac{1}{2}$ 53. T
54. T 55. F $\left(\frac{d_1}{d_2}\right)^2$ 56. T 57. T

Fill in the Blanks

1. 0.054 m	**2.** 71 MPa	**3.** 23 KN – m	**4.** Length	**5.** 80
6. 40 m	**7.** p	**8.** Changes	**9.** Shear	
10. 2 (Two)	**11.** Cubic	**12.** Support	**13.** $\frac{4}{3} q_{av}$	**14.** Shear
15. 1.0 m	**16.** More	**17.** Triangular	**18.** Two	**19.** Zero
20. 16	**21.** 1 : m^4	**22.** 80	**23.** Resilience	**24.** Proof
25. Modulus	**26.** Parabolic	**27.** $\frac{wl}{2}$	**28.** $\frac{wL^2}{9\sqrt{3}}$	**29.** $\frac{L}{\sqrt{3}}$
30. Bottom	**31.** Bending	**32.** Neutral axis	**33.** Zero	
34. $\frac{\text{Greater}}{\text{more}}$	**35.** Brittle	**36.** Ductile	**37.** Buckling	**38.** 32
39. $\frac{1}{1600}$	**40.** Shear	**41.** Columns	**42.** 1 to 2%	
43. 20 mm	**44.** 4 (Four)	**45.** Yield point	**46.** Toughness	
47. Ultimate	**48.** Stretch	**49.** 45°	**50.** Double	
51. Maximum	**52.** d^4	**53.** One-fourth		

SECTION 20

Structural Analysis

Definitions – 37

Multiple Choice Questions – 102

True & False – 52

Fill in the Blanks – 51

DEFINITIONS

Arch An arch is a curved vertical structure spanning an open space underneath it. Arches support load above them. Arches transfer load to abutments at **springing** points. Hinges may be provided at these points. The top most point is called the **crown** which sometimes also has a hinge. There are three types of arches– the two-hinged arch, three-hinged arch and hingeless fixed arch.

Braced Frames: It is construction method that uses diagonal members to resist lateral loads, such as wind and seismic pressures. These are economical and common methods of building structural frames. These are robust and use mainly steel. They are simple to analyse.

Cantilever Beam: It is a structural element that extends horizontally and is supported on only one end. The **cantilever** extends beyond the support point. The best example of its use is in balconies.

Castigliano's First Theorem: In a linearly elastic structure, partial derivative of the strain energy (total internal energy) with respect to a load is equal to the deflection of the point where the load is acting, the deflection being measured in the direction of load. Mathematically $\frac{dU}{d P_i} = \Delta_i$ and $\frac{dU}{d M_j} = \theta_j$ where, U = Total strain energy P_1 and M_j are loads; δ_i and θ_j are deflections.

Castigliano's Second Theorem: The first partial derivative of the total internal energy in a structure with respect to the force applied at any point is equal to the deflection at the point of application of that force in the direction of its line of action.

Conjugate Beam: It is an imaginary beam of same span as the origin beam loaded with $\frac{M}{EI}$ diagram of the original beam, such that the Shear force and bending moment at a section will represent the rotation and deflection at that section in the original beam.

Degree of Static Indeterminacy (DSI): It is the number of redundant forces in the structure. Redundant forces are the forces that cannot be found by writing and solving only the equations of equilibrium. Static indeterminancy is of two types-External and Internal indeterminancy. External indeterminancy is related to the support system of the structure whereas internal indeterminancy is related to internal stress in the members of structure.

Dynamic Analysis: It is the assessment of structures subjected to loads which vary with time, the examples are wind force, seismic forces etc.

Finite Element Method (FEM): It is a numerical technique used to obtain approximate solutions to complex structural problems by dividing the structure into smaller, manageable elements.

Fixed End: Such end cannot have any linear or rotational moment. At the fixed end, the support can develop not only resisting force in any direction but also the resisting moment.

Framed Structure: It refers to a structure composed of a system of members like beams, columns, and braces, that support loads and provide stability, typically arranged like in a grid system.

Free-End: At the free end a structure can have linear or rotational freedom in any direction and hence no **reaction** is developed e.g. Free end of a cantilever.

Hinged Ends: At such ends it is assumed that the member cannot have linear motion in any direction but can rotate freely along the support point, i.e., the end is pinned. In such cases the support can develop resisting force in any direction and not the resisting moment.

Idealization of Structures: For convenience it is necessary to idealize any structure. Normally their centroidal axes represent the members. There may be difference between clear spans and centre-to-centre spans ordinarily used in analysis. These differences can be ignored unless the cross-sectional dimensions are large.

Influence Line: For a given function, such as reaction, axial force, shear force, or bending moment, is a graph (diagram) that shows the variation of that function at any given point on a structure due to application of unit load at any point on the structure.

Kinematic Indeterminancy of a Structure: It is also referred as **degree of freedom** (DOF) for a structure, is the number of unrestrained components, of joint displacements. In simple terms, DOF of the structure is the number of directions the structure can be moved freely. For example, in a two-dimensional structure; each joint has 3 possible degrees of freedom – one in horizontal, one vertical and one rotation.

Linear and Non-Linear Systems: A system is called a linear system if its material has linear stress-strain relationship and a small deflection. A system will be treated as a non-linear system if its material does not have linear stress-strain relationship and its deformation is large.

Load Path: The route that loads take through a structure, from the point of application to the ground (or supports).

Maxwell's Theorem of Reciprocal Deflection: It states, "Displacement at a point A due to a load at B will be same as displacement of point B due to the same load acting at point A. The displacement is measured in the direction of the loads."

Point Load (Concentrated Load): Load applied to a very small area of or a specific point inside a structure. They are commonly applied to columns and beams.

Propped Cantilever: It is a type of structural element or beam that is supported at one end (the cantilever end) and restrained or propped at the other end. The cantilever end is free to deflect vertically but cannot rotate. The propped end is typically supported by a vertical column or a wall. It is indeterminate to the first degree.

Roller Support: At such support beam is free to move along the support and can rotate freely. Hence there is no reaction along the support and the resisting moment is zero. In other words, there will be a reaction only in the direction normal to the support.

Shear Modulus: The shear modulus is one of several quantities for measuring the **stiffness** of materials. In material science shear modulus (G) is defined as the ratio of stear stress to shear strain.

Static Analysis: The evaluation of structures under static loads, which do not change with time.

Statically Determinate and Indeterminate Structures: A structural system which can be analysed by using equations of statical equilibrium only is called as **statically determinate structure**, examples are beams, trusses with ends simply supported, one end hinged and other on rollers and cantilevers. A structure which cannot be analysed by using equilibrium equation only is called **Statically Indeterminate Structures**, examples are fixed beams, continuous beams, propped cantilevers. Indeterminate structures are also called **redundant structures**.

Stiffness (K): Moment (M) required to rotate an end by unit angle (θ, 1 radian) when rotation is permitted at the end, is called stiffness of the beam, $K = \frac{M}{\theta}$.

Strain Energy: When an external load acts on a structure, the structure undergoes deformation and hence, the work is done. To resist these external forces, the internal forces develop gradually from zero to their final value is stored in the structure as strain energy and it helps the structure to spring back to the original shape and size, whenever the external loads are removed.

Stress Concentration: The location in a material (structure) where stress is significantly higher than the average due to geometric discontinuity or load application. Example: Sharp corners.

Structural Analysis Software: While it is not always possible to set up a manual process for Finite Element Analysis (FEA) so there is a vast number of Structural Analysis **software's** available for complex problems. It is less time taking, more efficient and normally accurate method. However needs expert handling.

Structural Design Basics: All the component members are to be arranged in a manner such that they transmit their self weight and other superimposed loads to foundation or supporting structure by economical means to satisfy requirements of architecture and structural stability.

Temperature Stresses: Changes in temperature cause a change in the length of a member. In redundant frames, a change in length of any member gives rise to forces in all other members. When free expansion is not possiblc, compressive force may develope in the members.

$$\text{Temp. Stress} = \alpha\, TE : E = \text{Elastic Modulus}$$

$$T = \text{Temp. rise}$$

$$a = \text{Coefficient of thermal expansion}$$

Three-hinged Arch: In three hinged arches two hinges are at the support and one at the crown. It is statically **determinate** structure though four reaction occur at the supports but an additional independent fourth moment equation can be obtained for the hinge at the crown.

Truss: It is a rigid structural framework made up of straight members that are connected at their ends to form triangles. The members are designed to carry only tension or compression along their length, and the connections are like hinges that transfer loads between members. They are used in buildings, bridges, cranes, aircrafts.

Two-hinged Arch: In two-hinged arch, there are four unknown reactions, but there are only three equations of equilibrium available. Hence the degree of statical indeterminacy is **One** for two hinged arch. The fourth equation is written considering horizontal reaction as the redundant and is evaluated by the method of least work.

Uniformly Distributed Load (UDL): It is the load in engineering whose **magnitude remains uniform** throughout the length (span) or loaded with same load per unit length.

Unit Load Method: It is a technique that helps to quantify displacements and rotations of the equilibrium configuration, that is the shape of the structure after it has managed to equilibrate the applied loads. It is also known as Virtual Work Load method and is used to find the deflection and slope at a point on structure with the help of law of conservation.

Vibrational Analysis: It includes the study of natural frequencies and mode shape of any structure to find its response to the dynamic loading, examples : machine foundation.

Virtual Work (δw): It is imaginary work done when a particle subject to a number of forces is moved a small imaginary distance. It may also be defined as the total work done by the applied forces and the inertial forces of a mechanical system as it moves through a set of virtual displacements. When considering forces applied to a body in static equilibrium, the principle of least action requires the virtual wok of these forces to be zero.

MULTIPLE CHOICE QUESTIONS

1. How many independent equations have to be satisfied for static equilibrium of a plane structure?
   (a) 1 (b) 2 (c) 3 (d) 6
2. If there are (m) unknown member forces, (r) unknown reaction components and (j) number of joints, then the degree of static indeterminacy of a pin-jointed plane frame is given by
   (a) $m + r + 2j$ (b) $m - r + 2j$ (c) $m + r - 2j$ (d) $m + r - 2j$
3. The degree of static indeterminacy of a rigid-jointed plane frame having 15 members, 3 reaction components and 14 joints is
   (a) 2 (b) 3 (c) 4 (d) 6
4. The degree of static indeterminacy of a pin-jointed space frame is given by
   (a) $m + r - 2j$ (b) $m + r - 3j$ (c) $m + r + 3j$ (d) $3m + r - 3j$
5. The degree of kinematic indeterminacy of a pin-jointed plane frame is given by
   (a) $2j - r$ (b) $j - 2r$ (c) $3j - r$ (d) $2j + r$
6. A pin-jointed plane frame becomes unstable if
   (a) $(m + r) = 2j$ (b) $(m + r) < 2j$ (c) $(m + r) > 2j$ (d) None of these
   (where m = Number of members, r = Reaction components and j = Number of joints)
7. Degree of Kinematic indeterminacy of the beam shown
   (a) 2 (b) 6 (c) 8 (d) 10
8. A rigid-jointed plane frame is stable and statically determinate, if
   (a) $(m + r) = 2j$ (b) $(m + r) = 3j$
   (c) $(3m + r) = 3j$ (d) $(m + 3r) = 3j$
9. The rigid-jointed frame shown here is

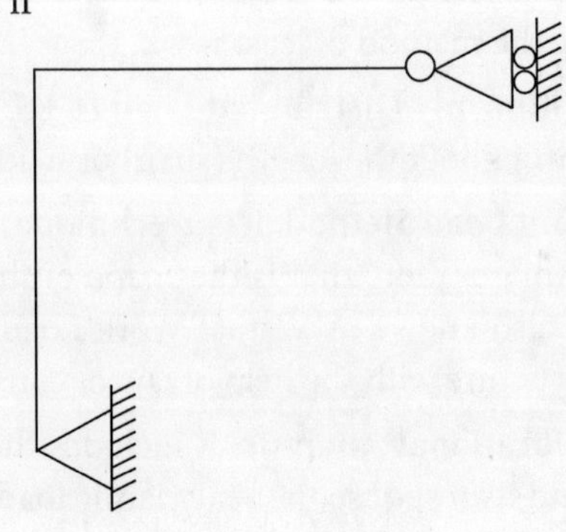

   (a) Unstable
   (b) Stable & determinate
   (c) Stable but indeterminate by 1 degree
   (d) Stable but indeterminate by 2 degrees

10. Castigliano's I-theorem is applicable
(a) For statically determinate structures only
(b) When system behaves elastically
(c) Only when principle of superposition is valid
(d) None of these

11. The portal frame as shown here is statically indeterminate to the

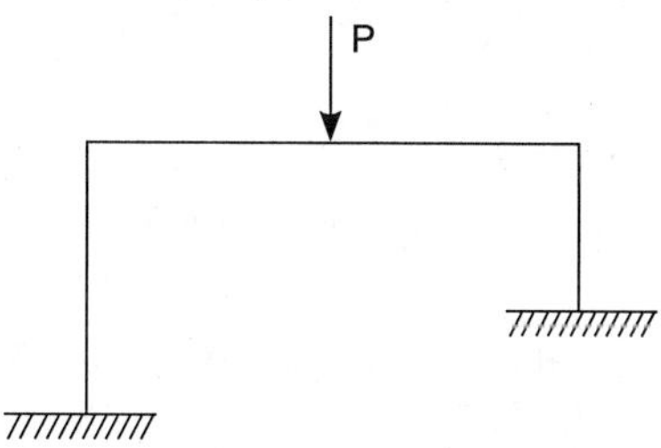

(a) I-degree (b) II-degree (c) III-degree (d) None of these

12. A perfect plane frame having (n) number of members and (j) number of joints should satisfy the relation
(a) $n < 2j - 3$ (b) $n = 2j - 3$ (c) $n > 2j - 3$ (d) $n = 3 - 2j$

13. The number of independent equations to be satisfied for static equilibrium in a space structure is
(a) 2 (b) 3 (c) 4 (d) 6

14. Neglecting axial deformation, the Kinematic indeterminacy of the structure here is

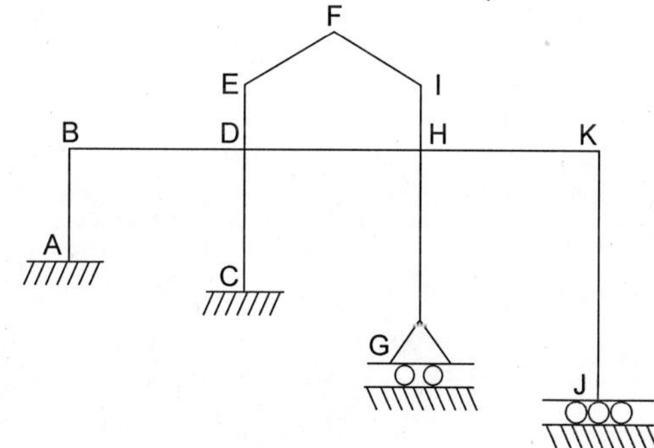

(a) 12 (b) 14 (c) 20 (d) 22

15. Which of the following beams is unstable?

(a) 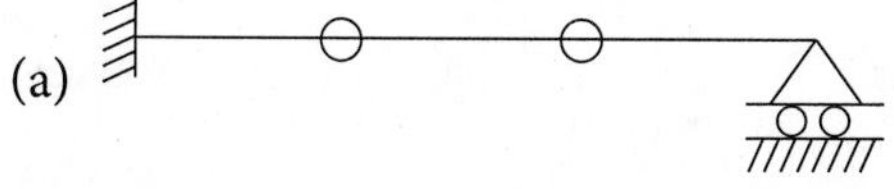
(b) 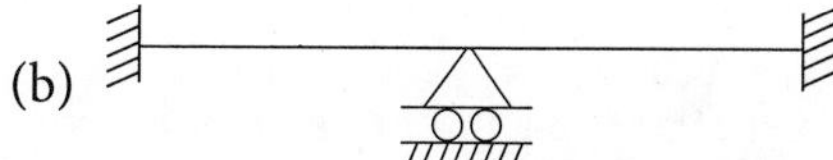
(c) 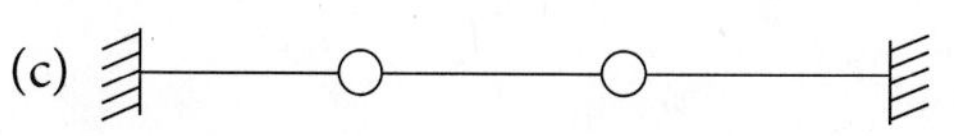
(d) 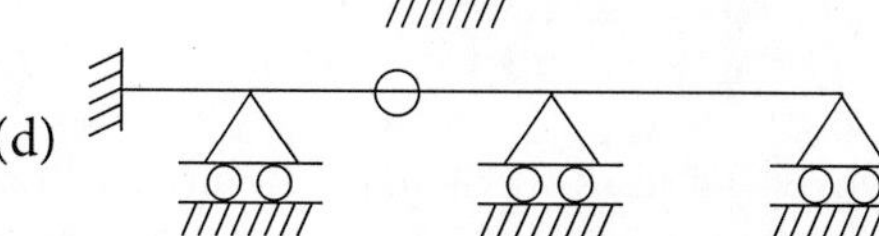

16. In moment distribution method the sum of distribution factors of all the members meeting at any point is always
(a) Zero (b) Less than 1 (c) 1 (d) More than 1

17. Considering the beam as axially rigid, the degree of freedom of the plane frame shown here is

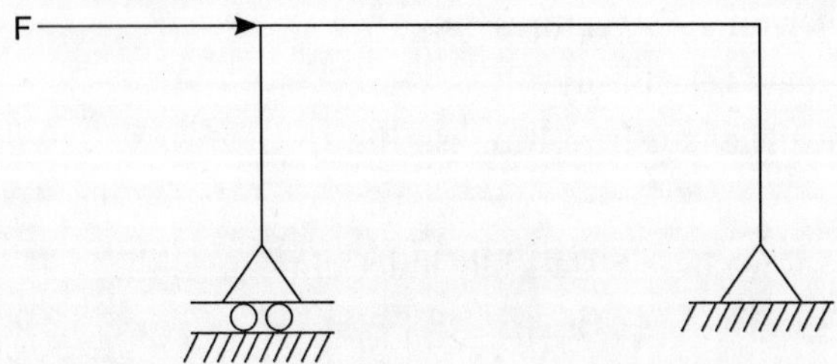

(a) 9 (b) 8 (c) 7 (d) 6

18. The kinematic indeterminacy of single bay portal frame fixed at base is

(a) 1 (b) 2 (c) 3 (d) Zero

19. The beam fixed at ends and subjected to lateral loads only is statically indeterminate of degree

(a) 1 (b) 2 (c) 3 (d) 4

20. The kinematic indeterminacy of the plane truss shown is

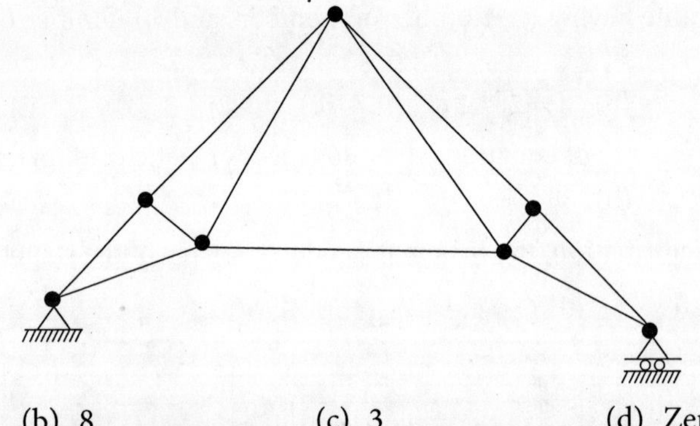

(a) 11 (b) 8 (c) 3 (d) Zero

21. The degree of kinematic indeterminacy of the pin-joined frame as shown here is

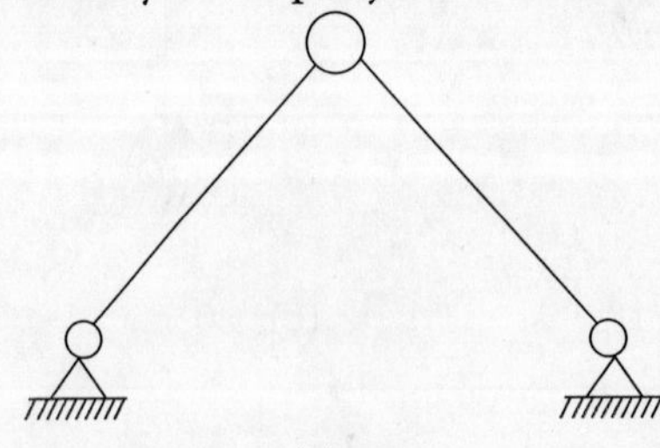

(a) 0 (b) 1 (c) 2 (d) 4

22. The carryover factor in a prismatic member whose far end is fixed is

(a) 0 (b) $\frac{1}{2}$ (c) $\frac{3}{4}$ (d) 1

23. A carryover factor in a prismatic member whose far end is hinged is

(a) Zero (b) $\frac{1}{2}$ (c) cc (d) 1

24. The degree of kinematic indeterminacy of the rigid frame with clamped ends at A and D as shown here

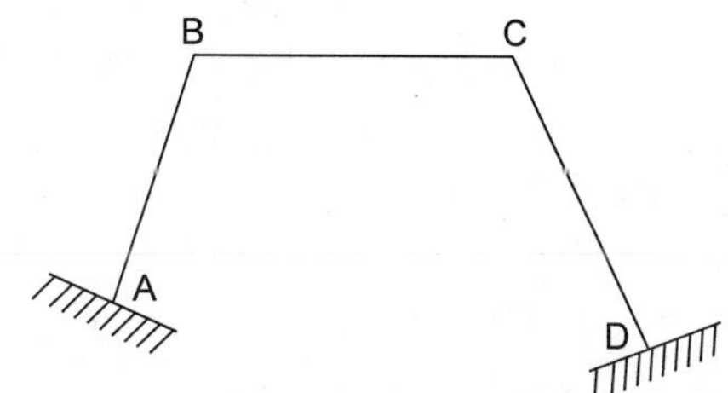

(a) 4 (b) 3 (c) 2 (d) Zero

25. Influence line for redundant structures can be obtained by

(a) Castgliano's theorem (b) Muller Breslau Principle

(c) Unit load theorem (d) M-B reciprocal Theorem

26. The ordinates of influence line diagram for Bending Moment always have the dimensions of

(a) Force (b) Length (c) Force × Length (d) $\frac{\text{Force}}{\text{Length}}$

27. Bending Moment at any section in a conjugate beam gives in the actual beam

(a) Slope (b) Curvature (c) Deflection (d) Bending Moment

Common data for Questions 28, 29 and 30.

A beam PQRS is 18 m long and is simply supported at points Q and R 10 m apart. Overhangs PQ and RS are 3 m and 5 m respectively. A train of two point loads of 150 KN and 100 KN, 5m apart, crosses, this beam from left to right with 100 KN load leading.

28. The maximum sagging moment under the 150 KN load anywhere is

(a) 500 KN-m (b) 450 KN-m (c) 400 KN-m (d) 375 KN-m

29. During the passage of loads, the maximum and the minimum reactions at support R in KN are respectively

(a) 300 and – 30 (b) 300 and – 25 (c) 225 and – 30 (d) 225 and – 25

30. The maximum hogging moment in the beam anywhere, is

(a) 300 KN-m (b) 450 KN-m (c) 500 KN-m (d) 750 KN-m

31. Consider the beam ABCD and the influence line as shown here. The influence line pertains to

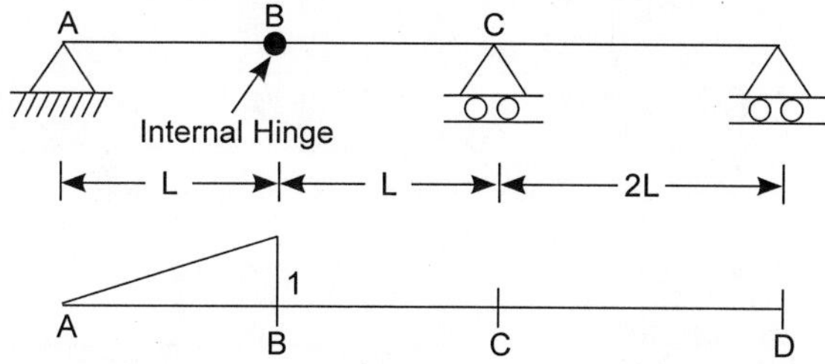

(a) Reaction at A, R_A (b) Shear Force at B, V_B

(c) Shear Force left of C, V_C^- (d) Shear Force right of C, V_C^+

32. The influence line for shear at section x (F_x) at a distance of 4 m from the left supported girder AB is shown here

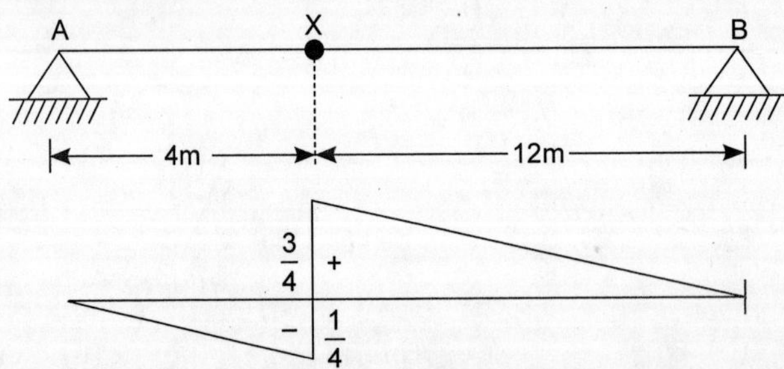

The Shear Force at section x due to uniformly distributed dead load of intensity $\frac{2\text{ t}}{\text{m}}$ covering the entire span will be

(a) 8 t (b) 4 t (c) 2 t (d) 1 t

33. A single rolling load 8 KN rolls along a girder of 15 m span. The absolute maximum bending moment will be, (in KN – m)

(a) 8 (b) 15 (c) 30 (d) 60

34. Principle of superposition is applicable under

(a) Deflection is a linear function of applied force

(b) Material obeys Hooke's Law

(c) The action of applied forces will be affected by small deformations of the structure

(d) None of these

35. The maximum BM due to a train of wheel loads on simply supported girder

(a) Always occurs at the centre of span (b) Always occurs under a wheel load

(c) Never occurs under a wheel load (d) None of these

36. Casligliano's second theorem can be used to compute deflections

(a) In statically determinate structures only (b) For any type of structures

(c) For beams and frames only (d) At any point under the load only

37. The correct Shear Force diagram for the beam shown is

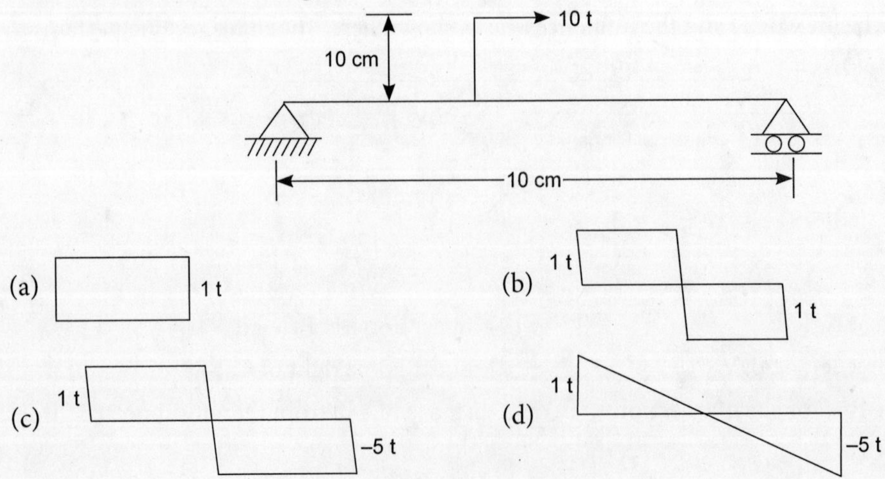

38. The reaction at support A of the propped cantilever beam shown here is

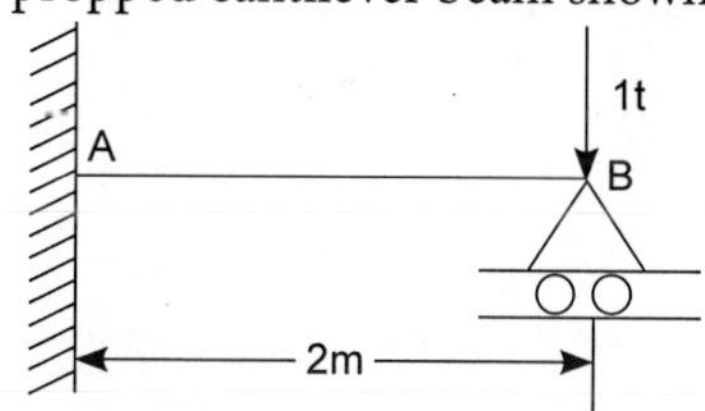

(a) 0 (b) 1 t (c) 0.5 t (d) 2 t

39. For a 2-hinged arch, if one of the supports settles down vertically then the horizontal thrust

(a) Increases (b) Decreases (c) Remains same (d) Becomes zero

40. In the 2-hinged parabolic arch as shown, the maximum bending moment is equal to

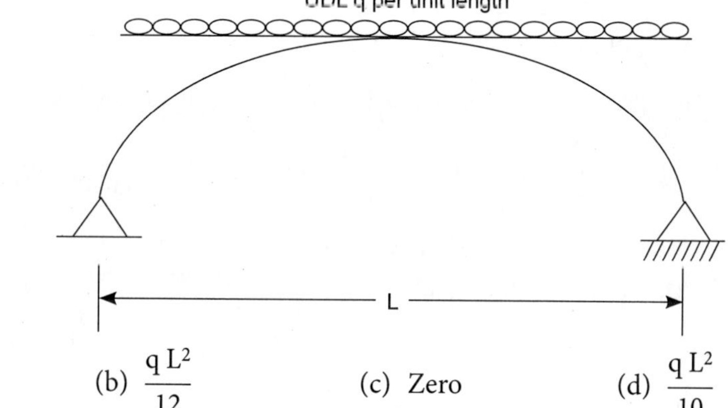

(a) $\frac{qL^2}{8}$ (b) $\frac{qL^2}{12}$ (c) Zero (d) $\frac{qL^2}{10}$

41. In a 2-hinged arch an increase in temperature induces

(a) No bending moment in the arch rib (b) Uniform BM in the arch rib

(c) Maximum bending at the crown (d) Maximum BM at the crown

42. A 3-hinged arch is subjected to a point load at crown. The magnitude of horizontal reaction at A will be

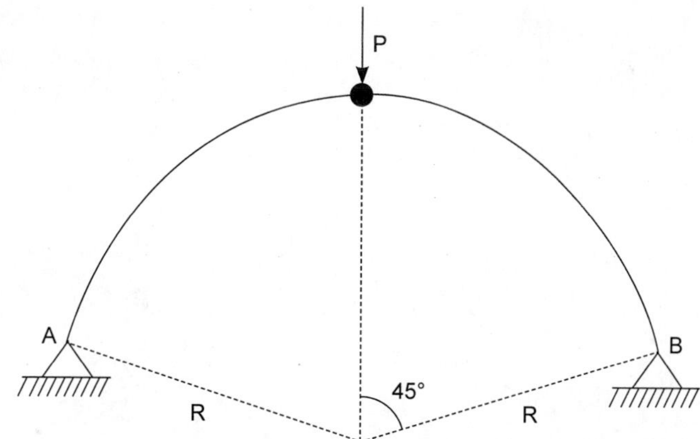

(a) $\frac{P}{2(\sqrt{2}-1)}$ (b) $\frac{P}{2}$ (c) $\sqrt{2}-1$ (d) $\frac{P}{2\sqrt{2}}$

43. The influence line for horizontal thrust of a two-hinged parabolic arch of span (L) and rise (h) will be shown in

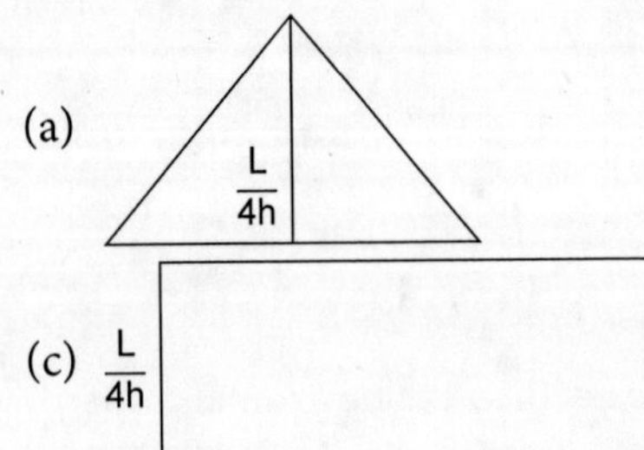

(b) $\frac{25}{128}\frac{L}{H}$

(d) None of these

44. The horizontal thrust due to rise in temperature in a semi-circular two-hinged arch of radius R is proportional to

(a) R (b) R^2 (c) $\frac{1}{R}$ (d) $\frac{1}{R^2}$

45. A 3-hinged semi-circular arch of Radius R carries a Uniformly distributed load w per unit run over the whole span. The horizontal thrust is

(a) wR (b) $\frac{wR}{2}$ (c) $\frac{4\ wR}{3\ \pi}$ (d) $\frac{2\ wR}{3\ \pi}$

46. An arch is used in long-span bridges because there is

(a) Considerable reduction in horizontal thrust
(b) Considerable reduction in Bending moment
(c) Reduction in shear and Bending moment
(d) No change in shear or Bending moment

47. Force in the member BC of the truss shown here is

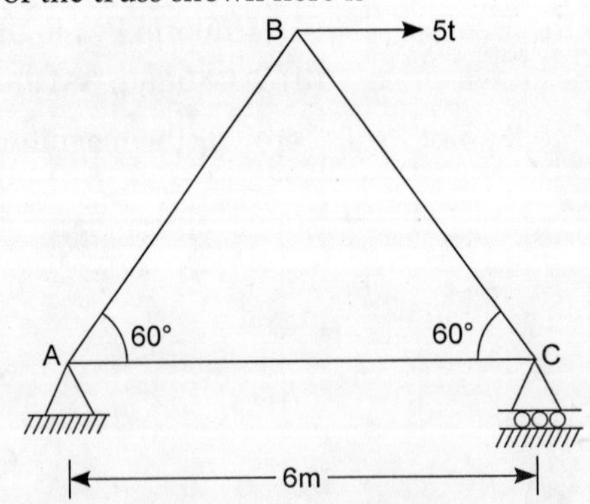

(a) 5t (tensile) (b) Zero
(c) 2.88 t (compressive) (d) 5 t (compressive)

48. The reaction component (locus) of a 2-hinged semi-circular arch is

(a) Straight line (b) Parabola (c) Circle (d) Hyperbola

49. Masonry arches resist the external loads by

(a) Normal Thrust and Shear (b) Normal Thrust, radial Shear and moment
(c) Bending moment and radial shear (d) Thrust only

50. A symmetrical 2-Hinged parabolic arch when subjected to a UDL on entire horizontal span is subjected to

(a) Radial shear only
(b) Normal thrust alone
(c) Normal thrust and BM
(d) Normal thrust, BM and radial shear

51. The degree of static indeterminacy of structure shown is

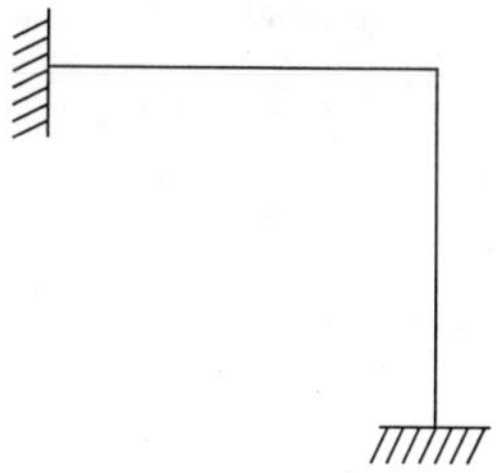

(a) 1 (b) 2 (c) 3 (d) 4

52. The degree of kinematic indeterminacy of the beam shown is

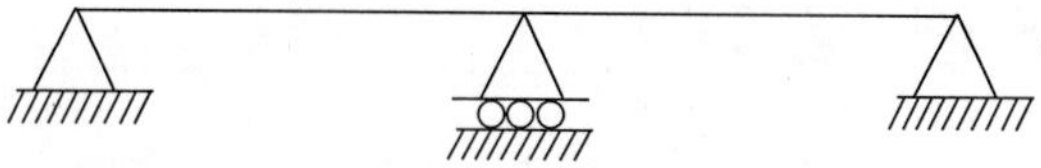

(a) 2 (b) 3 (c) 4 (d) 5

53. Due to uniformly distributed load, maximum reaction at support B for the beam shown here will be when

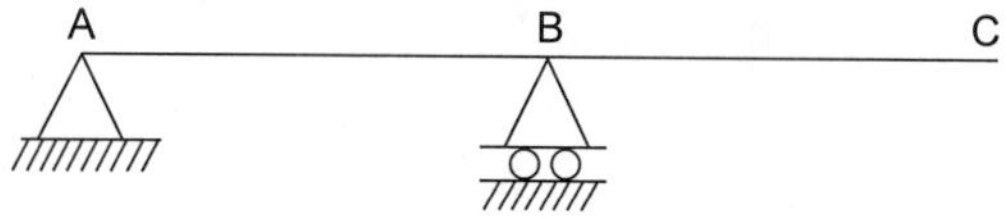

A B C

(a) Only AB is loaded
(b) Only BC is loaded
(c) Entire span is loaded
(d) None of the above

54. For a single point load W moving on a symmetrical 3-Hunged parabolic arch of span (L), the maximum sagging moment occurs at a distance x from ends

(a) 0.211 L (b) 0.25 L (c) 0.234 L (d) 0.5 L

55. For approximate analysis of building frames under vertical loads, the point of inflection is assumed at

(a) Centre of each beam
(b) Centre of each column
(c) One-tenth of the span length from each end of beam
(d) Both (a) and (b)

56. Due to some point load anywhere on a fixed beam, the maximum free bending moment is M. The sum of fixed end moment is

(a) M (b) 1.5 M (c) 2.0 M (d) 3.0 M

57. A 2-hinged arch (semi-circular) of radius (R) carries a concentrated load (w) at the crown, the horizontal thrust will be

(a) $\frac{w}{2\pi}$ (b) c (c) $\frac{2w}{3\pi}$ (d) $\frac{4w}{3\pi}$

58. A steel frame as shown here has joint E as rigid, the rotational stiffness of the frame at E is given by (EI is constant)

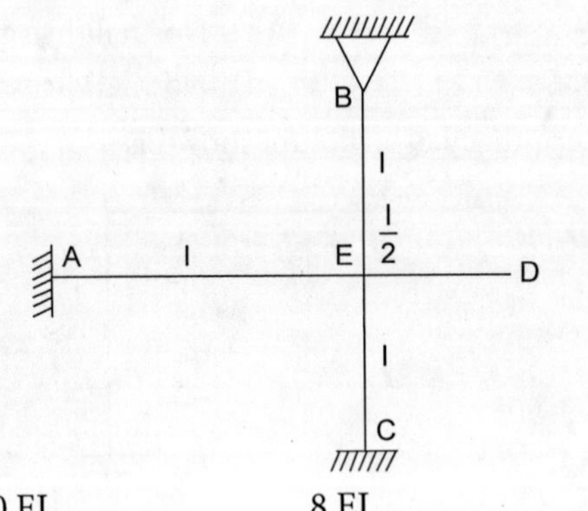

(a) $\frac{11\ EI}{l}$ (b) $\frac{10\ EI}{l}$ (c) $\frac{8\ EI}{l}$ (d) $\frac{6\ EI}{l}$

59. For the portal frame shown here, the final end moments are

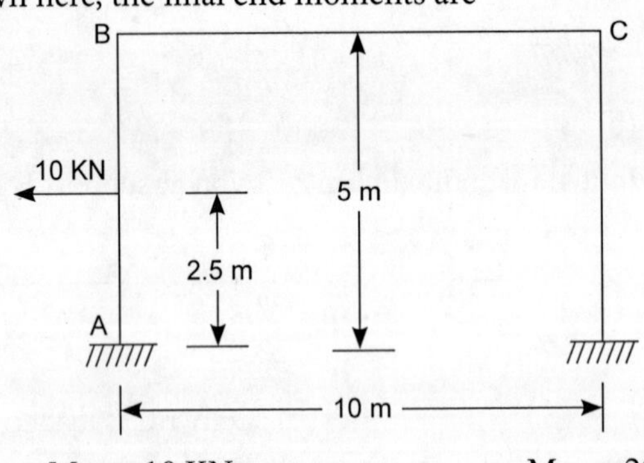

M_{AB} = 15 KN – m M_{BA} = 10 KN – m M_{CD} = 20 KN – m

The end moment M_{DC} will be

(a) 10 KN – m (b) 20 KN – m (c) 30 KN – m (d) 40 KN – m

60. A fixed beam of uniform section is carrying a point load at its mid-span. If the moment of inertia of the middle half length is reduced to half its previous value, then the fixed end moment will

(a) Increase (b) Decrease (c) No change (d) Change in direction

61. The rigid portal frame shown here will not have any side sway, (I_1 is moment of Inertia of column and I_2 is moment of Inertia of beam cross-section)

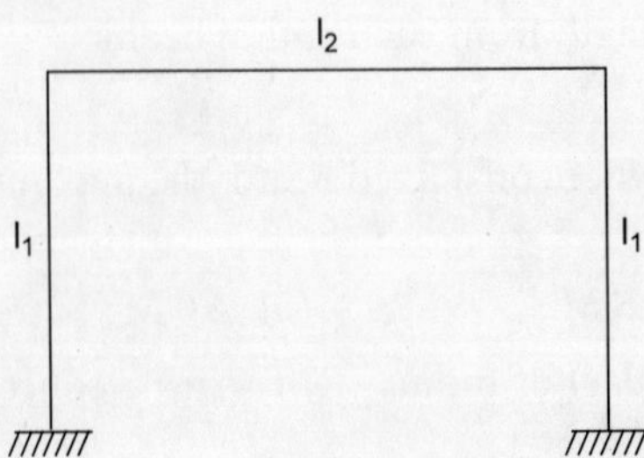

(a) Subjected to vertical loading only (b) $I_2 - 2I_1$

(c) Loaded in any manner (d) Loading is symmetrical about central line

62. A symmetrical parabolic arch of span 20 m and rise 5 m is hinged at the springings. It has a UDL of $\frac{2\text{ t}}{\text{m}}$ run over the spam. The horizontal thrust in tonnes at each springing is

(a) 8 (b) 16 (c) 20 (d) Zero

63. A rigid cantilever frame ABC is fixed at O. It carries a load (P) at A as shown here. Neglecting axial deformation, the vertical deflection of the point A is given by

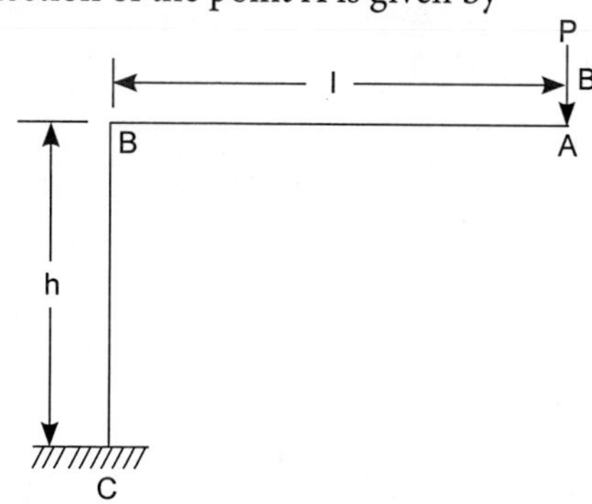

(a) $\frac{P\,l\,h^2}{2\,EI}$ (b) $\frac{P\,l^2\,h}{3\,EI}$ (c) $\frac{P\,l^2\,(3h+1)}{3\,EI}$ (d) $\frac{P\,l^2\,(2h+1)}{2\,EI}$

64. In the displacement method of structural analysis, the basic unknowns are

(a) Displacement (b) Force

(c) Displacement and Forces (d) None of these

65. The width of the analogous column in the method of column analogy is

(a) $\frac{2}{EI}$ (b) $\frac{1}{EI}$ (c) $\frac{1}{2\,EI}$ (d) $\frac{1}{4\,EI}$

66. The Strain Energy of a structure due to bending is given by

(a) $\int \frac{M^2 dx}{EI}$ (b) $\frac{1}{2}\int \frac{2M^2 dx}{EI}$ (c) $\int \frac{M^2 dx}{EI}$ (d) $\frac{1}{3}\int \frac{M^2 dx}{EI}$

67. In the framed structure here, the forces in AB and BC are

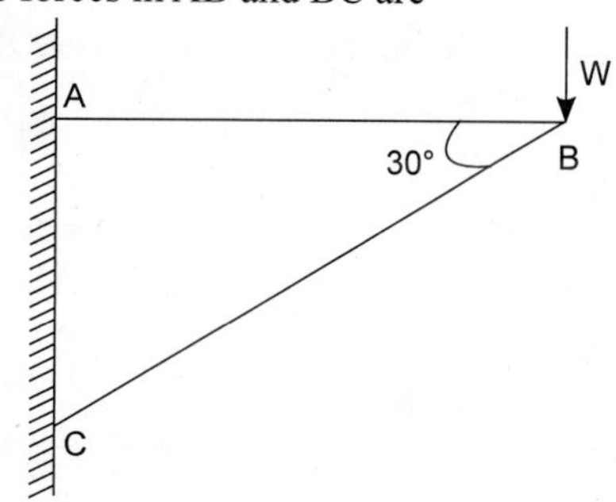

(a) $\sqrt{3}$ W (Tensile) and 2W (Compressive)

(b) 2W (Tensile) and $\sqrt{3}$ W (Compressive)

(c) $2\sqrt{3}$ W (Tensile) and $2\sqrt{3}$ W (Compressive)

(d) None of these

68. The span(s) to be loaded uniformly for maximum positive (upward) reaction at support P, as shown here, is (are)

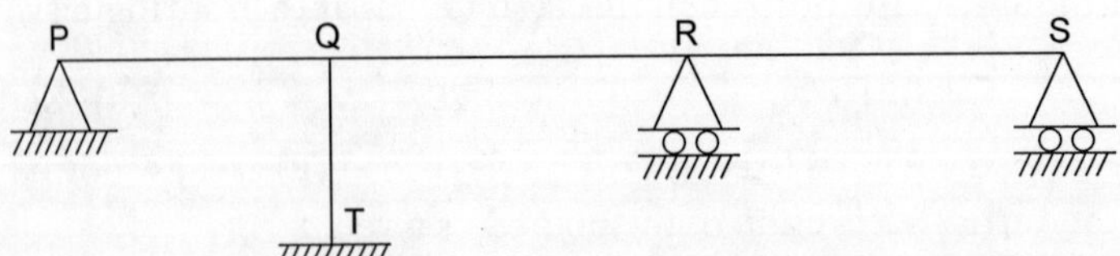

(a) PQ only (b) PQ and QR (c) QR and RS (d) PQ and RS

69. Influence line for redundant structure can be found out by

(a) Castgliano's theorem
(b) Muller-Breslau Principle
(c) Unit Load theorem
(d) Maxwell-Bette's reciprocal theorem

70. The influence diagram shown here is for which member?

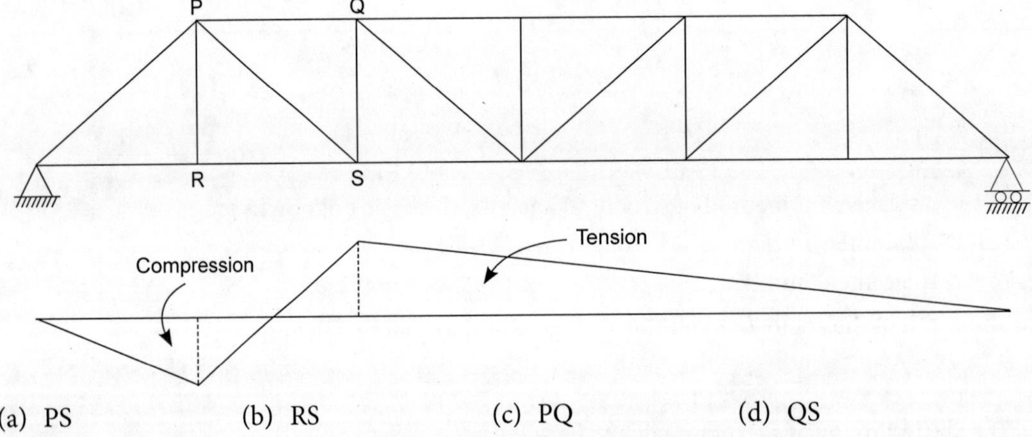

(a) PS (b) RS (c) PQ (d) QS

71. The fixed end moment of uniform beam of span L and fixed at the ends, subjected to a central point load P is

(a) $\frac{PL}{2}$ (b) $\frac{PL}{8}$ (c) $\frac{PL}{12}$ (d) $\frac{PL}{16}$

72. The magnitude of the bending moment at the fixed support of the beam is

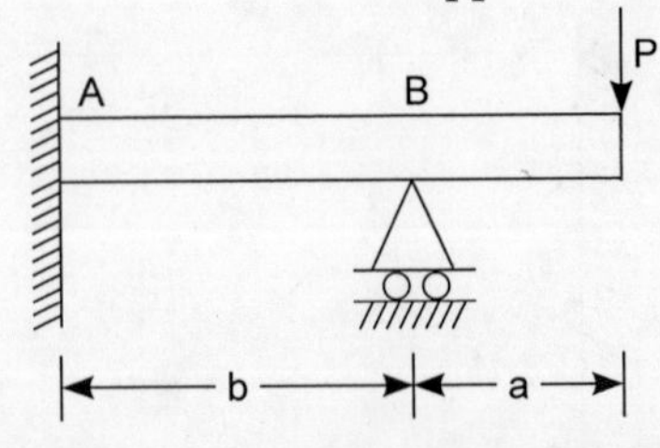

(a) P . a (b) $\frac{P . a}{2}$ (c) P . b (d) P (a + b)

73. A two span continuous beam having equal spans each of length L is subjected to UDL (w) per unit length. The beam has constant flexural rigidity. The reaction at the middle support is

(a) w L (b) c? (c) $\frac{5 wL}{4}$ (d) $\frac{wL^2}{16}$

74. In the above problem the BM at the middle support is

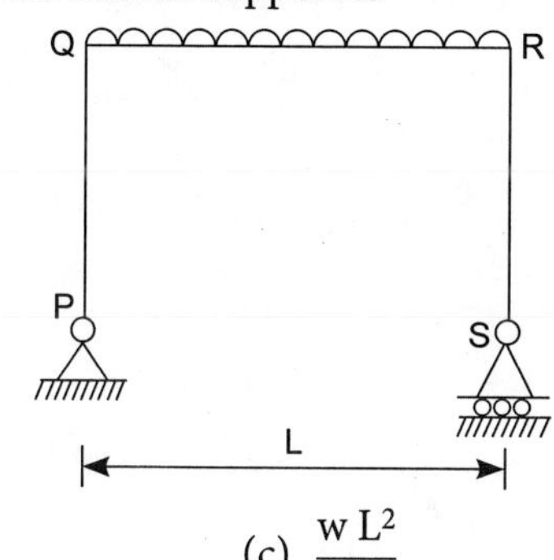

(a) $\frac{w L^2}{4}$ (b) $\frac{w L^2}{8}$ (c) $\frac{w L^2}{12}$ (d) $\frac{w L^2}{16}$

75. A UDL of $\frac{60 \text{ KN}}{\text{m}}$ run of length 5m on a girder of span 16 m. What is the maximum positive shear force at a section 6 m from the left end?

(a) 140.625 KN (b) 65.625 KN (c) 90.625 KN (d) 45.625 KN

76. For the semi-circular 2-hinged arch, a moment of 50 t – cm is applied at B, produces a displacement of 0.5 cm at A. If a concentrated load is applied at A, the rotation at B in the arch will be

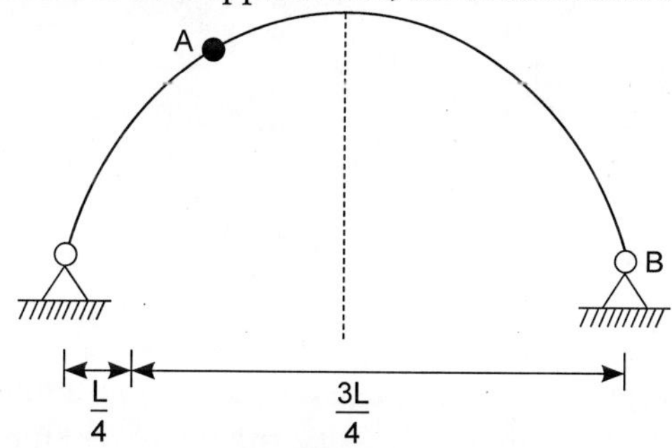

(a) 0.1 (b) 0.001 (c) 0.01 (d) 0.0001

77. A 3-hinged arch of span 30 m and rise 6m is subjected to a rise of temperature of 40° C. The change in the rise of arch (Take $\alpha = 12 \times 10^{-6}$ per °C) is

(a) 20.9 mm (b) 30.9 mm (c) 10.9 mm (d) 40.9 mm

78. The figure below shows a portal frame with one end fixed and other hinged. The ratio of the fixed end moments $\frac{M_{BA}}{M_{CD}}$ due to side sway will be

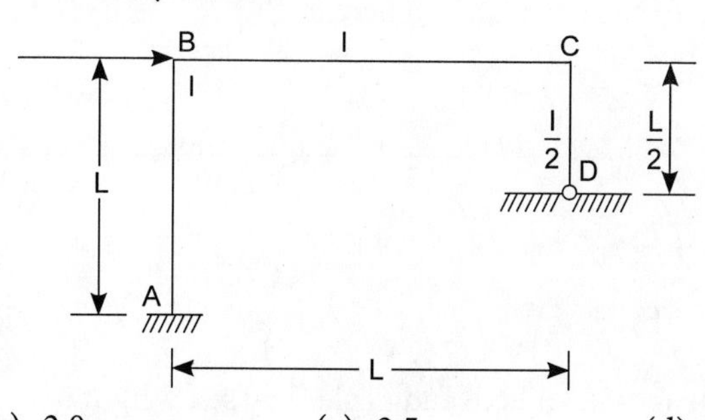

(a) 1.0 (b) 2.0 (c) 2.5 (d) 3.0

79. The fixed end moment M_A for the beam shown in the given figure is

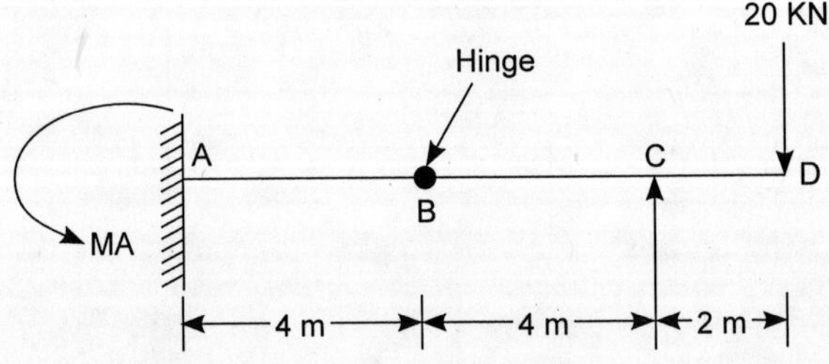

(a) + 40 KN – m (b) – 40 KN – m (c) + 80 KN – m (d) – 80 KN – m

80. Members AB and BC in the figure here are identical. Due to moment 2M applied at B, the value of axial force in the member AB will be

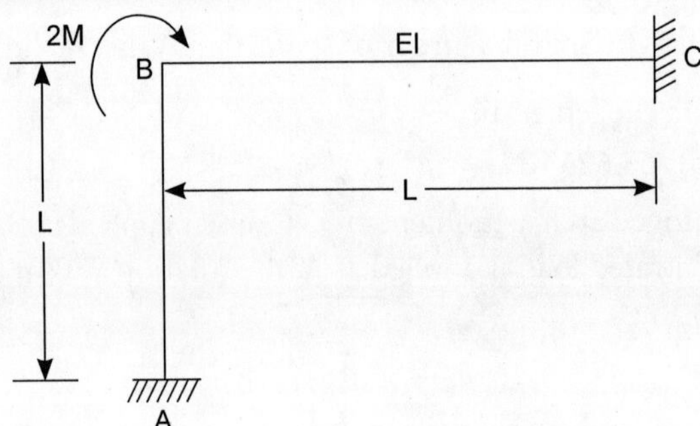

(a) $\frac{M}{L}$ (Compressive) (b) $\frac{M}{L}$ (Tension)

(c) $\frac{1.5\,M}{L}$ (Compressive) (d) $\frac{1.5\,M}{L}$ (Tension)

81. a uniform simply supported beam is subjected to a clockwise moment M at the left end. What moment will be required at the other end so that the rotation of the right end is zero?

(a) 2 M (b) M (c) $\frac{M}{2}$ (d) $\frac{M}{3}$

82. If a point load acting at the mid-span of a fixed beam of uniform section produces fixed end moments of 60 KN – m, then the same load spread uniformly over the entire span will produce fixed end moments equal to

(a) 20 KN – m (b) 30 KN – m (c) 40 KN – m (d) 45 KN – m

83. Carryover factor C_{AB} for the beam shown here is

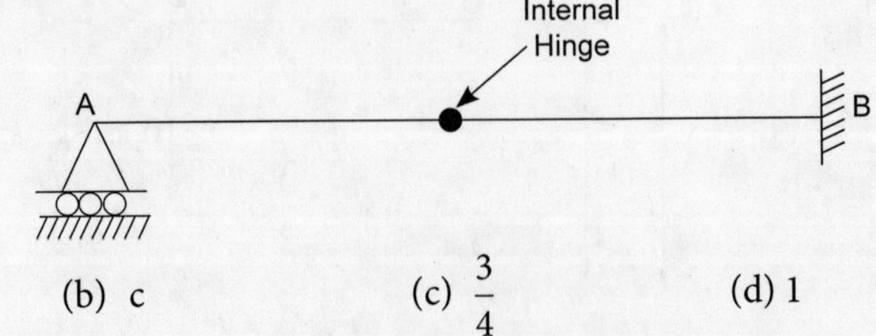

(a) $\frac{1}{4}$ (b) c (c) $\frac{3}{4}$ (d) 1

84. Consider the planar truss shown here and neglecting self weight of members, what will be number of zero-force members in the truss under the action of the load P

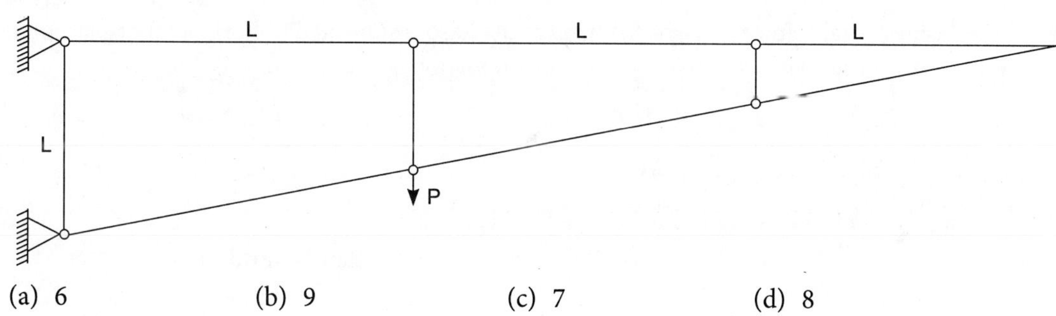

(a) 6 (b) 9 (c) 7 (d) 8

85. For the truss shown here, the member PQ is short by 3 mm. The magnitude of vertical displacement is

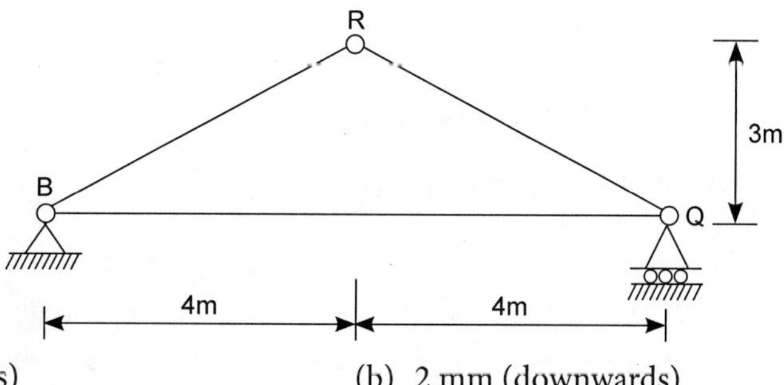

(a) 2 mm (upwards) (b) 2 mm (downwards)
(c) 3 mm (upwards) (d) 3 mm (downwards)

86. The ratio of the stiffness of a beam at the near end when the far end is hinged to the stiffness of the beam at the near end when the far end is fixed is

(a) $\frac{1}{2}$ (b) $\frac{3}{4}$ (c) 1 (d) $\frac{4}{3}$

87. In a linear structural element

(a) Stiffness is directly proportional to flexibility
(b) Stiffness is inversely proportional to flexibility
(c) Stiffness is equal to flexibility
(d) Stiffness and flexibility are not related

88. The portal frame shown here is subjected to uniformly distributed load w per unit length. The Bending Moment in the beam at joint Q is

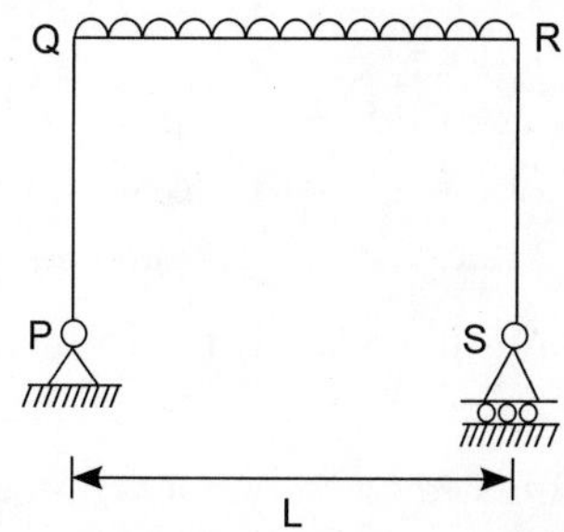

(a) Zero (b) $\frac{wL^2}{24}$ (Hogging) (c) $\frac{wL^2}{12}$ (Hogging) (d) $\frac{wL^2}{8}$ (Sagging)

89. A single bay portal frame of height (h) fixed at the base is subjected to horizontal displacement of (Δ) at the top. The base moment developed is proportional to

(a) $\frac{1}{h}$ (b) $\frac{1}{h^2}$ (c) $\frac{1}{h^3}$ (d) None of these

(Note : All members are prismatic)

90. For analysis of multistoreyed frame best method is

(a) Strain Energy Method (b) Muller Breslau Method
(c) Kani's Method (d) Complimentary Energy Method

91. The moment required to rotate the near end of a prismatic beam through unit angle, without translation is given by

(a) $\frac{EI}{L}$ (b) $\frac{2\,EI}{L}$ (c) $\frac{3\,EI}{L}$ (d) $\frac{4\,EI}{L}$

(Note : EI is flexural rigidity, L is span and far end is fixed)

92. The deflection at any point of a perfect frame can be obtained by applying a Unit Load at the joint in

(a) Vertical direction (b) Horizontal direction
(c) Inclined direction (d) In the direction of required deflection

93. The flexibility matrix corresponding to coordinates 1 and 2 for the structure shown here is

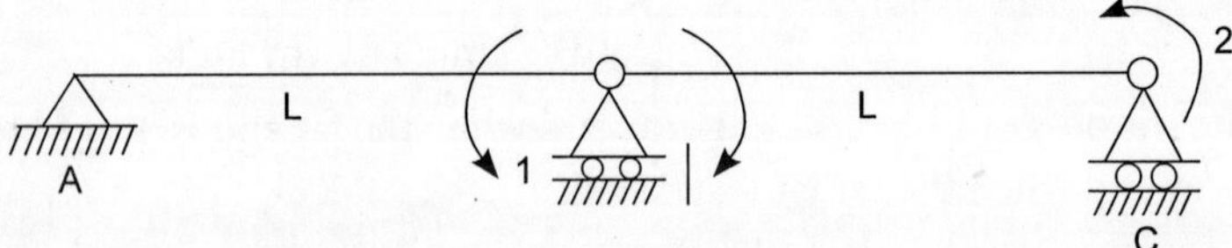

(a) $\frac{L}{3\,EI}\begin{bmatrix}4 & 1\\1 & 2\end{bmatrix}$ (b) $\frac{L}{2\,EI}\begin{bmatrix}3 & 2\\2 & 4\end{bmatrix}$ (c) $\frac{L}{6\,EI}\begin{bmatrix}4 & 1\\1 & 2\end{bmatrix}$ (d) $\frac{L}{3\,EI}\begin{bmatrix}3 & 2\\2 & 4\end{bmatrix}$

94. When a unit rotation is given at end A of the given beam, the moment produced at the end B will be

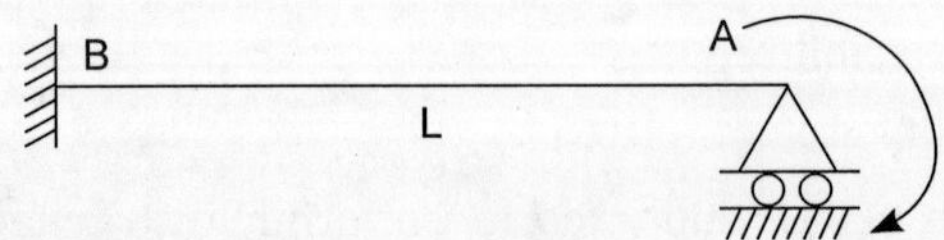

(a) Zero (b) $\frac{2\,EI}{L}$ (c) $\frac{4\,EI}{L}$ (d) $\frac{6\,EI}{L}$

95. Which of the following methods of structural analysis is a Force Method?

(a) Slope Deflection (b) Column Analogy
(c) Moment distribution (d) None of these

96. Consider the following assumptions in the analysis of a plane truss

1. Individual members are straight
2. Individual members are connected by frictionless hinge
3. The loads and reactions act only at joints

(a) 1 and 2 are valid (b) 1 and 3 are valid (c) 2 and 3 are valid (d) All are valid

97. Two bars AO and BO are of uniform area (A) and are hinged at (O) as shown here, A load (w) is applied at O, the vertical deflection of O will be

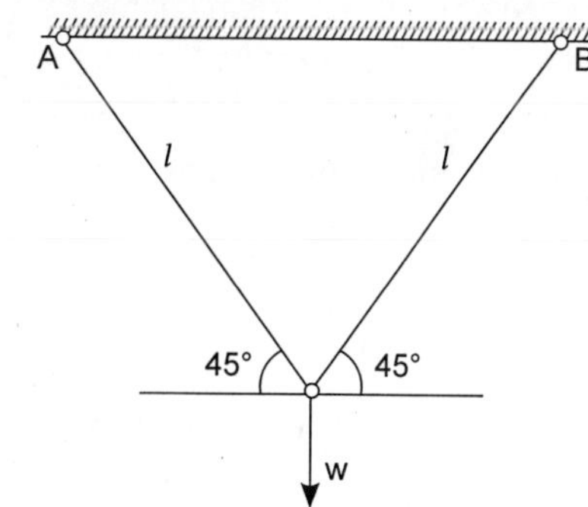

(a) $\frac{wl}{AE}$ (b) $\frac{wl}{2\ AE}$ (c) $\frac{w^2 l}{AE}$ (d) $\frac{w^2 l}{2AE}$

98. In structural analysis, the moment of Inertia (I) of a beam's cross-section is important because it

(a) Determines the beam's ability to resist Shear Force

(b) Determines the beam's ability to resist Bending

(c) Determines the beam's ability to resist Axial Force

(d) Determines the beam's ability to resist Torsion

99. The principal of superposition is applicable when

(a) The structure is linear elastic (b) The material follows Hooke's Law

(c) Deformations are small (d) All these

100. In a statically determinate planar frame, if one of the horizontal members is removed, how will it affect the structures stability?

(a) No effect on stability (b) Remain stable but flexible

(c) Become more rigid (d) Will become unstable

101. Consider the beam PQ as shown, fixed at P and hinged at Q and subjected to load P. The static and kinematic degrees of indeterminacy, respectively are

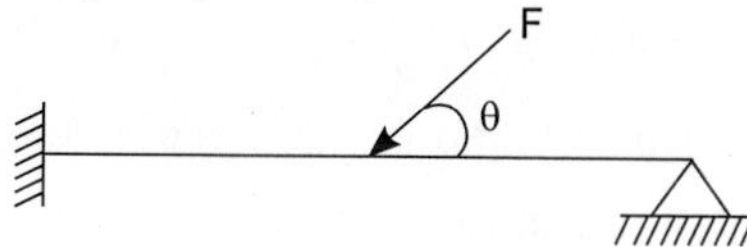

(a) 2 and 1 (b) 2 and 0 (c) 1 and 2 (d) 2 and 2

102. The degree of Kinematic indeterminacy in the given pin-joined frame is

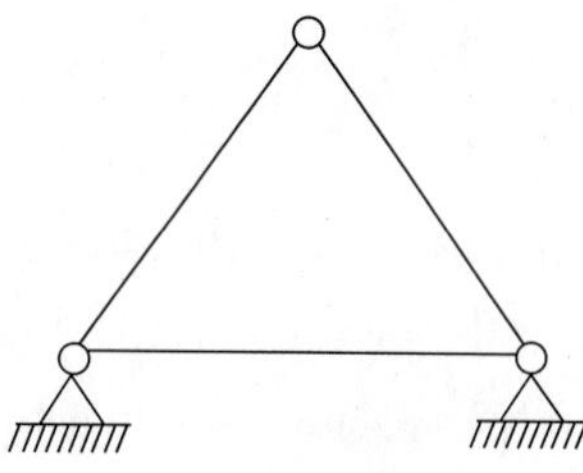

(a) 1 (b) 0 (c) 2 (d) 4

TRUE OR FALSE QUESTIONS

1. The number of independent displacement components at each joint of a rigid-jointed space frame is six. (T or F)
2. Moment distribution method is displacement method in structural analysis. (T or F)
3. Influence line for horizontal thrust in a 2-hinged parabolic arch is rectangle. (T or F)
4. The flexibility method is suited for analysing statically determinate structure. (T or F)
5. In beams or other linear members Shear Force is algebraic sum of forces perpendicular to the axis of structure. (T or F)
6. In a statically indeterminate structure, the internal forces can be determined using only the equation of equilibrium. (T or F)
7. The stiffness matrix method approach is used to analyse structures by breaking them down into elements and producing stiffness matrix. (T or F)
8. the Moment Distribution method (Hardy cross method) is alternative procedure for analysis of indeterminate structures. (T or F)
9. A truss is considered unstable if the number of members is equal to (2n – 3), where (n) is number of joints. (T or F)
10. In the analysis of beams, the deflection is maximum where B.M. is zero. (T or F)
11. Influence lines are graphical representations that show how reaction at a support or the internal force in a member varies as a unit load moves across the structure. (T or F)
12. For statically determinate structures, the number of unknown forces must be greater than the number of equilibrium equations available. (T or F)
13. In structural analysis, Shear Force, and Bending Moment diagrams are used to determine internal forces in beams and frames. (T or F)
14. The Flexibility method of structural analysis is same as Force method. (T or F)
15. Statically indeterminate structures cost more in terms of material. (T or F)
16. Carryover moment at the end B, due to moment M at the end A for the beam will be. (T or F)

17. Portals and sway bracing in bridge trusses provide additional stability. (T or F)
18. At the end of the bridge truss the roller provided is to allow thermal expansion. (T or F)
19. A truss which undergoes rigid body translation for an arbitrary load is classified as statically unstable. (T or F)

20. The absolute maximum BM in a simply supported beam (span 20 m) due to a moving UDL of $\frac{4\ t}{m}$ spanning over 5 m is 87.5 t – m at midpoint. (T or F)

21. Influence Line Diagram (ILD) for Shear Force at the fixed end of a cantilever and SFD due to unit load at the free end are same. (T or F)

22. The degree of static indeterminacy of a rigid-jointed plane frame having 15 members, 3 reaction components and 14 joints is Eight (8). (T or F)

23. If in a pin-jointed plane frame, number of members is (m) and number of reaction components is (r) and number of joint (j) it will be unstable, if m + r = 2j. (T or F)

24. The principle of superposition is not applicable when the material does not obey Hooke's law. (T or F)

25. The three moments equation is applicable only when there is no discontinuity like hinge within the span. (T or F)

26. Column analogy method is an example of displacement method. (T or F)

27. The degree of static indeterminacy of the beam is 3 (three). (T or F)

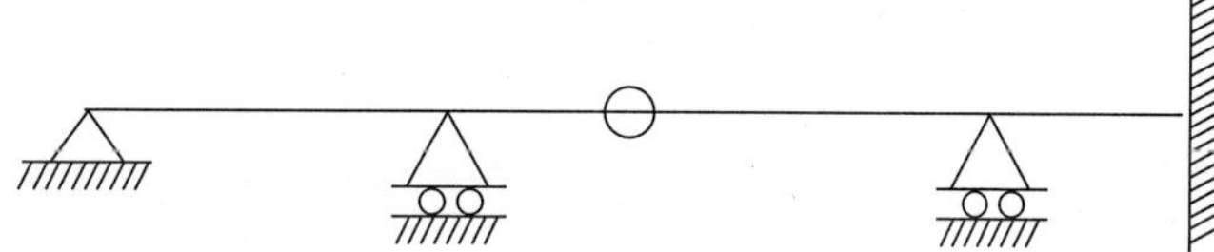

28. The degree of Kinematic indeterminacy of the pin-jointed space frame, if (j) is number of joints and (r) reaction components is 3 j – r. (T or F)

29. When a Uniformly Distributed Load (UDL) longer than the span of girder moves the maximum BM at mid-section will occur when UDL occupies whole span. (T or F)

30. The ordinates of influence line diagram for Bending moment always have the dimensions of Force. (T or F)

31. A fixed beam of span (L) is carrying a point load (P) at mid span. If Moment of Inertia of the middle half length is twice that of remaining length, then the fixed end moment is $\frac{5\ PL}{48}$. (T or F)

32. The Bending Moment Diagram for the structure is shown here true or false. (T or F)

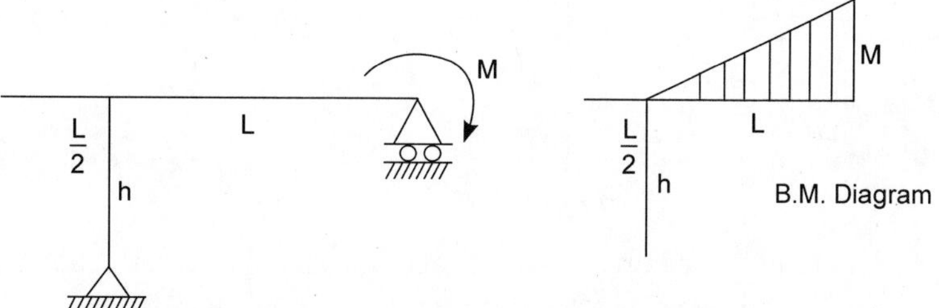

33. A perfect frame is that which is composed of numbers just sufficient to keep it in equilibrium when loaded without change in shape. It satisfies the condition, n = 2j – 1. (T or F)

34. Stiffness (k) is the force required to produce unit deformation. (T or F)

35. In the cantilever beam the moment to be applied at free end for zero vertical deflection at that point is 18 KN – m. (T or F)

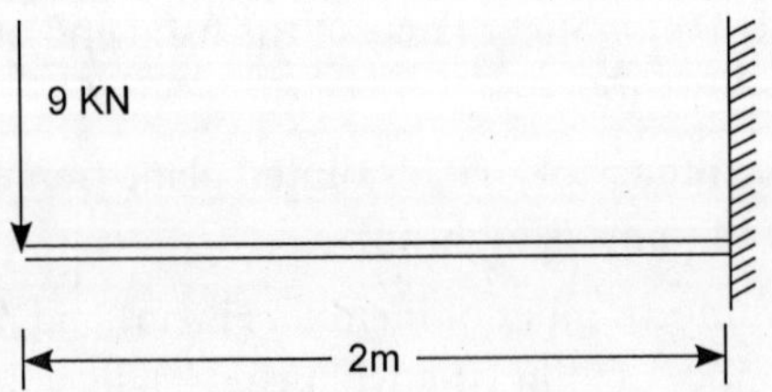

36. In the propped cantilever beam carrying UDL of $\frac{w\ N}{m}$, the reaction at support is $\frac{3}{4}$ wL. (T or F)

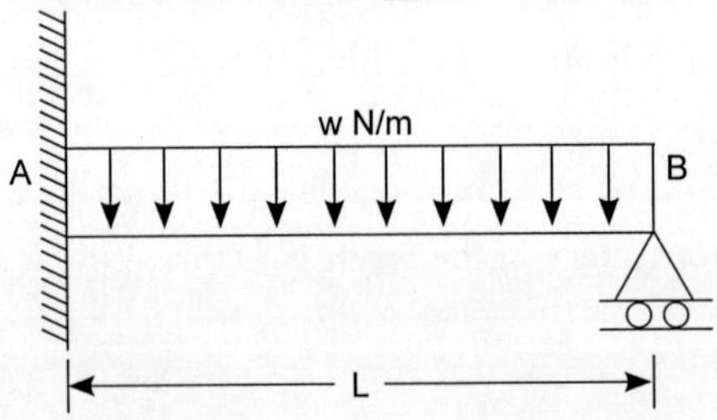

37. A cantilever beam of span L is subjected to a downward load of 800 KN uniformly distributed over its length and a concentrated load P (upwards) at the free end. For vertical displacement to be zero at free end the P required is 800 KN. (T or F)

38. Strain energy method is appropriate for finding deflection of linear structures and deflection of non-linear structures also. (T or F)

39. Kani's method of analysis is appropriate for analysis of multistoreyed buildings. (T or F)

40. Double overhanging structures are statically indeterminate. (T or F)

41. R.C.C. can be used in making trusses. (T or F)

42. Truss members carry bending load. (T or F)

43. If a structure has 10 joints, then the number of joints where Equilibrium equations should be concurrently satisfied is 10. (T or F)

44. This structure is unstable. (T or F)

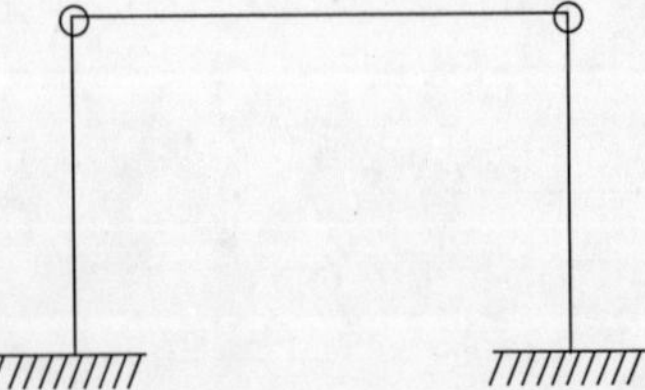

45. A fixed support in structure can resist vertical and horizontal forces, but not moments. (T or F)

46. In slope-deflections method, the primary unknowns are the rotation and translation at the joints. (T or F)

47. The stiffness matrix method is used only for analysis of trusses. (T or F)

48. In truss, all joints are assumed to be pinned, which means they are able to resist moments. (T or F)

49. In a truss subjected to external loads, all members are in pure tension or pure compression. (T or F)

50. The influence line for BM at section $X(M_X)$ as shown here of a simply supported girder AB. (T or F)

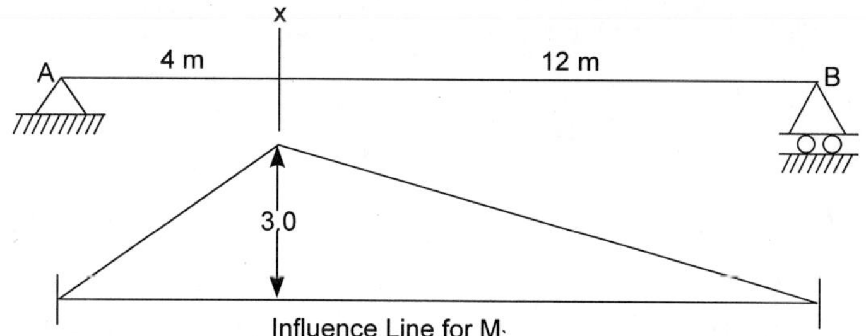

A uniformly distributed load of $\frac{2\text{ T}}{\text{M}}$ longer than the span crosses the girder A to B. The maximum Bending moment at X will be 96 t – m.

51. The influence line for shear at section $X(F_X)$ as shown here of a simply supported girder AB (T or F)

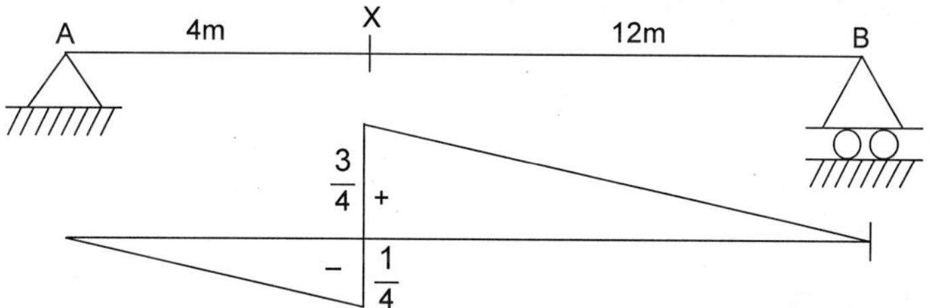

The shear force at section X due to a uniformly distributed lead load of intensity $\frac{2\text{ t}}{\text{M}}$ covering entire span will be 8 t.

52. A simply supported beam with an overhang is traversed by a unit concentrated moment for left to right. (T or F)

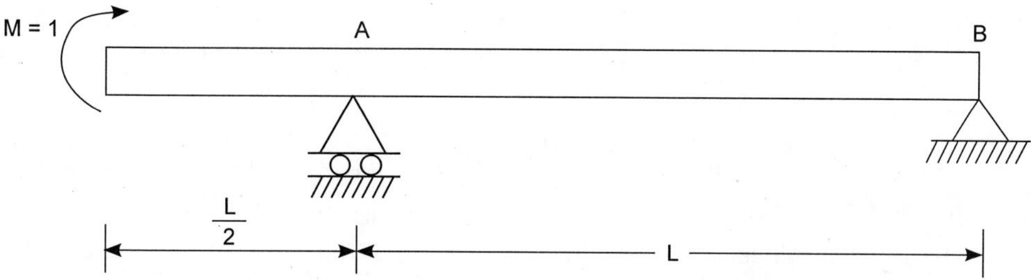

The influence line for reaction at B is given by

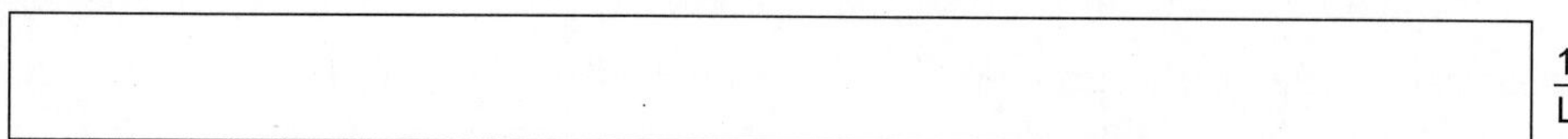

FILL IN THE BLANKS

1. Truss members carry .... load.
2. The carryover moment at B is ....

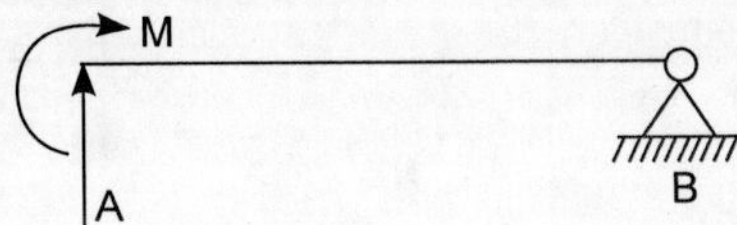

3. The number of unknown internal forces in each member of a rigid jointed plane frame is ....
4. the degree of static indeterminacy upto which column analogy method can be used is ....?
5. The strain energy of a structure due to bending is given by formula ....
6. the fixed support in a real beam becomes in the conjugate beam a ....
7. The deformation of a spring produced by a unit load is called ....
8. A cantilever beam AB fixed at A and carrying a load W at free end B is found to deflect δ at the mid-point of AB. The deflection of B due to a load of c at the mid-point will be ....
9. The portal frame is statically indeterminate to the .... degree.

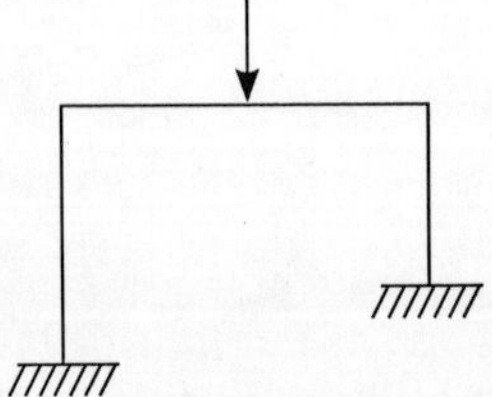

10. The Kinematic indeterminacy of the given beam is ....

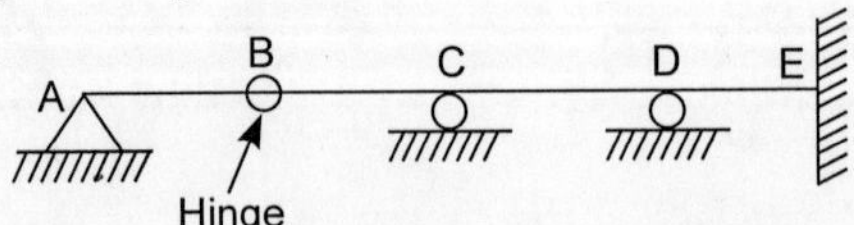

11. A symmetrical 3-hinged parabolic arch of span (L) and rise (h) is hinged at the springs and crown. It is subjected to UDL over the whole span. The BM at a section $\frac{L}{4}$ from left support is ....
12. A 3-hinged arch (semi-circular) of radius (R) carries a UDL of (w) per unit Length over the whole span The horizontal thrust is ....
13. A fixed beam AB is subjected to a triangular load varying from zero at the end A to (w) at the end B varying linearly. The ratio of fixed end moment at B to A will be ....
14. The portal method of structural analysis is generally suitable for .... rise buildings.
15. A fixed beam of uniform section is carrying a point load at mid span. The moment of Inertia of the middle half length is now reduced to half, then the fixed end moment will ....

16. In the given truss zero force will be induced in the member ....

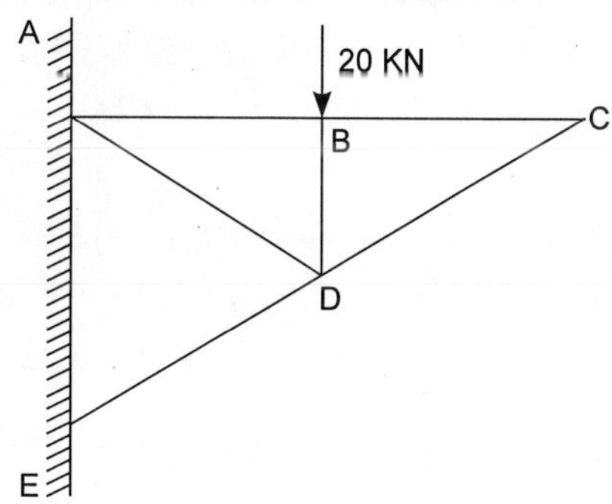

17. The forces in members AB and BC will be .... and ....

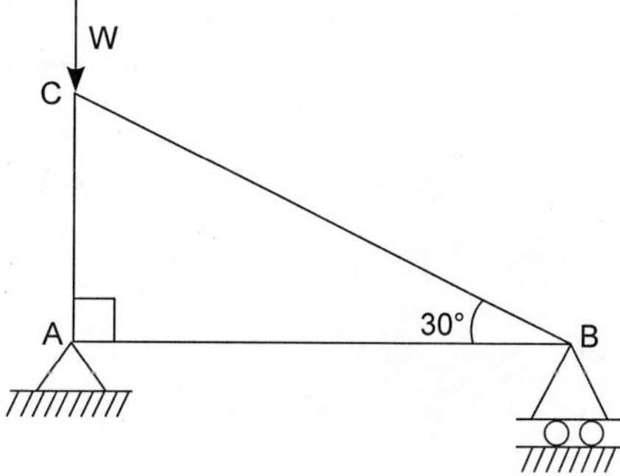

18. The force in the member AB of the truss will be ....

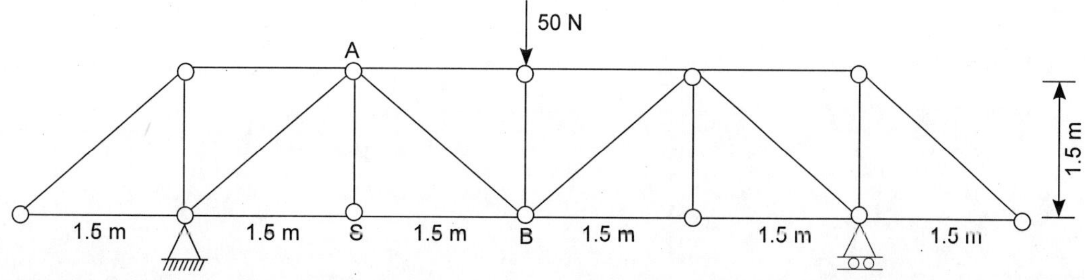

19. The maximum BM at the left quarter point of a simple beam due to crossing of UDL of length shorler than span in the direction left to right, would occur after the load had just crossed the section by .... of its length.

20. A symmetrical 2-hinged parabolic arch rib has a span of 32 m between abutment pins at the same level and a central rise of 5 m. When a rolling load of 100 KN crosses the span, the maximum horizontal thrust at the hinges will be .... KN.

21. The distribution factors at joint B for the members BA and BC respectively are ....

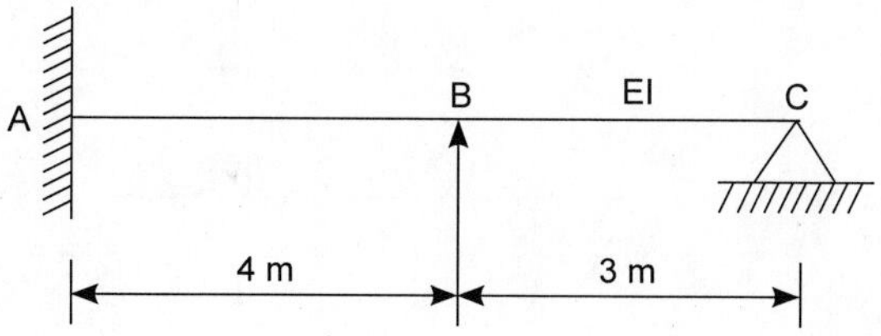

22. A rigid-jointed plane frame as shown here will sway to ....

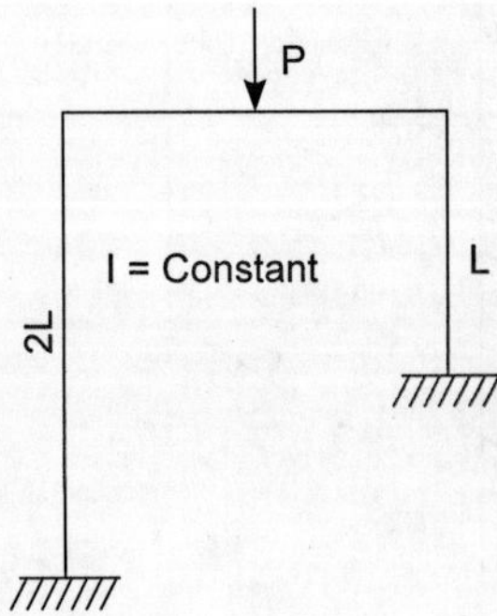

23. A two span continuous beam ABC is continuous over support B. Span AB and $\frac{2t}{m}$ over both the spans. EI is constant for entire beam. The fixed end moment at B in span BA or BC will be ....

24. In the simply supported truss, the force in the member BE is ....

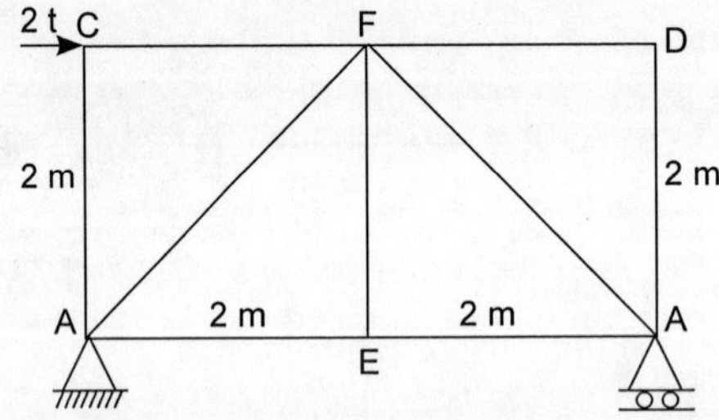

25. In the portal frame shown, the value of P is ....

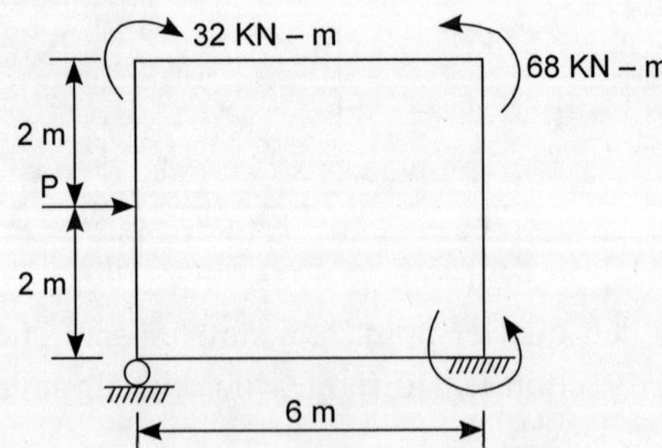

26. Force in the member BC of the truss shown will be ....

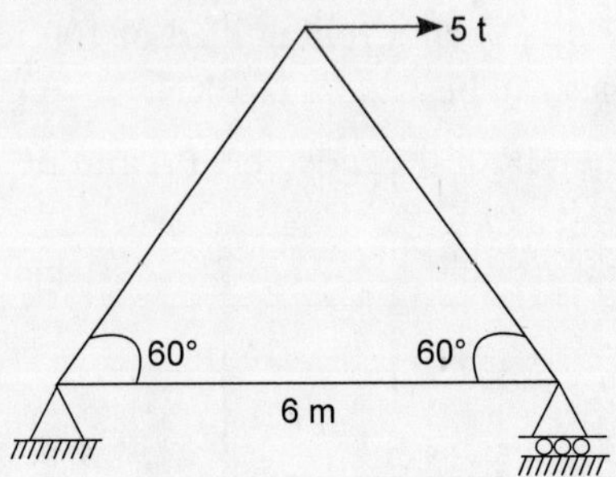

27. The fixed end moment for the beam shown here will be ....

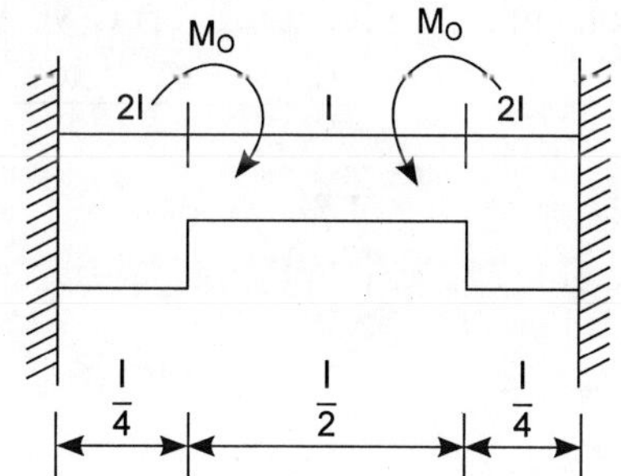

28. Unequal settlement of supports of statically determinate structure develop no ....

29. The number of independent displacement modes for the frame shown are ....

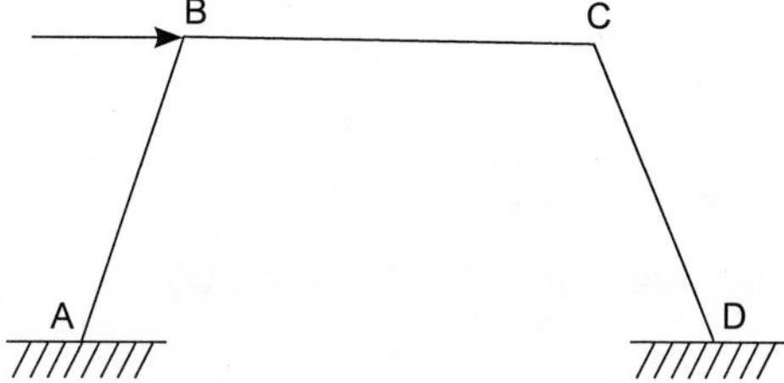

30. The force induced in the vertical member CD of the symmetrical plane truss as shown is ....

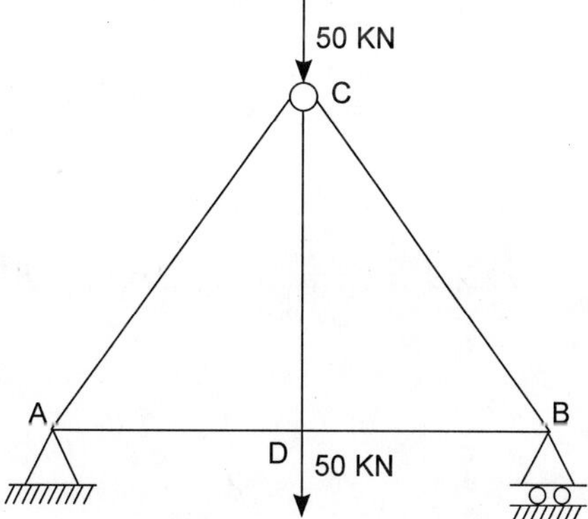

31. In the framed structure the force in the member is ....

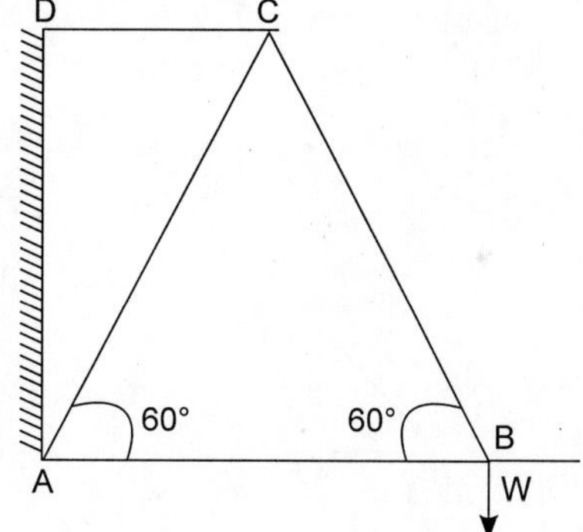

32. Consider the beam ABCD, for a moving concentrated load of 50 KN on the beam, the magnitude of the maximum bending moment (KN – m) at support C will be ....

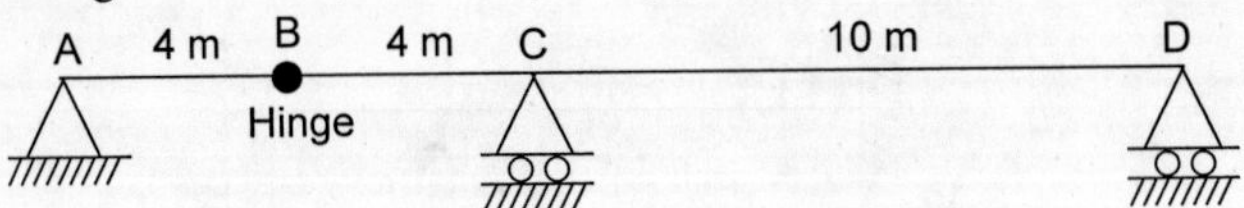

33. The maximum Bending Moment generated in the planar elastic arch is ....

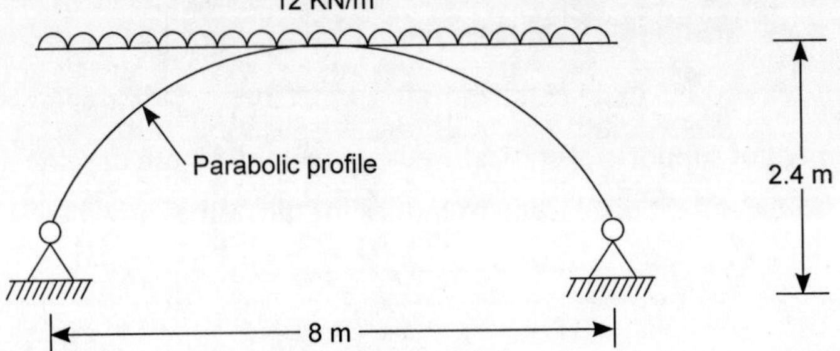

34. The unit load method used in structural analysis is derived from .... theorem.

35. For the beam shown here, the support moment M is ....

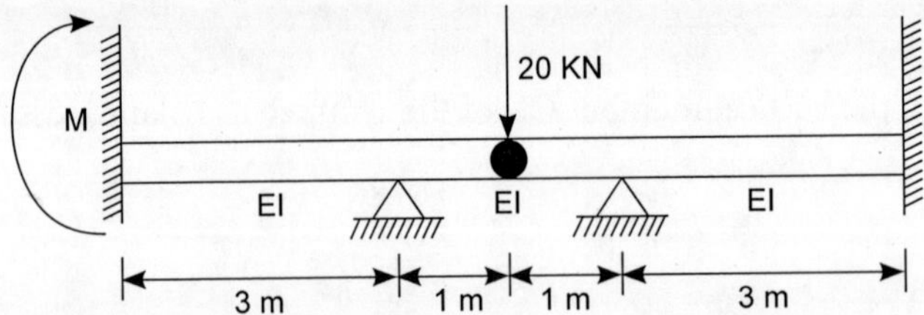

36. The force in the member QR is.................

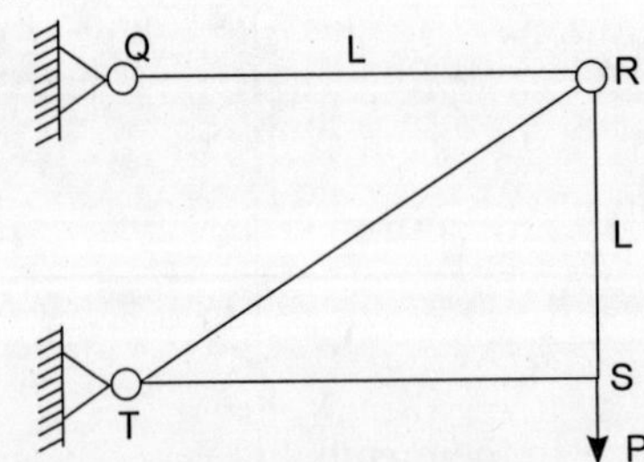

37. A right triangular truss made of members having equal cross-sectional area of 1550 mm² and Young's Modulus of 2×10^5 Pa. The horizontal deflection of the joint Q will be ....

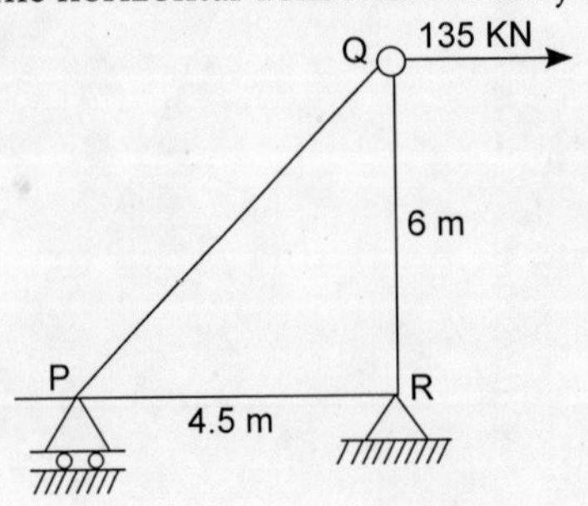

38. In flexibility method the unknown quantities are Force, Moment whereas in stiffness method the unknown quanties are ....

39. Flexibility of structures may be defined as the displacement caused for unit force and stiffness is defined as the force caused for .... displacement.

40. For a linear elastic frame, if stiffness matrix is doubled with respect to existing stiffness matrix, the deflection of the resulting frame will be .... the existing value.

41. The effective length of a column of length (L) fixed against rotation and translation at one end is ....

42. On a simply supported beam of length (L) carrying a point load (W) in the mid-span, the deflection under the load is given by ....

43. A simply supported beam with gradually varying load from zero at B and (W) at A. The shear Force at B will be ....

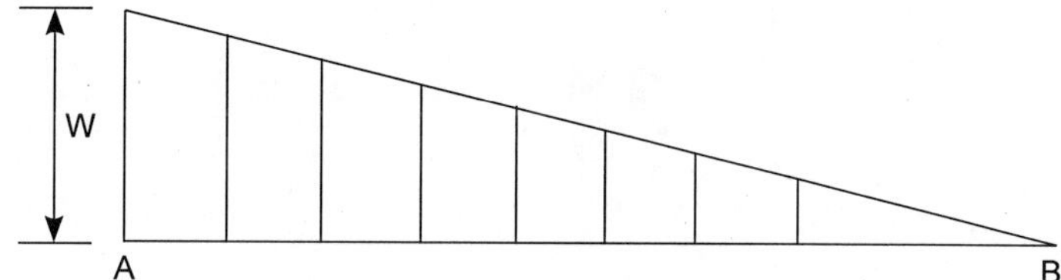

44. For the beam in Q. 43 the Shear force at A will be ....

45. For the beam in Q. 43, the maximum bending moment will be ....

46. When shear force in a beam at a point is zero, the Bending Moment will be.................. .

47. The number of simultaneous equations to be solved in the slope-deflection method is equal to the degree of Kinematic ....

48. The order of flexibility matrix for structures is equal to number of .... forces.

49. For a linear elastic structural system, minimization of potential energy Yields .... conditions.

50. A prismatic beam PQR of flexural rigidity EI = 1×10^4 KNm^2 is subjected to a moment of 180 KN m at Q. The rotation at Q in radians will be ....

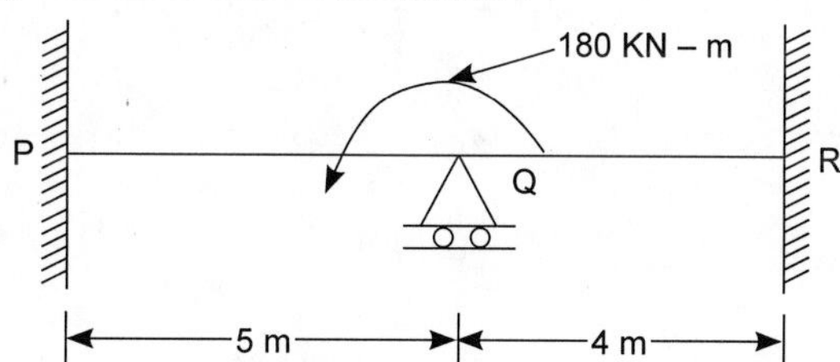

51. The value of M in the beam ABC shown here such that joint does not rotate. The support reaction at B is equal to ....

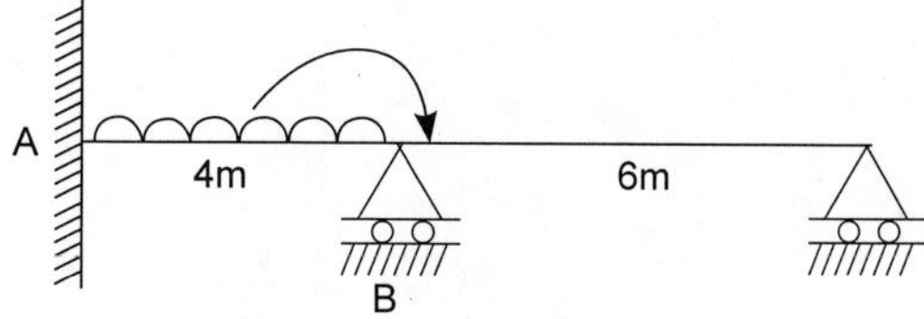

ANSWERS

Multiple Choice Questions

1. c 2. c 3. d 4. b 5. a 6. b 7. d 8. b 9. b

10. c 11. c 12. b 13. d 14. b 15. a 16. c 17. b 18. c

19. b 20. a 21. c 22. b 23. a 24. b 25. b 26. b 27. c

28. c 29. a 30. d 31. b 32. a 33. c 34. a 35. b 36. b

37. a 38. a 39. c 40. c 41. c 42. a 43. b 44. d 45. b

46. b 47. d 48. a 49. d 50. b 51. c 52. d 53. c 54. a

55. c 56. a 57. b 58. a 59. b 60. a 61. d 62. c 63. c

64. a 65. b 66. b 67. a 68. d 69. b 70. a 71. b 72. b

73. c 74. b 75. a 76. a 77. a 78. a 79. b 80. d 81. c

82. c 83. d 84. d 86. b 85. a 87. b 88. a 89. b 90. c

91. d 92. d 93. c 94. b 95. b 96. d 97. a 98. b 99. d

100. d 101. a 102. c

True & False

1. T 2. T 3. F Cubic 4. F Indeterminate

5. T 6. F 7. T 8. T

9. F It is for stable determinate truss 10. F Slope of deflection curve is zero

11. T 12. F Must be equal 13. T 14. T

15. F Costs Less 16. F Will be zero 17. T 18. T

19. F Geometrically unstable 20. T 21. T

22. F Is 6 (Six) 23. F $(m + r) < 2j$ 24. T 25. T

26. F Not an example 27. 2 (Two) 28. T 29. T

30. F Of Length 31. T 32. T 33. F $n = 2j - 3$

34. T 35. F 12 KN – m 36. F $\frac{3}{8}$ wL 37. F 300 KN

38. T 39. T 40. T 41. F Only metal bars

42. F Only axial loads 43. T 44. F It is stable

45. F Consists all three 46. T

47. F It is used for various structures 48. F They can resist only axial force

49. T 50. F 48 t – m 51. T 52. T

Fill in the Blanks

1. Axial
2. Zero
3. 3 (Three)
4. 3 (Three)
5. $\frac{1}{2}\int\frac{M^2dx}{EI}$
6. Free End
7. Flexibility
8. 2δ
9. Third
10. Nine (9)
11. Zero
12. $\frac{wR}{2}$
13. $\frac{3}{2}$
14. Low Rise
15. Increase
16. BC
17. Zero and zero
18. $25\sqrt{2}$ KN (Tension)
19. $\frac{3}{4}$ of its length
20. 125 KN
21. 0.5 and 0.5
22. Left
23. 9 t – m
24. 1.0 t
25. 50 KN
26. 5 t (compression)
27. $\frac{2}{3}M_0$
28. Reaction
29. 1 (One)
30. 50 (KN) (Tension)
31. $\frac{2w}{\sqrt{3}}$ (Tension)
32. – 200 KN – m
33. 96 KN – m
34. Castigliano's
35. 5 KN – m
36. P units
37. 15.68 mm
38. Deflection Rotation
39. Unit
40. Ilalf
41. 2 L
42. $\frac{wL^3}{48\,EI}$
43. $\frac{wL}{6}$
44. $\frac{wL}{3}$
45. $\frac{wL^2}{9\sqrt{3}}$
46. Maximum
47. Indeterminacy
48. Redundant
49. Compatibility
50. $\theta_Q = 0.01$ rad
51. 40 KN – m_1

SECTION 21

Surveying

Definitions – 52

Multiple Choice Questions – 101

True & False – 101

Fill in the Blanks – 101

DEFINITIONS

Accuracy and Precision: Precision is the degree of perfection used in the instruments, methods and the observations. Accuracy is the degree of perfection obtained.

Alidade: It is a kind of ruler with a sight line attached and is used on plane table for bisecting objects, drawing rays and direction lines. It has tow **vanes** at the ends. One of the sight vanes is provided with a narrow rectangular slit, while the other is provided with a central vertical wire or hair. Telescopic alidade are used for accuracy and long range.

Astronomical Survey: It offers to the surveyor means of determining the **absolute** location of any point or the absolute location and direction of any line on the surface of earth. This consists in observations to the heavenly bodies such as sun or any fixed star.

Azimuth: It is the angle between North, measured clockwise around the observer's horizon, and a celestial body (sun, moon etc.). It determines the direction of celestial body. Also defined as an arc of the horizon measured between a fixed direction (like true north) and the vertical circle passing through an object (celestial body).

Cadastral Survey: It is meant to the fixing of property lines, the calculation of land area, or the transfer of land property from one owner to another. They are also made to fix the boundaries of municipalities, of state and Federal Jurisdiction.

Chain: Chains are formed of straight links of galvanised mild steel wire bent into rings at the ends and joined each other by three small circular or oval wire rings. The length of a link is the distance between the centres of two consecutive middle rings, while the length of the chain is measured from outside of one brass handle to the outside of the other handle.

Chaining: It is a term which is used to denote measuring distance either with the help of a chain or a tape and is the most accurate method of making direct measurements. For work of ordinary precision, a chain can be used, but for higher precision a tape or special bar can be used. The distances determined by chaining form the basis of all surveying.

Change Point (or Turning Point): It is the location on which both Fore Sight (or minus sight) to know the elevation of that point and Back Sight (or Plus Sight) on the same point from new location of the Dumpy Level to know the height of line of sight.

Closing Error: When a closed traverse is plotted according to the field measurements, the end point of the traverse may not coincide with the starting point, owing to the errors in field measurements. Such error is known as closing error. Various methods are used to adjust closing error such that starting and end points of traverse coincide.

Contour: It is an imaginary line on the ground joining the points of equal elevation. **Contour Line** is a line drawn on the map representing a contour. From a contour map nature of the ground can be easily guessed.

Contour Interval: The vertical distance between any two consecutive contours is called contour interval. The **contour interval** is kept constant for a contour plan, otherwise general appearance of the map will be misleading. The horizontal distance between two points on two consecutive contours is known as **horizontal equivalent.**

Diagonal Scales: These are used in engineering to read lengths with higher accuracy as it represents a unit into three different multiples in meters, centimeters and millimeters. It plays an important role in engineering drawing.

Dumpy Level: It is an instrument used in surveying to measure difference in reduced level (height) of any location on the ground relative to a reference point. It is simple to use and gives almost accurate results. They are robust and cost effective. Main components are telescope, bubble tube, compass, vertical spindle, foot screws and tripod. Dumpy level identifies point in the horizontal plane only.

Fathometer: It is a device that uses **echo-sounding** to measure the depth of the ocean or any other deep water body. It may indicate the depth visually or indicate graphically. It contains– Transmitting and receiving Oscillator, Recording Unit and transmitter.

Geodetic Surveying: In this type of surveying the **shape** of the earth is taken into account. All lines in the surface are curved lines and the triangles are spherical triangles. All geodic surveys include work of larger magnitude and high degree of precision.

Geodimeter: It is an instrument which works based on the propagation of modulated light waves. The instrument is suitable for night time observations and requires a prism system at the end of the line for reflecting the waves. It gives reasonably better results.

Gunter's Chain or Surveyor's chains: This chain is 66 feet long and consists of 100 links. It was adopted for convenience in land measurement since 10 square chains are equal to 1 acre. Also 10 Gumter's chains are equal to 1 furlong.

Hydrographic (Marine) Survey: It deals with bodies of water. The work includes measurement of discharge of streams, making topographical survey of shores and banks, finding depth of water and water surface level variations.

Instrumental Error: Error may arise due to faulty adjustment of the instrument used in survey. This may be due to bad or no maintenance of the instruments.

Levelling: It is a branch of surveying used to find elevations of places with respect to a given or assumed datum and also to establish points at a given elevation or at different elevations with respect to a datum.

Levelling Survey Terms:

(i) **Station:** It is that point where the level rod or staff is held and whose elevation is to be found.

(ii) **Height of Instrument (HI):** It is the elevation of line of sight with respect to the assumed datum. It does not mean the height of telescope above the ground where the level stands.

(iii) **Back Sight (BS):** It is the reading on staff held at a point of **known elevation** to ascertain the amount by which the line of sight is above that point. The object of back sighting is, therefore, to ascertain the height of the line of sight.

(iv) **Fore Sight (FS):** It is a sight (reading) taken on the staff held at the point of **unknown elevation** to ascertain the amount by which the point is below the line of sight, thus obtaining the elevation at the desired location.

Local Attraction: Sometimes the magnetic needle may get attracted and prevented from indicating the true magnetic meridian when it is in the proximity to certain magnetic substances. Local Attraction is the term used to denote any such influence. Some of the sources of local acceleration are : magnetite in ground, Electric wires, steel structures etc.

Meridian: It is an imaginary circle or closed curve on the surface of globe that lies in a plane passing through the poles. **Prime meridian** is the line of zero degree longitude. It is geographical reference passing through Greenwich (London).

Metric Chain: After the introduction of metric units in India, the metric chains are widely used. Metric chains are generally available in lengths of 5 m, 10 m, 20 m, and 30 m. IS : 1492-1970 covers the requirements of metric surveying chains.

Mines' Surveying: Mine surveying is the practice of determining the relative position of points on or beneath the surface of earth of direct or indirect measurements of distance, direction and elevation. Due to limitations of space, specially designed instruments are used. Short sights, darkness and steep slopes need special efforts.

Natural Errors: These may be due to variation in natural phenomena such as temperature, humidity, gravity, refraction, wind and magnetic declination.

Odometer: It registers the number of revolutions of a wheel. It is fitted to the **wheel** which is rolled along the line whose length is required. The number of revolutions recorded is multiplied by the circumference of the wheel. The well known **Speedometer** is based on this principle.

Optical Square: It is a hand instrument used to lay off right angles that are multiples of 90° or 45°. They normally comprise of two optical glass penta prisms in a sturdy housing. Mainly they are used for setting out right angles.

Passometer: It is an instrument shaped like a watch and is carried in pocket or attached to one leg. The mechanism of the instrument is operated by motion of the body and it automatically registers the number of paces. The number of paces registered by the **passometer** is multiplied by the average length of one step to get the approximate distance.

Pedometer: It is a device similar to **Passometer** except that it can be adjusted to the length of the pace of the person carrying it, it directly registers the total distance covered by any number of paces.

Personal Error: Error may arise due to lack of perfection of human sight in observing or in handling of instruments. 'Ihese are common in readings by theodolite or holding of staff or in fixing temporary bench marks etc.

Photogrammetry: It is the science and art of obtaining accurate measurements by use of photographs. The photographs taken either from the air or from station on the ground. These are termed as aerial photogrammetry and terrestrial photogrammetry respectively.

Plane Surveying: It is the type of surveying in which the mean surface of the earth is considered as a plane and the spheroidal shape is neglected. The level line is considered as straight and all plumb lines are considered parallel. It may be noted that the length of an arc 12 km. long lying in the earth's surface is only 1 cm, greater than the subtended chord. Also difference between sum of the angles in plane triangle compared to sum of those in spherical triangle is only one second for a triangle having area of 195 km^2 on earth's surface.

Plane Table: If consists of a smooth table surface mounted on a sturdy base. The connection between the table top and the base permits one to level the table precisely, using bubble levels in a horizontal plane. The base, a tripod is designed to support the table over a specific point on the ground.

Planimeter: It is an instrument which measures area of plan of **any shape** accurately.

Reconnaissance Survey : It is a vital initial step in planning and executing any project. It provides a broad overview of site conditions, identify key features, constraints, and opportunities that can influence the project feasibility and design.

Remote Sensing: It is the process of detecting and monitoring the physical characteristics of an area by measuring its reflected and emitted radiation at a distance (typically from satellite or aircraft). Special cameras collect remotely sensed images, which help **Sense** things about the Earth.

Representative Fraction (RF): One unit of length on the plan represents some number of same units of length on the ground, such as $\frac{1}{100}$ or $\frac{1}{1000}$ etc. This ratio of map distance to the corresponding ground distance is independent of units of measurements and is called **Representative Fraction.** (Example : For a scale 1 cm to 20 m R.F. = $\frac{1}{20 \times 100} = \frac{1}{2000}$).

Shrunk Scale: Due to changes in temperature the sheet on which drawing is made shrinks. It becomes essential to find the shrunk scale of the drawing (plan). First **shrinkage factor** has to be found out which is the ratio of the **shrunk length** to the actual length. (Shrunk Scale = Shrinkage factor × original scale).

Sight Distance: It is the unobstructed visibility available to drivers along a roadway. It refers to the maximum distance a driver can see ahead, providing essential reaction time to avoid potential hazard. For any given value of the difference in the tangential grades, the length of vertical curve must be large enough to provide minimum required sight distance throughout the vertical curve.

Super-elevation: It is the banking of the roadway such that the outside edge of pavement is higher than the inside edge. The use of super-elevation allows a vehicle to travel through a curve more safely and at a higher speed than otherwise be possible. Also known as **cant.**

Tacheometry: It is a system of **rapid surveying**, by which the horizontal and vertical positions of points on Earth's surface relative to one another are determined using a **tacheometer** which is a form of theodolite. It is used without a chain or tape for distance measurement or any levelling instrument.

Tapes: These are used for more accurate measurements. They are made of Cloth or Linen, metal or invar. **Invar tapes** are made of alloy of nickel (36%) and steel. They have very low coefficient of expansion. Invar tapes are used where high degree of accuracy is required.

Tellurometer: It is an instrument which determines distance by measuring the roundtrip travel time of reflected microwaves. Though not accurate it is portable and easy to use.

Theodolite: It is a versatile surveying instrument which can measure precisely horizontal and vertical angles. It has wide applicability in surveying such as– laying of horizontal angles, locating points on line, establishing grades, finding difference in elevation, setting of curves etc.

Topographical Survey: It consists of horizontal and vertical location of certain location by linear and angular measurements and is made to determine the natural features of a country such as rivers, lakes, woods, hills etc. and man-made features like roads, railways, dams, canals, villages and towns.

Traverse Surveying: It is the type of survey in which a number of connected survey lines form the framework and the directions and lengths of the survey lines are measured. When the lines form a circuit which ends at the starting point, it is known as a **closed traverse**. If the circuit ends elsewhere, it is said to be an open traverse.

Trigonometric Levelling: It is the process of determining the difference of elevations of stations from observed vertical angles and known distances.

True Value and Measured Value: The difference between measured value and true value of the quantity measured is called **true error**, which is never known since true value is of the quantity is never known. In survey **limit of error** is prescribed by the competent authority depending upon the nature and purpose of survey.

Tunnel Surveying: The basic procedure of tunnel surveying is to align the centre line in the ground and transfer it to the tunnel. Levelling of surface and internal (underground) tunnel is also done.

Vertical Curve: A vertical curve is used to join two intersecting grade lines of rail-roads, highways or other routes to smooth out the changes in vertical motion. It contributes to the safety and comfort to the travellers. Circular or parabolic curves are used as vertical curves.

Zenith and Nadir: The zenith (z) is the point on the upper portion of the celestial sphere marked by plumb line above the observer. It is thus the point on celestial sphere immediately above the observer's station. The **Nadir** (z′) is the point on the lower portion of the celestial sphere vertically below the observer's station.

MULTIPLE CHOICE QUESTIONS

1. In which of the following type of survey only linear measurements are taken?
   (a) Chain survey (b) Contouring (c) Dumpy level (d) Theodolite survey
2. In chain survey which of the following is not used?
   (a) Pegs (b) Arrows (c) 20 m chain (d) Dumpy level
3. Which is the last step in chain surveying?
   (a) Fixing (b) Marking
   (c) Running survey lines (d) Reconnaissance
4. Gunter's chain has how many links?
   (a) 500 (b) 50 (c) 100 (d) 1000
5. Chain surveying uses the principle of
   (a) Traversing (b) Chaining (c) Ranging (d) Triangulation
6. Chain measurements are recorded in
   (a) Record book (b) Field book (c) Chain book (d) Survey book
7. Survey stations are marked with
   (a) Poles (b) Peg (c) Rod (d) Arrow
8. The position of an object can be fixed more accurately by
   (a) Cross staff (b) Optical square (c) Oblique offset (d) Perpendicular offset
9. For a well-conditioned triangle none of the angles should be less than
   (a) 30° (b) 20° (c) 45° (d) 60°

10. A tie line is to
    (a) Check accuracy of work
    (b) To take details of nearby objects
    (c) To take offsets for details
    (d) All these
11. Permissible Error in chaining for measurement with chain on rough or hilly ground is
    (a) 1 in 100 (b) 1 in 250 (c) 1 in 500 (d) 1 in 1000
12. The correction for sag is
    (a) Always additive
    (b) Always subtractive
    (c) Always zero
    (d) Sometimes additive and sometimes subtractive
13. The maximum tolerance in a 20 m chain is
    (a) ± 2 mm (b) ± 3 mm (c) ± 5 mm (d) ± 8 mm
14. When a chain is found to be short on testing, it can be adjusted by
    (a) Straightening the links
    (b) Adding small circular links
    (c) Rejecting the chain
    (d) Replacing the bent links
15. A metallic tape is made of
    (a) Steel (b) Invar (c) Linen (d) Cloth & Wire
16. The angle of intersection of the two plane mirrors of an optical square is
    (a) 30° (b) 45° (c) 60° (d) 90°
17. The length of a chain is measured from
    (a) Centre of one handle to the centre of other handle
    (b) Outside of one handle to outside of other handle
    (c) Outside of one handle to inside of other handle
    (d) Inside of one handle to inside of other handle
18. In which of the following type of surveying only linear measurements are made?
    (a) Theodolite survey (b) Contouring (c) Chain survey (d) Dumpy level survey
19. Compass surveying is not good for
    (a) Small areas (b) Undulating area (c) Crowded area (d) Near TV-towers
20. Which of the following is an indirect method of surveying?
    (a) Contouring (b) Chain survey (c) Tacheometry (d) All these
21. The type of survey in which Earth's Curvature is also considered is called?
    (a) Geodetic survey
    (b) Plane survey
    (c) Preliminary survey
    (d) Geographical survey
22. The survey carried out to delineate natural features like hills, rivers, forests and manmade features, like towns, roads, buildings, villages, transmission lines and canals is classified as
    (a) Engineering survey
    (b) Geological survey
    (c) Land survey
    (d) Topographical survey
23. The plan of a map was photocopied to a reduced size such that a line originally 100 mm measures 90 mm. The original scale was 1 : 1000 the revised scale is
    (a) 1 : 900 (b) 1 : 1111 (c) 1 : 1121 (d) 1 : 1221

24. The plan of a survey plotted to a scale 10 m : 1 cm is reduced in such a way that a line originally 10 cm long now measures as 9 cm and the area of the reduced plan is measured as 81 cm^2. The actual area of survey is

(a) 10,000 m^2 (b) 6561 m^2 (c) 1000 m^2 (d) 656 m^2

25. The true bearing of a line AB is 269° 30′, the Azimuth of the line is

(a) 0° 30′ (b) 89° 30′ (c) 90° 30′ (d) 269° 30′

26. A nautical mile equals to how many meters?

(a) 1954 (b) 1852 (c) 2000 (d) 1000

27. Different grades are joined together by a

(a) Compound curve (b) Transition curve (c) Reverse curve (d) Vertical curve

28. The process of a location of intermediate points on a survey line is

(a) Aligning (b) Extending (c) Ranging (d) Offsetting

29. The longest of survey line is called

(a) Base line (b) First line (c) Tie line (d) Main survey line

30. The principle of chain survey is

(a) Traverse (b) Triangulation (c) Traversing (d) Chaining

31. Where a plane table set up for preparing a road map?

(a) Middle of the road (b) Left side of road

(c) Right side of the road (d) Along any side of the road

32. The instrument used for accurate centering in plane table survey is

(a) Spirit level (b) Alidade (c) Plumbing fork (d) Trough compass

33. Longitudes are measured from 0° to

(a) 180° E (b) 180° W (c) 180° E or W (d) 360° E

34. Which of the following methods of plane table surveying is used to locate the position of an inaccessible point?

(a) Radiation (b) Intersection (c) Traversing (d) Resection

35. Which type of survey facilitates field observations and plotting on sheet simultaneously?

(a) Plane table (b) Chain (c) Compass (d) None of these

36. The first temporary adjustment to be done after setting of plane table at any station will be

(a) Centering (b) Levelling (c) Orientation (d) Sighting

37. In plane table survey Lehman's method is also known as

(a) Trial & Error method (b) Mechanical method

(c) Graphical method (d) Measurement

38. In plane table survey which instrument is used to sight an object?

(a) Compass (b) U-fork (c) Alidade (d) Plumb

39. The Rise and Fall method provides a complete check on

(a) Back sight (b) Intermediate sight

(c) Fore sight (d) All these

40. The required slope correction for a length of 50 m along a gradient in 1 in 40 is

(a) 15.625 mm (b) 18.625 mm (c) 20.625 mm (d) 13.625 mm

41. The side of a square land was measured as 150 m with error 0.05 m, the corresponding error in the computed area of land is

(a) 5 m^2 (b) 10 m^2 (c) 15 m^2 (d) 20 m^2

42. While levelling in an undulating terrain, it is preferable to set up the instrument on?

(a) Top of summit (b) A flat location

(c) Bottom of a valley (d) One side of slope

43. The sensitiveness of a bubble tube in a level would decreases if

(a) The radius of curvature of internal surface of the tube is increased

(b) The diameter of the tube is increased

(c) The length of vapour bubble is increased

(d) The viscosity of liquid is increased

44. Which of the following statement of false?

(a) Plumb line is along the line of gravity

(b) Mean Sea Level (MSL) used as reference for horizontal control

(c) Geoid is equipotential surface of gravity

(d) MSL is simplification of Geoid

45. The distance between the instrument station and staff station is 1 km, the correction for refraction in (m) is

(a) 0.0673 (b) 0.0785 (c) 0.0112 (d) 0.0121

46. The staff reading at a distance of 80 m from a level with bubble at the centre is 1.52 m and, when the bubble is moved 5 divisions away from centre, the reading is 1.6m, the angular value of bubble's one division is

(a) 20.62 sec. (b) 41.25 sec. (c) 14.53 sec. (d) 25.05 sec.

47. Two points P and Q located on a map have the following co-ordinates.

	P	Q
Latitude (L)	+ 40 m	+20 m
Departure (D)	– 20 m	+30 m

The length of PQ is ....

(a) 53.85 m (b) 34.89 m (c) 34.98 m (d) 12.89 m

48. The reduced bearing of a 10 m long line is N 30° E, the departure of the line is

(a) 10 m (b) 8.66 m (c) 7.52 m (d) 5.00 m

49. The magnetic bearing of a line AB is S 45° E and declination is 5° W, the true bearing of AB, is

(a) S 45° E (b) S 40° E (c) S 50° E (d) S 50° W

50. The latitude and departure of a line AB are +78 m and –45.1 m respectively. The Whole Circle Bearing (WCB) of AB is

(a) 30° (b) 150° (c) 210° (d) 330°

51. In quadrantal bearing system, bearing of a line varies between

(a) 0° and 360° (b) 0° and 180° (c) 0° and 90° (d) 0° N to 90° S

52. In a closed loop traverse of total 1 km length the closing error in departure and latitude are 0.3 m and 0.4 m respectively. The relative precision of this traverse will be

(a) 1 : 5000 (b) 1 : 4000 (c) 1 : 3000 (d) 1 : 2000

53. The following measurements were made during testing a levelling instrument

Instrument at	Staff reading at	
	P_1	Q_1
P	2.8 m	1.7 m
Q	2.7 m	1.8 m

P_1 is close to P and Q_1 is close to Q. If the reduced level of station P is 100.0 m. The reduced level of station Q is

(a) 99.0 m (b) 100.0 m (c) 101.0 m (d) 102.0 m

54. Which of the following errors can be eliminated by reciprocal measurements in differential levelling

(I) Error due to earth's curvature (II) Error due to atmospheric refraction

(a) Both I and II (b) I only (c) II only (d) Neither I nor II

55. Curvature correction to a staff reading in a differential levelling survey is

(a) Always subtractive (b) Always zero

(c) Always additive (d) Dependent on latitude

56. The staff reading taken on a workshop floor using a level is 0.645 m. The inverted staff reading taken to the bottom of a beam is 2.96 m. The reduced level of the floor is 40.5 m. The reduced level in (m) of the bottom of the beam is

(a) 44.105 (b) 43.46 (c) 42.815 (d) 41.125

57. The relationship between the length (l) and radius (r) to an ideal transition curve is given by

(a) $l \propto r$ (b) $l \propto r^2$ (c) $l \propto \frac{1}{r}$ (d) $l \propto \frac{1}{r^2}$

58. A circular curve of radius R connects two straights with deflection angle of 60°, the tangent length is

(a) 0.577 R (b) 1.55 R (c) 1.732 R (d) 3.464 R

59. The co-ordinates of A are 100 N and 200 E and those of C are 100 S and 200 E, then the length AC is

(a) 400 (b) 200 (c) 244 (d) 282

60. Among the following instruments which is used in plane table survey for measuring horizontal and vertical distances directly

(a) Telescopic alidade (b) Clinometer (c) Tacheometer (d) Plain alidade

61. The linear error in a 50 m long traverse is 0.01 m. The angular error (in seconds) considering equal precision will be

(a) 81 (b) 41 (c) 21.5 (d) Zero

62. Mean Sea Level (MSL) at any place is the average datum of hourly tide heights observed over a period of nearly

(a) 5 years (b) 10 years (c) 20 years (d) 50 years

63. If parallax difference between the top and bottom of a tree is measured as 1.32 mm on a sterio-pair of photos taken at 3000 m above ground and the average photobase is 66 mm, then the height of the tree will be

(a) 60 m (b) 45.49 m (c) 39.50 m (d) 23.51 m

64. The sides of a rectangle are (120 ± 0.05) m and (180 ± 0.06) m. The probable error in the area will be

(a) ± 16.8 m^2 (b) 12.35 m^2 (c) ± 16.2 m^2 (d) ± 11.35 m^2

65. If the bearing of a line AB is N 60° 30′ and that of BC is 122° of a closed traverse ABCDE, then the measure of the interior angle B is

(a) 154° 00′ (b) 240° 30′ (c) 122° 00′ (d) 118° 30′

66. The distance between two Bench Marks is 1 km. If during levelling, the total error due to collimation, curvature and refraction is + 0.120 m, then magnitude of collimation error is

(a) 0.00527 m (b) 0.0527 m (c) 0.527 m (d) 0.673 m

67. A circle of radius 7 m has a standard error of 0.02 m on the radius. The standard error of its area is

(a) 0.04 m^2 (b) 0.14 m^2 (c) 0.28 m^2 (d) 0.88 m^2

68. The Fore Bearing (FB) and Back Bearing (BB) of lines PQ and QR have been measured as

Line	FB	BB
PQ	50° 0′	235° 0′
QR	125° 30′	309° 30′

The correct value of the interior angle is

(a) 109° 30′ (b) 105° 00′ (c) 250° 00′ (d) 255° 00′

69. The azimuths of the two tangents to a circular curve of radius 100 m are due north and due east, then the area bounded by the two tangents and the circular curve will be

(a) 7857 m^2 (b) 5000 m^2 (c) 3143 m^2 (d) 2143 m^2

70. A 300 m long line lying at an elevation of 450 m measures 10 cm on vertical photograph. The focal length of the camera is 21 cm. The scale of the photograph for the area having an elevation of 1000 m will be

(a) 1 : 27381 (b) 1 : 25008 (c) 1 : 20606 (d) 1 : 30421

71. Setting out a curve using two theodolite methods involves with

(a) Linear measurements only
(b) Angular measurements only
(c) One linear and one angular measurement
(d) One linear and two angular measurements

72. The method of Orientation used, when the plane table occupies a position not yet located on the map is called as

(a) Traversing (b) Radiation (c) Levelling (d) Resection

73. The size of a plane table is
(a) 750 mm × 900 mm (b) 600 mm × 750 mm
(c) 450 mm × 600 mm (d) 300 mm × 450 mm

74. The angle between the plane of the equator and plane of the ecliptic, is known as obliquity of the ecliptic and its value is
(a) 22° 30′ (b) 23° 27′ (c) 23° 30′ (d) 24° 0′

75. For a tacheometer the additive and multiplying constants are respectively
(a) 0 and 100 (b) 100 and 0 (c) 0 and 0 (d) 100 and 100

76. Overturning of vehicles on a curve can be avoided by using
(a) Compound curve (b) Vertical curve (c) Reverse curve (d) Transition curve

77. Thc Polaris rcmains bclow horizon at
(a) 10° N Latitude (b) 50° N Latitude (c) Equator (d) 5° S Latitude

78. The great circle whose plane is perpendicular to the axis of rotation of earth, is called
(a) Equator (b) Terrestrial equator
(c) 0° latitude (d) All these

79. For a line XY,
(a) The forbearing of XY and backbearing of XY differ by 180°
(b) The forbearing of XY and backbearing of YX differ by 180°
(c) Both a and b correct
(d) None is correct

80. A 'level line' is a
(a) Horizontal line
(b) Line parallel to the mean spheriodal surface of earth
(c) Line passing through the centre of cross-hairs and centre of eye piece
(d) Line passing through objective lens and eye piece of a level

81. If the horizontal distance between the staff point and the point of observation is (d), then the error due to curvature of earth is proportional to
(a) d (b) $\frac{1}{d}$ (c) d^2 (d) $\frac{1}{d^2}$

82. If length (L) of a transition curve and (R) is the radius of circular curve, then shift of the curve is directly proportional to
(a) R and $\frac{1}{L^2}$ (b) $\frac{1}{R^2}$ and L^2 (c) $\frac{1}{R^2}$ and L (d) R^2 and $\frac{1}{L}$

83. If an upgrade of 1.5% is followed by a downgrade of 0.5% and rate of change of grade is 0.2% per 20 m chain, then the length of vertical curve is
(a) 200 m (b) 100 m (c) 300 m (d) 400 m

84. The combined correction for curvature and refraction for a distance of 1400 m is
(a) 0.021 m (b) 0.094 m (c) 0.132 m (d) 0.153 m

85. The representative fraction $\frac{1}{2500}$ means that the scale is 1 cm equal to

(a) 0.25 m (b) 2.5 m (c) 25m (d) 2.5 km

86. The difference between the sum of the angles of a plane triangle and the sum of the angles of a spherical triangle is one second when the triangle on earth surface has an area about

(a) 105 km^2 (b) 135 km^2 (c) 159 km^2 (d) 195 km^2

87. The size of a theodolite is specified by

(a) Telescope's length (b) Diameter of vertical circle
(c) Diameter of lower plate (d) Diameter of upper plate

88. The cross hairs in a survey telescope are placed

(a) Midway between eye piece and objective lens
(b) Much closer to eye piece
(c) Much closer to objective lense
(d) Anywhere between two lenses

89. Benchmark is established by

(a) Hypsometry (b) Barometric levelling
(c) Spirit levelling (d) Trignometric levelling

90. Intersection method of detailed plotting is most suitable for areas

(a) Forest (b) Hilly (c) Plains (d) Urban

91. The instrument used for accurate centering in plane table survey is

(a) Spirit level (b) Alidade (c) Trough compass (d) Plumbing Fork

92. The number of horizontal cross-wires in a stadia diaphram is

(a) One (b) Two (c) Three (d) Four

93. If intercept on a vertical staff is observed as 0.75 m from a tacheometer, the horizontal distance between tacheometer and staff is

(a) 75 m (b) 50 m (c) 25 m (d) 7.5 m

94. Different grades are joined together by a

(a) Compound curve (b) Transition curve
(c) Reverse curve (d) Vertical curve

95. A level instrument at a height of 1.32 m has been placed at a station (R L 112.565). The instrumental reads –2.835 m on staff held at the bottom of a bridge deck. The RL of the bottom of the bridge deck is

(a) 116.72 m (b) 116.08 m (c) 114.08 m (d) 111.05 m

96. Tilt of a staff in tacheometer survey increases the intercept if it is

(a) Away from telescope pointing uphill
(b) Towards the telescope pointing uphill
(c) Away from telescope pointing downhill
(d) None of these

97. The lengths and bearings of a closed traverse PQRSP are given below, x and y will be

Line	Length (m)	Bearing (WCB)
PQ	200	0°
QR	1000	45°
RS	907	180°
SP	x	y

(a) 207 m, 270° (b) 707 m, 270° (c) 707 m, 180° (d) 907 m, 270°

98. Which of the following is a type of Bench Mark?

(a) GTS Bench Mark (b) Permanent BM (c) Temporary BM (d) All these

99. Two straight lines intersect at an angle of 60°. The radius of a curve joining the two straight lines is 600 m. The length of the long chord and mid-ordinates in (m) of the curve are

(a) 80.4, 600.0 (b) 600.0, 80.4 (c) 600.0, 39.89 (d) 49.89, 300.0

100. If the probable error in single observation is $\pm$ 0.04 m, and that of mean is $\pm$ 0.01 m, then number of observations is

(a) 4 (b) 10 (c) 16 (d) 64

101. Three point problem can be solved by the method

(a) Tracing paper (b) Bessels method (c) Lehman's method (d) All of these

TRUE OR FALSE QUESTIONS

1. A Gunter's chain is 100 feet long. (T or F)
2. A passometer counts steps. (T or F)
3. Heliograph is a type of Sun signal used in triangulation work. (T or F)
4. The quadrental bearings of the lines AB and CA are S 30° E and S 70° E the included angle CBA is 320°. (T or F)
5. The survey stations must be mutually visible. (T or F)
6. Check lines or proof lines are the lines which are run to check accuracy of work. (T or F)
7. On a change point both Fore and Back sights are taken. (T or F)
8. Cyanometer is used to measure elevation. (T or F)
9. The correction for sag is always additive. (T or F)
10. The representative fraction of $\frac{1}{2500}$ means that scale is 1 cm equal to 25 m. (T or F)
11. For a well-condition triangle in surveying no angle should be less than 30°. (T or F)
12. For a line AB the Fore bearing of AB and back bearing of BA differ by 180°. (T or F)
13. Detailed plotting is done by Radiation method. (T or F)

14. The size of a theodolite is specified by the length of its telescope. (T or F)
15. The instrument used for accurate centering in plane table survey is alidade. (T or F)
16. The correction for sag is always subtractive. (T or F)
17. A negative declination shows the magnetic meridian is to the East of the true meridian. (T or F)
18. The cross hairs in the surveying telescope are placed midway between eye piece and objective lense. (T or F)
19. For locating inaccessible points intersection method is applied. (T or F)
20. In Geodetic survey curvature of the Earth is neglected. (T or F)
21. At the end of each chain during chaining the pins used are called Arrows. (T or F)
22. In surveying a well conditioned triangle is which has none of the angles less than 30° or more than 120°. (T or F)
23. If the staff at a levelling station is not held vertical, the observation will be more than true R.L. (T or F)
24. A series of closely spaced contour lines represent steep slope. (T or F)
25. An aerial photograph may be assumed as Parallel projection. (T or F)
26. With a simple vernier reading can be taken in one direction only. (T or F)
27. The direction of a true meridian at a station is invariable. (T or F)
28. The lines of earth's magnetic field run from East to West. (T or F)
29. The axis of the bubble tube must be perpendicular to the vertical. (T or F)
30. The line of collimation is parallel to the horizontal axis. (T or F)
31. A dull image is obtained by a telescope, if the magnification is low. (T or F)
32. For accuracy the length of Back sight and Fore sight are kept unequal. (T or F)
33. The reduced level of plane of collimation is equal to Height of the Instrument. (T or F)
34. The levelling should start at a temporary B.M. and end at a permanent Bench Mark. (T or F)
35. In plane table survey, field work is recorded in a field book to be plotted later. (T or F)
36. The points on a contour have same elevation. (T or F)
37. In tacheometric method of setting a curve no chain or tape is used. (T or F)
38. The magnitude of superelevation depends upon the speed on curve and radius of curve. (T or F)
39. The area of any irregular figure can be measured by planimeter. (T or F)
40. The diurnal variation of magnetic needle is more near the equator. (T or F)
41. In high latitudes, the diurnal variation of magnetic needle is greater. (T or F)
42. Tri-angulation station selected close to the main station for avoiding intervening obstruction is called Eccentric station. (T or F)
43. Error due to carelessness of an observer is called compensating error. (T or F)
44. When a chain used is too long then the measured distance is more than its true distance. (T or F)
45. The two point and three point problem are methods of resection and orientation. (T or F)

46. Proof lines in chain survey are required to indicate accuracy of survey work. (T or F)

47. If δ is declination and λ latitude of any place, the azimuth will be $\frac{\sin \delta}{\sin \lambda}$. (T or F)

48. Detailed plotting is generally done by radiation. (T or F)

49. The size of a plane table is 300 mm × 450 mm. (T or F)

50. The angle between the plane of equator and plane of the ecliptic is 23° 27′. (T or F)

51. The plane of ecliptic is imaginary plane containing Earth's orbit around the Sun. (T or F)

52. The Principal line is the line joining the Principal point and Isocentre. (T or F)

53. A series of closely spaced contour lines represent plane surface. (T or F)

54. Benchmark is established by Trigonometrical levelling. (T or F)

55. Latitudes south of equator are taken as negative. (T or F)

56. Longitudes east of Greenwich are taken as negative. (T or F)

57. Level line is curved while horizontal line is straight line. (T or F)

58. Contour interval is inversely proportional to the scale of the map. (T or F)

59. The rise and fall method is less accurate than Height of Instrument method. (T or F)

60. The smaller horizontal angle between the true meridian and a survey line is called, Azimuth. (T or F)

61. The latitude (λ) of place and the altitude (α) of the pole are related by, $\lambda = \alpha$. (T or F)

62. Homologous points are corresponding points on the ground and photograph. (T or F)

63. Limiting gradient for locating the base line on evenly – sloping ground is 1 in 6. (T or F)

64. Correction for parallax is always positive. (T or F)

65. The great circle passing through the zenith, nadir and poles is known as vertical circle. (T or F)

66. The angle of intersection of the two plane mirrors of an optical square is 90°. (T or F)

67. Number of links in a 30 m metric chain is 100. (T or F)

68. The horizontal angle between the true meridian at place is called declination. (T or F)

69. Theodolite besides horizontal angle can also measure vertical angles. (T or F)

70. A level line is a horizontal line. (T or F)

71. With the rise of temperature the sensitivity of a bubble tube decreases. (T or F)

72. Contour interval is inversely proportional to the scale of the map. (T or F)

73. In a vertical cliff contour lines coincide. (T or F)

74. A compound curve is combination of two curves bending in reverse direction. (T or F)

75. The R.F. of the scale is the ratio of the distance on the map to the distance on the ground. (T or F)

76. 10 divisions of vernier scale are equal to 11 divisions of main scale of each 1 mm. The least count is 0.01. (T or F)

77. In geodetic survey, the surface of the earth is considered inclined. (T or F)

78. One cm on a map represents 10 m on the ground, the RF = $\frac{1}{100}$. (T or F)

79. The bearing of a line in the direction of the progress of survey is called Back Bearing. (T or F)
80. The bearing observed with a prismatic compass is whole circle Bearing. (T or F)
81. At equator, the value of dip is zero. (T or F)
82. An angle between the inclined line of sight and the horizontal is called deflection angle. (T or F)
83. The contour interval throughout the survey should be constant. (T or F)
84. The most likely error in plane table survey is due to error in sighting. (T or F)
85. Rivers are obstacles in chaining survey. (T or F)
86. For any line AB the Fore Bearing and Back Bearings of AB differ by 180°. (T or F)
87. On a turning point only backsight is taken. (T or F)
88. The sensitivity of a bubble tube can be increased by reducing tube's diameter. (T or F)
89. If the horizontal distance between staff point and the point of observation is x, then the error due to curvature of earth is proportional to d^2. (T or F)
90. Normal size of the plane table is 60 cm × 75 cm. (T or F)
91. The intercept on a vertical staff is observed as 0.75 m, from a tacheometer, the horizontal distance between tacheometer station and staff station is 75 m. (T or F)
92. In a reverse curve, the super-elevation provided at the point of reverse curvature is zero. (T or F)
93. The angle subtended by the long chord of a simple circular curve at the centre is equal to (180° – Angle of deflection). (T or F)
94. If the true length of a line is 200 m. When measured with a 20 m tape, the length is 200.80. The correct length of the tape is 20.08 m. (T or F)
95. The best method to estimate area of an irregular curved boundary is Simpson's method. (T or F)
96. The scale for an area plan is 1 cm = 10 m. The plan has shrunk such that a line now measure 14.5 cm of originally 15 cm. The shrunk scale gives 1 cm equal to 10.34 m. (T or F)
97. The tangent distance for a circular curve of length 150 m joining two straights meeting at an external deflection angle of 30° is 76.76 m. (T or F)
98. The minimum number of satellites needed for a GPS to determine its position precisely is 3. (T or F)
99. A rectangular area of 12 km² is represented by 1.5 cm × 0.5 cm rectangle on the map, the scale of the map is 1 : 400000. (T or F)
100. The bearing observed with a prismatic compass is Quadrantal bearing. (T or F)
101. Compensating errors in chaining are proportional to the square root of length of the line. (T or F)

FILL IN THE BLANKS

1. The required slope correction for a length of 60 m along a gradient of 1 in 20 is ....
2. The length of a base line measured on ground at elevation of 300 m above MSL is 2250 m. The required correction to reduce to sea level length is, (Take radius of earth = 6370 kg) ....

3. The lengths and bearing of traverse PQR are

Segment	Length (m)	Bearing
PQ	40	80°
QR	50	10°
RS	30	210°

The length of line segment SP is ....

4. A line of levels has been run from a Bench Mark (23.47 m) and ends at another BM (23.50 m). The sum of the back sights is (16.26 m) and of foresights (16.29 m). The closing error is ....

5. In a levelling work sum of Back Sights (BS) and Fore Sights (FS) are 3.085 m and 5.645 m respectively. If the RL of starting station is 100 m. The R.L. of the Last Station is ....

6. A series of perpendicular offsets taken from a curved boundary wall to a straight survey line at 6 m interval are 1.22, 1.67, 2.04, 2.34, 2.14, 1.87 and 1.50 m. The area in m^2 bounded by the survey line, curved boundary wall, the first and the last offsets by Simpson's rule is ....

7. The length of a line measured with a 20 m chain was found to be 260 m. The true length of the line is, (if chain was 5 cm too long). ....

8. A field was surveyed by a chain and the area was found to be 51.53 Hectares. If the chain used was 0.8 percent too long. Correct area is ....

9. The distance between two station on a plan drawn at a scale 10 m to 1 cm was found to be 1286 m. Later it was found that the scale used was 20 m to 1 cm. The true distance is ....

10. An offset is laid out at 1° 30′ from its true direction on the field. The degree of accuracy with which the offset should be measured so that maximum displacement of the point on paper from both the sources is equal, is ....

11. If the shrinkage factor is $\frac{15}{16}$ and if the original scale is $\frac{1}{1500}$, the R.F. of shrunk scale will be ....

12. The quadrental bearing of lines AB and CA are 530° E and 570° E. The included angle is ....

13. Shrinkage Ratio is between shrink length and .... length.

14. The true length of a line is known to be 500 m. The line was again measured with a 20 m tape and found to be 502 m. What is the correct length of the 20 m tape .... ?

15. Clinometer is used to measure ....

16. The sum of the interior angles of a closed traverse of (n) sides is ....

17. Permanent point of reference whose elevation is know is called ....

18. Full form of GTS is ....

19. The imaginary line passing through the intersection of cross hairs and the optical centre of object glass is called ....

20. Full form of R.L. ....

21. Magnetic declination is the horizontal angle between the true meridian and .... meridian.

22. A line connecting the North and South poles is called ....
23. The first principle in surveying is from whole to ....
24. The first sight taken after setting up the level instrument is called ....
25. Full form of GPS? ....
26. Minimum number of satellites needs for a GPS to precisely determine the position is ....
27. Lateral measurements taken from the chain line are called ....
28. The angle or inclination of a survey line with respect to North-South direction is called ....
29. True meridian is same as .... meridian.
30. If magnetic bearing of a line is 134° 45′ and magnetic declination is 10° 15′, the True bearing is ....
31. The angle of dip at equator is ....
32. The angle of dip at Poles is ....
33. 150° 20′ WCB, in RB will be ....
34. 332° 40′ WCB, in RB will be ....
35. The bearing of a line is N 50° 25′ E what will be its WCB ....
36. The Fore Bearing of a line AB is N 28° W, the Back Bearing will be ....
37. The imaginary line passing through the optical centre of object and the point of intersection of cross-hairs is called the line of ....
38. The imaginary straight line passing through the optical centre of object, traversing the eye piece and entering the eye is known as line of ....
39. Last reading before shifting the instrument is called .... sight.
40. For finding the height of instrument at a new station the first reading is called .... sight.
41. Trough compass in plane table survey is used to orient the table in the direction ....
42. What is full form of GTS Bench Mark? ....
43. Imaginary line joining the points of equal elevations (RLs) is called ....
44. The theodolite in which telescope can be rotated horizontally through 180° is called .... theodolite.
45. The process of locating intermediate points in a straight line between two end points in a straight line is called ....
46. Gunter's Chain is 66 feet long, what is the length of Engineers' Chain ....
47. The Mean Sea Level (MSL) as adopted by the GTS is at .... Port.
48. In rough and difficult terrain where direct levelling and chaining is difficult which type of survey is taken ....
49. The accuracy of Tacheometry is .... %.
50. Moveable hair method in tacheometry is also known as ....
51. The number of horizontal cross-wires in a stadia diaphragm is ....
52. The point on the celestial sphere immediately above the Observer's Station is called ....

53. The point on the celestial sphere vertically below the Observer's Station is called ....
54. For ensuring verticality of staff we use ....
55. In chain survey for setting out right angles what instrument is used? ....
56. When two end stations are not inter-visible the ranging technique used is called? ....
57. Bearing is measured by a ....
58. Levelling rods are normally .... (m) long.
59. The rod reading taken on a point of known elevation is called ....
60. The Ratio of the map distance to the corresponding distance on ground is called ....
61. The preliminary inspection of area to be surveyed is called ....
62. Angle (or inclination of a survey line) with respect to North-South direction is called ....
63. Increasing value of circular contours from outside towards inside (centre) represents a ....
64. A true error is the difference between the true value of quantity minus its .... value.
65. In first order triangulation average triangle closure is less than .... second.
66. The measurement of depth below the water surface is called ....
67. Detailed plotting is generally done by ....
68. Marking of chaining pins are called ....
69. Full form of R.F. is Representative ....
70. For determining the feasibility and rough cost is called .... survey.
71. Setting the theodolite exactly over a station mark is called ....
72. Lines joining the points having same elevation above the datum are called ....
73. A series of closed contours on the map having higher values outside indicates a ....
74. Surveying in which the curvature of the earth is considered is called .... survey.
75. A diagonal scale is used to read .... dimensions.
76. A 50 m tape is held 2 m out of line. The true length will be ....
77. The cumulative errors that occur in chaining are proportional to .... of the line.
78. In an optical square, the angle between the first (incident ray) and the last (reflected) ray is ....
79. In a whole circle Bearing System S 25° 15′ E corresponds to ....
80. At the magnetic poles, the value of 'dip' is ....
81. The ratio of the focal length of the objective to that of an eye-piece of a telescope is called its ....
82. A fixed point of reference of known elevation is called ....
83. The height of instrument is equal to Bench Mark plus ....
84. The back sight of a station A is 0.855 and height of instrument is 550.265, the R.L. at A is ....
85. If R.F. is $\frac{1}{2500}$ then scale is ....

86. When the WCB of lines AB and AC are 115° and 41° respectively, then angle BAC ....

87. The capacity of a telescope of producing a sharp image is called ....

88. For contouring of hill areas best method is ....

89. For a chord of 60m, the mid ordinate for a circular curve of 50 m radius will be ....

90. For a line the Torebearing of AB and Backbearing of AB differ by ....

91. Latitude of a place is the angular distance from centre of ....

92. The maximum tolerance in a 20 m chain is .... mm.

93. Different grades are joined together by a .... curve.

94. Mistakes by the observes are also called .... Errors.

95. The shape of vertical curve generally provided is .... in shape.

96. An aircraft is flying at a ground speed of 200 km/hr. Focal length of camera is 200 mm. The ground distance covered is 2 km between exposure. The time interval between exposure is ....

97. If the standard meridian is 82° 30′ E and the standard time of longitude 90° E is known to be 8 H 30 m, the local mean time will be ....

98. Two points P and Q located on a map have the following co-ordinates

	P	Q
Latitude (L)	+ 40 m	+ 20 m
Departure (D)	– 20 m	+ 30 m

The Length PQ will be ....

99. The tangent distance for a circular curve of length 150 m joining two straights meeting at an external deflection angle of 30° is ....

100. In a reverse curve, the superelevation provided at the point of reverse curvature is ....

101. In spherical triangle any angle is less than ....

ANSWERS

Multiple Choice Questions

1. a	2. d	3. c	4. c	5. d	6. b	7. b	8. d	9. a
10. b	11. b	12. b	13. c	14. a	15. d	16. b	17. b	18. c
19. d	20. c	21. a	22. d	23. b	24. a	25. c	26. b	27. d
28. c	29. a	30. b	31. d	32. c	33. c	34. b	35. a	36. b
37. a	38. c	39. d	40. a	41. c	42. b	43. d	44. b	45. c
46. b	47. a	48. d	49. c	50. d	51. c	52. d	53. c	54. a
55. a	56. a	57. c	58. a	59. b	60. d	61. b	62. c	63. a
64. a	65. d	66. b	67. d	68. a	69. d	70. a	71. b	72. d
73. b	74. b	75. a	76. d	77. d	78. d	79. a	80. b	81. c
82. b	83. a	84. c	85. c	86. d	87. c	88. b	89. c	90. b
91. d	92. c	93. a	94. d	95. a	96. a	97. b	98. d	99. b
100. c	101. d							

True & False

1. F 66′	2. T	3. T	4. F 220°	5. T
6. T	7. T	8. F Blueness	9. F Negative	10. T
11. T	12. F	13. T	14. F Base Plate	
15. F Plumbing Fork	16. T	17. F West	18. F Near Eye Piece	19. T
20. F	21. T	22. T	23. F Less	24. T
25. F Central Projection		26. T	27. T	
28. F North-South	29. T	30. F Right angle	31. F High	32. T
33. T	34. F	35. F	36. T	37. T
38. T	39. T	40. F	41. T	42. T
43. F Mistake	44. F Less	45. T	46. T	
47. F $\frac{\cos \delta}{\cos \lambda}$	48. T	49. F 600 mm × 750 mm		50. T
51. T	52. T	53. F Steep	54. F Spirit Levelling	55. T
56. F Positive	57. T	58. T	59. F	60. T
61. T	62. T	63. F 1 in 12	64. F	
65. F Meridian	66. F 45°	67. F 150	68. T	69. T
70. F Normal to Plumb line		71. T	72. T	73. T
74. F Deviation curve		75. T	76. T	
77. F Curved	78. F $\frac{1}{1000}$	79. T	80. T	81. T

82. F Vertical	**83.** T	**84.** F Orientation	**85.** T	**86.** T
87. F Taken both BS FS		**88.** F Increase	**89.** T	**90.** T
91. T	**92.** T	**93.** F Equal to	**94.** F 19.92 m	**95.** T
96. T	**97.** T	**98.** F (4)	**99.** T	
100. F WCB	**101.** T			

Fill in the Blanks

1. 7.5 cm	**2.** –0.106 m	**3.** 44.79 m	**4.** 0.06 m
5. 97.44 m	**6.** 68.50 m	**7.** 260.65 m	**8.** 52.34 Ha
9. 643 m	**10.** 1 : 39	**11.** $\frac{1}{1600}$	**12.** 220°
13. Actual	**14.** 19.92 m	**15.** Angles	**16.** (2n – 4) 90°
17. Bench Mark	**18.** Great Trignometrical Survey		**19.** Line of Collimation
20. Reduced Level	**21.** Magnetic	**22.** Meridian	**23.** Part
24. Back sight	**25.** Global positioning system		**26.** 4 (Four)
27. Offsets	**28.** Bearing	**29.** Geographical	**30.** 145°
31. 0° (Zero)	**32.** 90°	**33.** S 28° 40′ E	**34.** N 27° 20′ W
35. I Quadrant 50° 25′		**36.** S 28° E	**37.** Collimation
38. Sight	**39.** Fore	**40.** Back	**41.** Magnetic North
42. Geodetic Triangulation Survey		**43.** Contour	**44.** Transit
45. Ranging	**46.** 100 feet	**47.** Bombay	**48.** Tacheometry
49. ± 0.5	**50.** Subtense	**51.** Three (3)	**52.** Zenith
53. Nadir	**54.** Plumb Bob	**55.** Optical Square	**56.** Reciprocal Ranging
57. Prismatic Compass		**58.** 3 m to 4 m	**59.** Back sight
60. RF Representative fraction		**61.** Reconnaissance	**62.** Bearing
63. Hill	**64.** Observed	**65.** 1″ (One)	**66.** Sounding
67. Radiation	**68.** Arrows	**69.** Fraction	**70.** Reconnaissance
71. Centring	**72.** Contours	**73.** Valley	**74.** Geodetic
75. Three (3)	**76.** 49.96 m	**77.** Length	**78.** 90°
79. 154° 45′	**80.** 90°	**81.** Magnification	**82.** Bench Mark
83. Back Sight	**84.** 549.41	**85.** 1 cm = 25 m	**86.** 74°
87. Definition	**88.** Tacheometry	**89.** 10 m	**90.** 180°
91. Earth	**92.** ± 0.5 mm	**93.** Vertical	**94.** Gross
95. Parabolic	**96.** 36 sec	**97.** 9 H	**98.** 53.85 m
99. 76.85	**100.** Zero	**101.** 180°	

SECTION 22

Transportation Engineering

DEFINITIONS

Aggregate: It is the broad category of coarse-to-medium grained particulate material used in construction, including sand, gravel, crushed stone, slag etc. They are composite material such as concrete and asphalt. Recycled material like blast furnace and steel furnace slag can be used as binding material instead of portland cement.

Axle Load: The weight exerted by a single axle of a vehicle on the pavement. Heavy axle loads (e.g., from trucks) can cause significant wear and tear, leading to pavement's damage such as cracking and rutting. Pavement design must account for cumulative effect of axle loads over the pavement's lifespan.

Berm in Highways: It is a small bank or raised strip of land that runs along the side of a highway, and serves several purposes– as noise barrier, helps in storm water management, and plantation. These can be made of variety of materials– gravel, crushed rock or soil.

Camber: It is a transverse slope on the road surface, meaning it slopes from the center crown towards the edges of the road. The intentional tilt acts like a drainage system, channeling rainwater away from the road surface and towards the shoulders. The maximum camber recommended by IRC is 4% for earthen roads, for concrete roads it should be between 1.7% and 2.0%.

Capacity: It is the maximum number of vehicles or passengers that a transportation facility can accommodate during specific time under prevailing conditions. It depends upon number and width of lanes, traffic control devices and roadway design.

Culvert: It is a structure that allows water to flow beneath the road, railway, or an embankment. It helps prevent flooding and erosion by managing stormwater and maintaining natural water flow. Various types are in use like– Pipe culvert, Box culvert and Arch culvers.

Dorry Abrasion Test: It is a laboratory test that measures the resistance of aggregates to abrasion and wear. It is used to assess the ability of aggregates to withstand the grinding and rubbing action that occurs during transportation or under traffic loads. The test is performed using Dorry abrasion machine.

Drainage System: A drainage system in place ensures removal of surface and sub-surface water from the road. Effective drainage system prevents water pooling, which can weaken the pavement and may lead to accidents. The components of drainage system include-Side-drains (parallel to road), cross-drain (beneath the road) and catch basin (collects water and discharge to appropriate place).

Embankment: It is a raised structure built to carry a roadway over low-lying areas. It provides the necessary elevation to maintain roadway alignment and ensure that roads remain above water level in flood-prone areas. Proper compaction and drainage design are critical for embankment stability.

Enoscope: An enoscope is a device used to measure the spot speed of vehicles passing through a particular point on the road. It can find- Average speed, Spot speed, Time mean speed and Root mean square speed. It is a portable device and can be installed easily.

Geometric Design: It deals with the physical dimensions and layout of transportation facility. The key elements include. **Land width** (typically 3.3 m to 3.7 m) for highways; shoulder width; curve radius and sight distance.

Grand Trunk Road (GT Road): It is a major road in Asia that connects the Indian subcontinent to Central Asia. It passes through Bangladesh, India, Pakistan and Afghanistan. The G.T. Road is about 3,650 km long. Gt Road starts from Teknaf (Near Dhaka in Bangladesh) and ends at Kabul (Afghanistan). It passes through Kolkata, Kanpur, Agra, Aligarh, Delhi, Amritsar in India and Lahore, Rawalpindi and Peshawar in Pakistan. It was built by Sher Shah Suri in 333 BC.

Hairpin Bend: It is a sharp turn in a road that makes a vehicle almost turn 180° to continue on the road. It is typically used when cycling up or down a steep hill. The bend is so sharp that is almost turns back in the opposite direction. It is also known as U-turn as a traffic control device.

Interchange: It is grade-separated junction designed to allow traffic to flow between intersecting roads without direct conflict. These are three types– Cloverleaf (for high-traffic); Diamond (Between major and minor roads) and Trumpet (T-junction or terminals). Interchanges improve safety and capacity by cutting vehicle stops and delays.

Intersection: It is defined as a **location** where two or more roads cross or join. Intersections can be controlled by signals, signs or roundabouts to manage traffic movement.

Lateral Friction in Highways: It refers to the friction force that acts sideways between a vehicle's tyres and the road surface, resisting the **Centrifugal force** acting on the vehicle when it is turning on a curve, essentially keeping it from sliding sideways; it is a crucial factor in highway design, with a typical design value for the coefficient of lateral friction being around 0.15. The lateral friction depends upon Road surface texture, Tire tread pattern and weather conditions (dry, wet, snow).

Level of Service (LOS): It is a qualitative measure used to evaluate traffic flow and operational conditions of a transportation facility. LOS ranges from A (Free Flow Condition) to F (Highly Congested Conditions). It also determines the travel time, traffic interruptions, comfort and convenience.

Longitudinal Slope of Road: It is the gradual rise or fall in the direction of vehicle movement. It is also known as the road gradient, which is the rate of change of the road's level. It helps in safety and comfort, better drainage (requires 0.3% slope). A minimum of 0.5% slope is provided in engineering design.

Median: It is the central part of a divided roadway, separating opposite traffic flows. It enhances safety by reducing head on collisions. Medians can be raised, depressed or flush (marked only by paints or barriers). Wide medians may also accommodate future expansion or turning lanes. They are also used for flower plantation.

Noise Barriers: For mitigation of traffic noise, noise-barriers are erected alongside roads in residential and sensitive areas (hospitals, schools). These structures can be built of concrete, wood or vegetation (green walls). These are essential for ensuring quality of life for nearby residents of noise source.

Parking Facility: It is designed for vehicle storage, on street parking, large parking lots and multi-level parking facility. Key aspects are– Capacity (availability of parking space); Layout (maximise space and easy maneuvering) and Types (Parallel, Angled or perpendicular parking). Proper parking facility helps in reducing urban congestion and improves accessibility.

Pavement: It is durable surface material laid down on an area intended to sustain vehicular or pedestrian traffic. These include flexible pavements (like of asphalt) and rigid pavements (of concrete).

Pavement Markings: These are painted lines and symbols on the road surface to guide drivers and pedestrians. The common pavement markings are– Centerlines for separating opposite traffic; Edge Lines indicate road boundary; Cross Walks are for pedestrian crossings. Reflective paints and thermoplastic materials are used to improve night time visibility.

Pedestrian Crossings: Pedestrian crossings are designated areas where pedestrians can safely cross roads. These include– At-Grade crossings or zebra crossings; Grade– separated crossings include overpasses or underpasses. These are essential where traffic is fast and has high foot-traffic.

Right-of-way (ROW): The area of land acquired for road construction and for its future development along its alignment is known as Right of Way. It is also known as **land width.** It is the legal land area for transportation facilities like– roads, railways or pipeline. It includes space for pavement, shoulders, sidewalks and utilities.

Roadway Lighting: Roadway lighting is essential for enhancing visibility for drivers, cyclists, pedestrians during low-light conditions. Roadway lighting helps in reducing accidents and improves security. These are positioned along the road. For large intersections high mast lighting is preferred. Reflective markings are used in order to increase the visibility of road features.

Rolling Terrain: It is a type of land where the natural slopes consistently rise above and fall below the road grade and where occasional steep slopes offer some restriction to normal horizontal and vertical roadway alignment. It can be gentle slopes or more rugged, and is often covered with vegetation of different kinds and size. As per cross slope terrains are classified as–

Terrain classification:	Plain	Rolling	Mountainous	Steep
% slope :	0 to 10	10 to 25	25 to 60	> 60

Shoulder: The strip of land adjacent to the travel lanes (Pavement) designed for emergency stops, breakdowns and providing lateral support to pavement. A set of adjoining lanes and shoulders is called roadway or carriageway.

Sight Distance: It is the distance over which a driver can see and respond to an object on the roadway. The two types sight distances include– **Stoping Sight Distance** (SSD), which is minimum distance required for a vehicle to stop safely, and **Passing Sight Distance** (PSD), it is the distance required to overtake a slower vehicle safely. Adequate sight distance is critical for safety and comfort.

Subgrade: The soil layer beneath the pavement system that provides support to pavement. It's strength and stability are critical to the performance of the pavement. Subsgrade is also called formation level. Soil, moorum, gravel and mixture of these and free from logs, stumps or roots is normally used subgrades.

Summit Curve: On a highway it is a vertical curve with an upward convexity, essentially the highest point of a hill on the road where the gradient changes from uphill to downhill, and is designed to provide adequate sight distance for drivers as they crest the hill. It is formed when two ascending gradients meet creating a smooth transition between them. For the summit curve circular curve is preferred.

Superelevation: It is defined as the raised outer edge which is provided in the road to counteract the centrifugal force for smooth riding on a curved path. As per Indian practice, superelevation (e) can be defined by the equation, $e = \frac{V^2}{127\,R}$, where V = velocity of vehicle in KMPH, R = Radius of curve in m.

Tack Coat: Also known as **bond coat** is a thin layer of liquid asphalt or asphalt emulsion that is applied to a surface to create a strong **bond** between layers. This bond is critical for the large-term performance of the road. Besides creating strong bond, track coat also seal porous layers and patches. It prevents water to enter substrate.

Terrain: Alternately **relief** or **topographical relief** refers to an area of land when you are considering its physical features. It is the dimension and shape of a given surface of land. This is usually expressed in terms of the elevation, slope and orientation of terrain features. Terrain affects surface water flow and distribution. Over large areas, it can affect weather and climate patterns.

Traffic Control Devices: These devices include road signs, traffic signals, and pavement markings that guide and regulate traffic flow. Examples are– Regulatory signs which indicate laws (like stop, yield); Warning signs (like school, sharp curve ahead) and Guide signs (like directions, distance). Use of these devices enhance safety and efficiency.

Traffic Signal Timing: It refers to the duration of green, yellow and red lights at intersections. Proper timing optimizes traffic flow, reduces congestion, and increases safety. Techniques involved are-Fixed-time control, Actuated control (time adjustment) and Adaptive control (Real-time adjustment based on traffic condition).

Traffic Volume Survey: It is defined as the procedure to determine the number of vehicles moving on the roads at a particular section during a particular time period. The time period can be taken in **minutes, hours** or **days** etc. Or, it can be defined as number of vehicles passing a specific point on roadway during a specific period, usually expressed as vehicles per hour (VPH) or Average Daily Traffic (ADT).

Transit Oriented Development (TOD): It is the development which means integrated urban places designed to bring people, activities, buildings, and public places together, with easy walking and cycling connection between them and near excellent transit service to the rest of the city.

Valley Curve: Also known as **sag curve** is the vertical curve provided when a descending gradient meets an ascending gradient or, a flat gradient to a descending gradient. It is designed to ensure a smooth and gradual change in slope. A cubic parabola is often used for valley curves, but other transition curves like spiral or lemniscates, can also be used for small angles of deviation.

Water Bound Macadam (WBM) Roads: are a type of road construction technique that uses crushed stone aggregates, water and other materials to create a durable road surface. Multiple layers of crushed stone aggregates are compacted and bound together with water and other materials (stone dust and soil). The layers are 7.5 cm to 10 cm thick. These are economical and reasonably strong for light traffic. The recommended camber in WBM road is 1 in 30 to 1 in 48.

Weaving Section: A weaving section is a part of a roadway where vehicles cross paths to merge or diverge, such as the area between an on-ramp and off-ramp. It requires careful design to ensure smooth traffic flow and minimize conflicts. Key parameters include weaving length and width.

MULTIPLE CHOICE QUESTIONS

1. The main mode of transport in India is
   (a) Railway (b) Air lines (c) Roadways (d) Shipping
2. The specifications for highway planning are supplied by
   (a) PWD (b) IRC (c) MVA (d) ORO

3. The part of road used by vehicular traffic is called
   (a) Carriage way (b) Shoulder (c) Express way (d) None
4. For a two lane road with raised Kerbs the width of carriageway according to IRC should be
   (a) 5.5 m (b) 6.5 m (c) 8.5 (d) 7.5 m
5. The roads connecting Capital Cities of States are called
   (a) National Highways (b) Expressways
   (c) State Highways (d) Major Highways
6. The approximate length of Grand Trunk Road is
   (a) 5200 km (b) 3650 km (c) 3300 km (d) 1200 km
7. The Headquarter of Indian Road Congress (IRS) is at?
   (a) New Delhi (b) Mumbai (c) Kolkata (d) Chennai
8. The minimum width of Village Road should be
   (a) 3.2 m (b) 2.25 m (c) 2.2 m (d) 2.45 m
9. Yellow and White colours on milestones indicate
   (a) National Highway (b) Rural Road
   (c) State Highway (d) Main District Road
10. As per IRC, the maximum length of truck plus trailer should be
    (a) 18 m (b) 12 m (c) 11 m (d) 8 m
11. The shape of camber for cement concrete pavements should be
    (a) Straight line (b) Elliptical (c) Parabolic (d) Combined
12. For Water Bound Macadam Roads in heavy rainfall areas the camber should be
    (a) 1 : 50 (b) 1 : 60 (c) 1 : 36 (d) 1 : 25
13. Camber in road is provided for
    (a) Counteracting Centrifugal Force (b) Effective drainage
    (c) Easy driving (d) Avoid accidents
14. The stopping sight distance depends upon
    (a) Driver's Reaction time (b) Speed of vehicle
    (c) Efficiency of brakes (d) All these
15. The ruling design speed on a National Highway (NH) in Km/Hour is
    (a) 120 (b) 80 (c) 100 (d) 60
16. IRC recommended maximum limit of superelevation for mixed traffic in a plain terrain as
    (a) 1 in 15 (b) 1 in 12.5 (c) 1 in 10 (d) Equal to camber
17. A terrain is classified as 'Rolling' if the cross slope of land is upto
    (a) Upto 10% (b) 10 to 25% (c) 25 to 60% (d) More than 60%
18. Reaction time of driver
    (a) Increases with speed (b) Decreases with speed
    (c) Does not depend on speed (d) None of these

19. Compared to a level surface on descending gradient the stopping sight distance is
(a) Less (b) Same
(c) Depends on speed (d) More

20. For calculating minimum ruling radius of curve in plain area, the design speed KMPH should be
(a) 20 (b) 12 (c) 8 (d) 16

21. The important factor considered in the design of summit curves on highways is
(a) Comfort to passengers (b) Sight distance
(c) Superelevation (d) Impact factor

22. The ideal form of curve for the summit curve is
(a) Spiral (b) Parabola (c) Circle (d) Lemniscate

23. On valley (sag) curves the available sight distance is determined on
(a) Design speed (b) Height of obstacle
(c) Height of driver's eye (d) Night-time driving condition

24. The value of camber recommended for cement concrete roads in areas of heavy rainfall is
(a) 1 in 50 (b) 1 in 40 (c) 1 in 33 (d) 1 in 25

25. The reaction time for calculating stopping distance may be assumed as
(a) 5 sec (b) 2.5 sec (c) 0.5 sec (d) 10 sec

26. The superelevation needed for a vehicle travelling at 60 KMPH on a curve of radius 128 m and on a surface with coefficient of friction of 0.15 is
(a) 0.07 (b) 0.015 (c) 0.022 (d) Zero

27. Width of carriageway for a single lane is recommended to be
(a) 7.5 m (b) 7.0 m (c) 3.75 m (d) 5.5 m

28. For a highway with design speed of 100 KMPH the safe overtaking sight distance is, (assume acceleration as $0.5 \frac{m}{s^2}$)
(a) 300 m (b) 750 m (c) 320 m (d) 470 m

29. The design value of lateral friction coefficient of highway is
(a) 1.5 (b) 0.5 (c) 0.35 (d) 0.15

30. Stopping Sight Distance (SSD) and frictional coefficient are
(a) Directly proportional to each other (b) Inversely proportional to each other
(c) Unrelated (d) Depends upon nature of pavement

31. The design speed for a National Highway is 100 Km/H, maximum permissible superelevation 0.10 and coefficient of lateral friction 0.15, the ruling minimum radius of horizontal curve on highway should be (m)
(a) 260 (b) 315 (c) 380 (d) 410

32. For a road (camber 3%), Design speed (80 KMPH), the minimum radius of the curve beyond which superelevation will not be needed
(a) 1680 m (b) 948 m (c) 406 m (d) 280 m

33. The coefficient of friction in the longitudinal direction of a Highway is 0.396. The braking distance for a car (speed 65 $\frac{Km}{H}$) is

(a) 87 m (b) 45 m (c) 42 m (d) 40 m

34. A vehicle negotiates a transition curve with uniform speed (V). If the radius of the horizontal curve and the allowable jerk are R and J, the minimum length of transition curve is

(a) $\frac{R^3}{VJ}$ (b) $\frac{J^3}{RV}$ (c) $\frac{V^2}{RJ}$ (d) $\frac{V^3}{RJ}$

35. If a 2-lane NH and 2-lane State Highway intersect at right angles, the number of potential conflict point at the intersection, assuming both roads are two ways is

(a) 11 (b) 17 (c) 24 (d) 32

36. The shape of the STOP Sign as per IRC : 67-2001 is

(a) Circular (b) Triangular (c) Octagonal (d) Rectangular

37. The minimum value of 15 minutes peak hour factor on a section of road is

(a) 0.1 (b) 0.20 (c) 0.25 (d) 0.33

38. On a highway section of speed $\left(v, \frac{km}{H}\right)$ density $\left(k, \frac{veh}{h}\right)$ relation is linear given by, $v = \left[80 - \frac{2}{3}k\right]$.

The capacity $\left(\frac{veh}{H}\right)$ of this section will be

(a) 1200 (b) 2400 (c) 4800 (d) 9600

39. A well-designed signalized intersection is one in which the

(a) Crossing conflicts increase

(b) Total delay minimized

(c) Cycle time is equal to the Sum of Red and Green times in all phases

(d) Cycle time equal to the Sum of Red and Yellow times in all phases

40. Stopping sight distance is always

(a) Less than overtakingsight distance (b) Equal to overtaking sight distance

(c) More than overtaking sight distance (d) None of these

41. With increase in speed of traffic stream, the minimum spacing of vehicles

(a) Increases (b) Decreases

(c) First decreases than increases (d) First increases than decreases

42. With increase in speed of traffic stream, maximum capacity of the lane

(a) Increases

(b) Decreases

(c) First increases, then decreases after reaching a maximum value at optimum speed

(d) First decreases and then increases after reaching a minimum value at optimum speed

43. If stopping distance and average length of a vehicle are 18 m and 6 m respectively, then the theoretical maximum capacity of traffic lane at a speed of 10 m/s as vehicles per hour is

(a) 1500 (b) 2000 (c) 2500 (d) 3000

44. 'Dead Slow' is a
(a) Warning sign (b) Regulatory sign (c) Informatory sign (d) None of these

45. Which of the following tests measure the toughness of road aggregates?
(a) Crushing strength (b) Abrasion (c) Impact (d) Shape

46. Los Angeles testing machine is used for
(a) Abrasion test (b) Impact test (c) Attrition test (d) Crushing strength test

47. If in Dorry abrasion test loss of weight is 18 g, then coefficient of hardness is
(a) 14 (b) 13 (c) 20 (d) 18

48. Traffic volume is defined as
(a) $\frac{\text{Traffic Density}}{\text{Traffic Speed}}$ (b) Traffic Density × Speed
(c) $\frac{\text{Traffic Speed}}{\text{Traffic Density}}$ (d) None of these

49. Which of the following is indicated by a warning sign?
(a) No parking (b) End of speed limit
(c) Overtaking prohibited (d) Level crossing

50. Where pedestrians are to cross the pavement, the pavement markings are
(a) Stop lines (b) Turn marking (c) Cross-walk lines (d) Lane lines

51. The term used for important or major roads of a county is
(a) Country road (b) Urban road (c) Highway (d) Royal road

52. The Central Public Works Department (CPWD) was originally formed by
(a) Lord William Bentincis (b) Lord Dalhousie
(c) Lord Mayo (d) Lord Rippon

53. Which of the following can be considered as a taking-off point in road planning in India?
(a) Bombay Plan (b) Delhi Plan (c) Lucknow Plan (d) Nagpur Plan

54. Office of the National CPWD Academy is located in
(a) Ghaziabad (b) Faridabad (c) Gurugram (d) Noida

55. The water-bound macadam road the recommended camber is
(a) 1 in 24 to 1 in 30 (b) 1 in 30 to 1 in 48
(c) 1 in 60 to 1 in 80 (d) 1 in 80 to 1 in 120

56. According to Indian Road Congress (IRC) the maximum width of a road vehicle is
(a) 1.85 m (b) 2.25 m (c) 2.45 m (d) 3.2 m

57. The highest point on road surface is called
(a) Crown (b) Camber (c) Gradient (d) Berm

58. The camber for hill roads in case of bituminous surface should be
(a) 3% (b) 2.5% (c) 3% (d) 3.5%

59. The slope of line joining the crown and the edge of the road surface is known as
(a) Cross fall (b) Cross slope (c) Camber (d) Any of these

60. The camber of road should be equal to .... the longitudinal gradient
(a) One half (b) Two times (c) Three times (d) Two-thirds of

61. According to IRC, the recommended width of shoulder is
(a) 4 m (b) 3 m (c) 2.5 m (d) 2 m

62. The centrifugal force on a car moving on a horizontal circular curve is proportional to
(a) $\frac{Wv}{gR^2}$ (b) $\frac{Wv^2}{gR^2}$ (c) $\frac{Wv}{gR}$ (d) $\frac{Wv^2}{gR}$,
where W = wt of vehicle, v = speed of vehicle, R = Radius of curve

63. The main advantage of super-elevation is
(a) To reduce stress on foundation
(b) To achieve higher speed
(c) To increase stability of fast moving vehicle when negotiating horizontal curve
(d) All these

64. To prevent overturning of Bullock cart on curve, maximum value of super-elevation as per IRC is
(a) 1 in 5 (b) 1 in 10 (c) 1 in 15 (d) 1 in 20

65. As per IRC 86 : 2018, extra wadening depends on
(a) Horizontal curve radius (b) Superelevation
(c) Number of Lanes (d) Longitudinal gradient

66. The type of transition curve generally provided on hill roads is
(a) Circular (b) Cubic parabola (c) Lemniscates (d) Spiral

67. The interface treatment provided to plug in the voids of pours surfaces and to bond loose particles in bituminous pavement is called
(a) Tack coat (b) Seal coat (c) Prime coat (d) Surface dressing

68. A road sign NO PARKING is a
(a) Warning Sign (b) Cautionary Sign
(c) Mandatory Sign (d) Informatory Sign

69. Cautionary signs are shaped as
(a) Circular (b) Triangle (c) Square (d) Rectangular

70. For 2-lane National Highway and State Highway Roadway width as per IRC is
(a) 12 m (b) 10 m (c) 15 m (d) 7.5 m

71. The ductility value of bitumen for suitability in road construction should not be less than
(a) 60 cm (b) 50 cm (c) 40 cm (d) 30 cm

72. Among the four stages of highway planning which of the following is not included?
(a) Financing (b) Route selection
(c) Collection of Material (d) Alignment and design

73. Maximum number of cities and towns are connected by which type of highways?
(a) State Highways (b) Village roads
(c) Major District Roads (d) National Highways

74. Which of the following is the longest International Highway?

(a) Karakoram (b) Australia 1 (c) Pan-America (d) Trans-Canada

75. Which of the following is not considered when designing Highways?

(a) Settlement (b) Cross-section (c) Sight distance (d) Level of service

76. Which of the following types of roads are most preferred for Highways?

(a) Unpaved (b) Bituminous (c) Cement concrete (d) Gravel

77. The Highway Development Work in India is undertaken by

(a) NHDP (b) State Govt. (c) Govt. of India (d) NHAI

78. The vertical alignment of a Highway includes

(a) Highway lighting (b) Design of valley curves & gradient

(c) Sight distance and traffic intersection (d) Pavement widening

79. The SSD for a design speed of 60 $\frac{Km}{Hr}$ for a 2-way traffic on a single lane road for $f = 0.37$, $t = 2.5$ sec. will be

(a) 80 m (b) 160 m (c) 140 m (d) 40 m

80. The design thickness of the CC-slab of important Highway with heavy traffic is

(a) 125 mm (b) 175 mm (c) 300 mm (d) 250 mm

81. The width of a pavement of a 2-lane National Highway should be

(a) 8.8 m (b) 3.0 m (c) 3.75 m (d) 7 m

82. In the absence of super-elevation on road along curves pot holes are likely to occur at

(a) Centre (b) Outer edge (c) Inner edge (d) All over

83. The distance measured along the centreline of a road over which a driver can see the opposite object on the road surface is called

(a) Sight distance (b) Visibility (c) Clear distance (d) None of these

84. Which of the following statement is correct?

(a) Steeper gradients are, sometimes provided to avoid deep excavation

(b) Steeper gradients are, sometimes provided to avoid long detours

(c) Exceptional gradient becomes necessary to avoid deep cutting

(d) All these

85. A transition curve cannot be a

(a) True spiral (b) Cubic spiral (c) Compound curve (d) Cubic parabola

86. As per IRC, the rate of change of radial acceleration C in $\frac{m}{S^3}$ for Highway varies as per relation

(a) $C = \frac{65 + V}{75}$ (b) $C = \frac{75}{65 + V}$ (c) $\frac{85 + V}{95}$ (d) $\frac{95}{85 + V}$

87. The length of the tangent of a simple curve having angle of deflection θ and Radius of curvature R, is equal to

(a) $R \sin \frac{\theta}{2}$ (b) $R \cos \frac{\theta}{2}$ (c) $R \tan \frac{\theta}{2}$ (d) $R \cot \frac{\theta}{2}$

88. In plains, the minimum length of transition curve (L) is

(a) $\frac{V^2}{R}$ (b) $\frac{V^2}{1.5\,R}$ (c) $\frac{2.7\,V^2}{R}$ (d) $\frac{V^2}{24\,R}$

89. As per IRC the safe overtaking sight distance at the vehicle speed of 80 KMPH is

(a) 235 m (b) 340 m (c) 640 m (d) 470 m

90. In detailed survey of a hill road, the interval for plotting the contours at sharp curves is generally

(a) 1 m (b) 2 m (c) 4 m (d) 6 m

91. A curve at the ridge of a hill having a convex shape is called a

(a) Valley curve (b) Summit curve (c) R-entrant curve (d) Salient curve

92. The width of a class 9 Roads is

(a) 2.45 m (b) 3.65 m (c) 4.9 m (d) 6 m

93. The bottom most layer of pavement is known as

(a) Sub-base course (b) Sub-grade (c) Wearing course (d) Base course

94. The thickness of the parapet wall, on the valley side of the roadway, is usually kept as

(a) 60 cm (b) 40 cm (c) 80 cm (d) 20 cm

95. The joints parallel to the centre-line of the road, are called

(a) Longitudinal (b) Transverse (c) Expansion (d) Stiff joints

96. Select the strength parameter of concrete used in design of plain jointed cement pavement from the following

(a) Tensile (b) Compressive (c) Flexural (d) Shear

97. The group index of a soil sub-grade is 7. The sub-grade soil is rated as

(a) Poor (b) Very poor (c) Good (d) Fair

98. Name the traffic survey data which is plotted by means of Desire lines?

(a) Accident (b) Classified volume
(c) Origin & Destination (d) Speed and delay

99. California Bearing Ratio (CBR)

(a) Is a measure of soil strength
(b) Procedure for design of flexible pavement
(c) Is method of soil identification
(d) Is measure to indicate the relative strengths of paving materials

100. A sub-grade soil sample was tested using standard CBR apparatus and the observation are

Load (kg)	Penetration (mm)
60.5	2.5
80.5	5.0

Assuming that load-penetration curve is convex throughout, the CBR value (%) of the sample is

(a) 6.5 (b) 5.5 (c) 4.4 (d) 3.9

101. For narrow roads the most economical lighting layout is

(a) Single side lighting
(b) Staggered system
(c) Central lighting system
(d) None of these

102. Driving vehicles on wet surfaced roads, is dangerous because it may

(a) Skid
(b) Slip
(c) Overturn
(d) All these

TRUE OR FALSE QUESTIONS

1. The importance of road is comparable to the veins in human body. (T or F)
2. A road connecting two towns is called highway. (T or F)
3. The central portion of a road for high speed vehicles is known as Express Way or motor way. (T or F)
4. The foundation of a road is also called soling or base. (T or F)
5. The strength and durability of a road depends upon its sub-grade. (T or F)
6. The camber of 1 in 30 means that 30 m wide road, the crown of the road will be 1.0 m above the edge of the road. (T or F)
7. A camber consisting of two straight slopes joining at the centre is called composite camber. (T or F)
8. Steeper camber may cause deterioration of the central portion of the road surface. (T or F)
9. The shape of the camber provided for cement concrete road is parabolic. (T or F)
10. In scanty rainfall areas the camber provided should be steeper. (T or F)
11. The overage Spacing between vehicles in a traffic stream is 50 m, then the density in $\frac{\text{veh}}{\text{km}}$ of the stream is 20 $\frac{\text{veh}}{\text{km}}$. (T or F)
12. Coefficient of subgrade modulus is expressed as c . (T or F)
13. The maximum spacing of contraction joints in rigid pavements is 2.5 m. (T or F)
14. The maximum thickness of expansion joint in rigid pavement is 25 mm. (T or F)
15. In rigid pavements expansion joints allow free expansion. (T or F)
16. A vehicle while passing from straight to curved path is under the influence of the weight of the vehicle and centrifugal force. (T or F)
17. Super-elevation should not be more than the camber. (T or F)
18. The super-elevation is directly proportional to the velocity of vehicle. (T or F)
19. Andhra Pradesh does not come under the Golden Quadrilateral National Highway. (T or F)
20. For cement concrete roads minimum gradient of 1 in 330 can be provided. (T or F)

21. Electronic sensors to warm dangerous situations are being considered for safety improvement. (T or F)
22. Tie bars in cement concrete pavements are at Expansion joints. (T or F)
23. 'Dead Slow' is a warning sign. (T or F)
24. Length of a vehicle affects extra width of pavement and minimum turning radius. (T or F)
25. With increase in speed of the traffic stream the minimum spacing of vehicles decreases. (T or F)
26. Thickness of bituminous carpet varies from 20 to 25 mm. (T or F)
27. The most raised portion of the pavement is called kerb. (T or F)
28. Enoscope measures spot speed. (T or F)
29. Expansion joints in concrete roads are provided at interval of 18 to 21 m. (T or F)
30. Drain provided parallel to roadway to intercept and divert water from hill slopes is known as cross drain. (T or F)
31. The wall which is necessary on the hill side of roadway to avoid slipping is known as retaining wall. (T or F)
32. On a hill road with ruling gradient of 1 : 20, a horizontal curve of 80 m radius is encountered, the compensated gradient will be 1 in 25. (T or F)
33. Ratio of the width of the car parking area required at kerb for 30° parking relative to 60° parking is approximately 2.0. (T or F)
34. A vehicle was stopped in 2 seconds by fully jamming the brakes. The skid marks measured 9.8 metres. The average skid resistance coefficient is 0.5. (T or F)
35. The general requirement in constructing a RCC road is to place a single layer of reinforcement near the bottom of slab. (T or F)
36. When two roads with two-lane, two way traffic, cross at an uncontrolled inter-section, the total number of potential major conflict points would be 32. (T or F)
37. The semi-official body which controls standards of roads, bridges in India is the Indian Road Congress (IRC). (T or F)
38. The length of National Highways (NH) as per III 20 years Lucknow road plan is given by $\frac{\text{Area of the country}}{75}$ or $\frac{A}{75}$. (T or F)
39. The width of any vehicle according to IRC should not be more than 2.44 m. (T or F)
40. For an ideal transition curve Length (L) should be directly proportional to radius (r). (T or F)
41. Maximum number of vehicles can be parked with 45° angle parking. (T or F)
42. For compaction of clayey soil Sheep Foot Rollers are most suitable. (T or F)
43. A traffic signal indicating the speed limit should be circular in shape with white background and red border. (T or F)
44. The percentage of free carbon in Bitumen is more than in Tar. (T or F)
45. The ductility of bitumen used in road construction should not be less than 30 cm. (T or F)
46. Bitumen is derived from destructive distillation of petroleum. (T or F)

47. The minimum ratio of radii of two circular curves of compound curve is normally kept as 2.0. (T or F)
48. For Water Bound Macadam Roads, binding material is stone dust. (T or F)
49. The headlight of vehicles should be such that its lower beam illuminates objects at a distance upto 50 m (T or F)
50. For calculating the sight distance the driver's eye above road surface, is assumed to be 120 cm. (T or F)
51. Floating gradient is provided at summit curves. (T or F)
52. Nagpur Road Plan recommended the use of Star and Circular pattern. (T or F)
53. For calculating stopping distance reaction time assumed as 5 sec. (T or F)
54. If an ascending gradient of c meets another of $\frac{1}{30}$, then angle of deviation is $\frac{1}{75}$. (T or F)
55. Camber on highway pavements takes care of centrifugal force. (T or F)
56. The main objective of Indian Road Congress is to promote scientific construction of Roads. (T or F)
57. The Indian Road Congress was established in the year 1950. (T or F)
58. NH-1 and NH-2 together make Grand Trunk Road. (T or F)
59. Golden Quadrilateral programme is the Development of Roads programme. (T or F)
60. The first 20-year Road Development Programme is also known as the Bombay Road Plan. (T or F)
61. The targeted road density of the Lucknow Road Plan is 32 km per 100 km^2. (T or F)
62. The minimum width of the 2-lane carriage way shall be 7.5 m with raised kerbs. (T or F)
63. The distance of rainwater inlets on road-sides should be 80 m to 100 m. (T or F)
64. A road bridge is classified a 'major bridge' if the span is between 60 m to 120 m. (T or F)
65. Yellow and White colours on milestones indicate 'State Highway'. (T or F)
66. Green and White strips on milestones indicate a 'State Highway'. (T or F)
67. A 'deck bridge' has its floor supported at the top of the superstructure. (T or F)
68. The skid number for highways should not be less than 25. (T or F)
69. For fast moving vehicles parabolic camber is preffered. (T or F)
70. Compared to level surface on descending gradient the stopping sight distance is less. (T or F)
71. The ruling design speed on a National Highway in plain terrain as per IRC is 120 KMPH. (T or F)
72. The lost time due to starting delay on a traffic signal is noted to be 3 sec, the actual green time is 25 sec and yellow time is 3 sec, the effective green time is 25 sec. (T or F)
73. The minimum ratio of radii of two circular curves of a compound curve is kept at 2.0. (T or F)
74. The superelevation on roads is directly proportional to velocity of vehicles. (T or F)
75. The first operation during the detailed survey of a hill road is fixing of Bench Mark. (T or F)
76. For long term of Road development, Bombay Plan (1958) recommended 50 years. (T or F)

77. In Water Bound Macadam roads, binding material is stone dust. (T or F)

78. On hill roads maximum superelevation should no exceed 15% (T or F)

79. Normal Traffic Growth is defined as "Increase in traffic volume, due to increase in transport vehicles". (T or F)

80. In mountainous regions the cross slope is 25% TO 60%. (T or F)

81. The head light of vehicles should be such that its lower beam illuminates objects at 10 m. (T or F)

82. The length of the sides of warning sign boards of roads is 45 cm. (T or F)

83. The basic capacity of Traffic Lane is defined as "Maximum Number of Passenger cars that can pass a given point during one hour under ideal conditions." (T or F)

84. Reinforcement in cement concrete slab of road pavement is generally not required. (T or F)

85. If x% is the gradient of an alignment and y% is the gradient after proper super-elevation along a curved part of highway, the differential grade along the curve is (x – y)%. (T or F)

86. In an ideal transition curve, the radius of curvature is directly proportional to the radius of main curve. (T or F)

87. The minimum permissible speed on a high speed road is decided on the basis of 15% percentile cumulative frequency. (T or F)

88. If (D) is degree of a curve, the percentage reduction in gradient will be 0.04 D. (T or F)

89. Length of vehicle does not affect the width of shoulders. (T or F)

90. Exceptional gradient is more than ruling gradient. (T or F)

91. Motor vehicle act was enacted in the year 1947. (T or F)

92. Absolute minimum sight distance required for stopping a vehicle with speed, 80 KMPH is 120 m. (T or F)

93. Macadam road are suitable for traffic load less than 3000 Tonnes. (T or F)

94. For maximum strength and durability of road minimum percentage of cement should be 30%. (T or F)

95. Minimum thickness of a layer of fine sand required to cutoff capillary rise of water completely should be 52 cm. (T or F)

96. While designing a hair-pin bend on highways, the minimum designed speed is 20 $\frac{\text{km}}{\text{Hr}}$. (T or F)

97. Floating gradients are generally provided at valley curves. (T or F)

98. A circular curve may be a Vertical curve. (T or F)

99. Compound curve is not a transition curve. (T or F)

100. A lamniscate fulfils the requirements of a true transition curve. (T or F)

101. If brakes of vehicles are effective, the vehicle running at a speed of 30km/H comes to stop in 12 m. (T or F)

102. If the difference is elevation between edge and crown of a 9 m wide pavement is 15 cm, the camber is 1 in 30.

FILL IN THE BLANKS

1. The Star and Grid pattern of road network was adopted in .... road plan.
2. The setback distance from the Centre-Line of the inner lane is .... m.
3. In One-lane one way homogenous traffic stream the observed average headway is 3 second. The flow $\left(\frac{\text{Vehicles}}{\text{Hours}}\right)$ in this traffic stream is .... $\frac{\text{Veh}}{\text{Hour}}$.
4. As per IRC the lateral coefficient of friction is ....
5. According to IRC the shoulder width should be ....
6. In what type of roads 3-4% camber is adopted .... ?
7. The instrument used to determine specific gravity of bitumen is ....
8. The thickness of cement concrete pavement is designed for load and .... stresses.
9. Full form of CBR ....
10. The usual width of parapet walls along Highways in Hilly regions should be .... cm.
11. Raising outer edge of a road with respect to inner edge is known as ....
12. Width of the shoulders of carriage way is generally kept as ....
13. Enoscopes are used to determine .... speed.
14. Minimum thickness of the base of a flexible pavement is .... cm.
15. Newly constructed pavement with ordinary portland cement attains its working strength after .... days.
16. When an up gradient of Highway meets a downgrade, the vertical curve provided is known as .... curve.
17. The wall made for stability of back filling portion of a road on the down hill side is known as ....
18. The rate of change of grade permitted along a vertical curve is (r) and total change in grade is g%, the length of curve provided is inversely proportional to ....
19. In case of a multi-lane road overtaking is generally permitted from ....
20. In case of a two-lane road overtaking is allowed only from .... side.
21. The pavement width of a road depends upon number of ....
22. For a vehicle moving with a speed of 80 KMPH. The brake reaction time in ordinary cases is ....
23. Reconnaissance is best done with the help of .... photographic survey.
24. The usual width of side drains along highways in hilly region is .... cm.
25. The number of vehicles moving in a specific direction that pass a given point during a specific unit of time is known as Traffic ....

26. Bottom-most component of a flexible pavement is ....
27. Round abouts are **not** recommended if the number of vehicles passing exceed ....
28. For comfortable travel on highways, the centrifugal ratio should **not** exceed ....
29. Hill roads seldom need ....
30. For poorly graded sub-grade soil, thickness of sub-base should be .... cm.
31. The distance moved by a vehicle during perception and brake reaction time is known as .... distance.
32. The width of formation of a road means the width of .... of embankments.
33. If (P) is number of vehicles per day in a particular year, (r) is the increase in traffic, the design for traffic after (n) years the formula used will be ....
34. On a pavement with parabolic camber, the angle of inclination of the vehicle will be more at the ....
35. Minimum radius of a simple circular curve deflecting through 5°, is ....
36. When hairpin bends are introduced, the minimum intervening distance is kept as .... m.
37. In a circular curve of radius 1400 m and speed 80 KMPH, no superelevation is required if camber is .... %.
38. For design of cement concrete pavement for corner loading, IRC recommends application of .... Formula.
39. Depth of reinforcement below the surface of CC pavement is generally kept .... cm.
40. Cumulative deformation of surface of road per horizontal distance is known as .... Index of road.
41. Degree of a road curve is defined as angle subtended at the centre by an arc of .... m.
42. The tangent length of a simple circular curve of radius (R) deflecting through θ° is ....
43. The thickness of broken line markings on multilane road for lanes is generally kept .... cm.
44. For a design speed of 80 KMPH, perception reaction time 3 sec. and coefficient of friction 0.5, the Stopping Sight Distance (SSD) will be ....
45. In urban areas, a median is provided to separate the opposite streams of traffic. As per IRC : 86-1983, the desirable minimum width (m) of the median is .... m.
46. Width of carriage way for single-lane is recommended as .... m.
47. The land width required across the alignment of road is .... of way.
48. Stopping sight distance on NH at a descending gradient of 2% is ....
49. Roads provided to give access to properties along an important Highway with controlled access are .... roads.
50. Desire lines are drawn based on origin and .... studies.
51. The Passenger Car Unit (PCU) for a car on urban road is ....
52. The type of 'Level Crossing' sign is a .... sign.
53. The most efficient traffic signal system is Flexible .... system.

54. The traffic study required for deciding speed limit for traffic regulation and control is .... speed.
55. When the pressure carried by a California Bearing Ratio (CBR) specimen at 2.5 mm penetration is $\frac{3.5\ N}{mm^2}$, the CBR of the soil is .... %.
56. In design of Highway expansion joints should be provided at .... m.
57. In design of Highway contraction joints are provided at .... m.
58. In Highway pavements Emulsions are mainly used in .... and maintenance.
59. Bituminous material has the properties of cementing and ....
60. Aggregate used in road construction should have Flakiness Index not more than .... %.
61. The recommended grade of Tar for grouting purpose is ....
62. For measuring toughness of road aggregate the recommended test is .... test.
63. The binder used in flexible pavements is ....
64. The function of a road base is to transmit load of the traffic from the .... to the subgrade.
65. The thickness of the road base should not be more than .... cm.
66. A vehicle while passing from straight to a curved path is under the influence of two forces, the weight of the vehicle and .... Force.
67. The equilibrium superelevation is given by the equation ....
68. The maximum rate of superelevation (e) is given by equation .... (v is speed of vehicle in KMPH and R is radius in m)
69. IRC recommended ruling gradient in plains as ....
70. According to IRC the ruling gradient in hilly areas should be ....
71. The Stopping Sight Distance (SSD) depends upon Reaction time, Breaking time and ....
72. The reaction time of driver .... with increase in speed.
73. The stopping sight distance is always .... than overtaking sight distance.
74. The stopping sight distance is the .... of reaction distance and braking distance.
75. Width of the rotary round should be equal to width of widest road plus the width of one ....
76. If the cross-slope of a country is 25% to 60%, the terrain is classified as ....
77. The normal land width for a major district road in open area is .... m.
78. On earth roads the camber preferred is 1 in 20 to ....
79. The convexity provided to the carriageway between crown and edges of the pavement is called ....
80. The width of the right of way in urban area, is kept between 24 m to 60 m in case of National and .... highways.
81. The full width of land acquired before finalising a highway alignment is known as .... of way.

82. Retaining walls are generally constructed with dry rubble stones with 60 cm top width and .... front batter.

83. In case of CC-pavements visibility during night is ....

84. Concrete pavement is recommended if the daily traffic load per lane exceeds .... tonnes.

85. If elevations along the road increase, the slope of the road in longitudinal direction is known as ....

86. Three points A, B and C, 500 m apart on a straight road have 500 m, 505 m and 510 m as their R.L., the road has a .... gradient.

87. In a braking test, a vehicle moving at 36 KMPH was stopped at a distance of 8 m. The skid resistance (friction coefficient) is ....

88. If present Average Daily Traffic (ADT) is 1500 vehicles and annual increase is 10%, the average future flow after 5 years will be .... vehicles.

89. Initially the G.T. road was between Calcutta and ....

90. The normal lane width for National as well as State Highway, in open areas should be .... m.

91. Excessive camber may cause .... of the fast moving vehicles towards the edges.

92. When cross-slope of a country is upto 10% the terrain is classified as ....

93. When each particle of aggregate is thinly coated with cement paste, the heterogenous solid thus formed is ....

94. Gradient resistance of moving vehicles along down slopes is .... $\frac{\text{kg}}{\text{tonne}}$.

95. Maximum number of vehicles that can pass a given point on a lane during one hour without creating unreasonable delay is known as .... capacity of lane.

96. In case of a hairpin bend of a highway minimum road way width at apex for a single lane highway must be .... m.

97. The maximum comfortable retardation applied to moving vehicles is .... $\frac{\text{m}}{\text{s}^2}$.

98. According to IRC the width of carriageway for 3-lanes pavements should be ....

99. The minimum cross fall of shoulders is kept at .... %.

100. Set-back distance is the distance between building line and road land ....

101. Horizontal curves on highways are also provided to break the .... of driving.

102. In retuning and breast walls, weep holes are provided .... cm vertical height and .... cm centre to centre horizontally.

103. Widening of roads on curves in hilly region is done on .... side.

104. The top height of a route marker above crown level is .... m.

105. The distance travelled by revolving the wheels of a vehicle more than its circumferential movement is the case of ....

ANSWERS

Multiple Choice Questions

1. c	**2.** b	**3.** a	**4.** d	**5.** c	**6.** b	**7.** a	**8.** d	**9.** a
10. a	**11.** a	**12.** c	**13.** b	**14.** d	**15.** c	**16.** a	**17.** b	**18.** b
19. d	**20.** d	**21.** b	**22.** c	**23.** d	**24.** a	**25.** b	**26.** a	**27.** c
28. b	**29.** d	**30.** b	**31.** b	**32.** b	**33.** c	**34.** d	**35.** c	**36.** c
37. c	**38.** b	**39.** b	**40.** a	**41.** a	**42.** c	**43.** a	**44.** b	**45.** c
46. a	**47.** a	**48.** b	**49.** d	**50.** c	**51.** c	**52.** b	**53.** d	**54.** a
55. b	**56.** b	**57.** a	**58.** b	**59.** d	**60.** a	**61.** c	**62.** d	**63.** d
64. c	**65.** a	**66.** d	**67.** c	**68.** c	**69.** b	**70.** a	**71.** b	**72.** a
73. d	**74.** c	**75.** a	**76.** b	**77.** d	**78.** b	**79.** b	**80.** c	**81.** d
82. b	**83.** a	**84.** d	**85.** c	**86.** b	**87.** c	**88.** c	**89.** d	**90.** b
91. d	**92.** d	**93.** b	**94.** a	**95.** a	**96.** c	**97.** a	**98.** c	**99.** d
100. c	**101.** a	**102.** b						

True & False

1. T	**2.** F Country road	**3.** T	**4.** T	**5.** T
6. F 0.5 m above	**7.** F Sloped camber	**8.** T	**9.** F Straight	**10.** F Flat
11. T	**12.** F $\frac{KN}{m^3}$	**13.** F 4.5 m	**14.** T	**15.** T
16. T	**17.** F Less	**18.** T	**19.** T	**20.** T
21. T	**22.** F Longitudinal	**23.** F Regulatory	**24.** T	
25. F Increase	**26.** T	**27.** F Crown	**28.** T	**29.** T
30. F Catch water drain		**31.** F Breast wall	**32.** T	**33.** F 0.5
34. T	**35.** F Near the top	**36.** T	**37.** T	**38.** F $\frac{A}{25}$
39. T	**40.** F Inversely proportional		**41.** F 90° angle	**42.** T
43. T	**44.** F Less	**45.** T	**46.** T	**47.** F 1.5
48. T	**49.** F 30 m	**50.** T	**51.** T	**52.** F
53. F 2.5 sec.	**54.** T $\left(\frac{1}{30}-\frac{1}{50}\right)$	**55.** F Drainage	**56.** T	
57. F 1934	**58.** T	**59.** T	**60.** F Nagpur	

61. F $\frac{82\ \text{Km}}{\text{Km}^3}$
62. T
63. F 30 m to 60 m
64. T
65. F (NH) National Highway
66. T
67. T
68. F Less than 55
69. T
70. F SSD is more
71. F100 KMPH
72. T
73. F Kept at 1.5
74. F Proportional to v^2
75. T
76. F 25 years
77. T
78. F 10%
79. T
80. T
81. F 30 m
82. T
83. T
84. T
85. F (y–x)%
86. F Inversely
87. T
88. T
89. T
90. F
91. F 1939
92. T
93. F 2000 Tonnes
94. T
95. F 80 cm
96. T
97. F Summit curves
98. F Parabolic
99. T
100. F
101. T
102. T

Fill in the Blanks

1. Nagpur
2. 9.77 m
3. 1200/Veh/Hr
4. 0.15
5. 2.5 m
6. Earthen Roads
7. Pyconometer
8. Temperature
9. California Bearing Ratio
10. 60 cm
11. Superelevation
12. 250 cm
13. Spot
14. 10 (Ten) cm
15. 42 days
16. Summit
17. Retaining wall
18. (r)
19. Both Right and Left
20. Right
21. Lanes
22. 2.5 sec.
23. Aerial
24. (Sixty) 60 cm
25. Volume
26. Sub-grade
27. 3000
28. 0.25
29. Camber
30. 30 cm
31. Lag
32. Top
33. $P(1 + r)^n$
34. Edges
35. 1719 m
36. 60 m
37. 2%
38. Pickers'
39. (Five) 5 cm
40. Raughness
41. (Twenty) 20 m
42. $R \tan \frac{\theta}{2}$
43. (Ten) 10 cm
44. 117 m
45. (Five) 5 m
46. 3.75 m
47. Right
48. 132 m
49. Frontage
50. Destination
51. (One) 1.0
52. Warning
53. Progressive
54. Spot
55. 50%
56. (Fifty) 50 m
57. (Ten) 10 m
58. Patching
59. Water Proofing
60. 25%
61. RT-5
62. Impact
63. Bitumen
64. Surfacing
65. (Thirty) 30 cm
66. Centrifugal
67. $\frac{v^2}{127\,R}$
68. $e = \frac{V^2}{225\,R}$
69. 1 in 30
70. 1 in 20
71. Speed of vehicle
72. Decreases
73. Less
74. Sum
75. Lane

76. Mountainous **77.** (Twenty five) 25 m **78.** 1 in 24

79. Camber **80.** State **81.** Right **82.** 1 : 3

83. High or Better **84.** 1250 Tonnes **85.** Gradient **86.** Positive **87.** 0.26

88. 8050 **89.** Lahore **90.** (Forty-five) 45 m **91.** Slip **92.** Plain

93. Concrete **94.** $\frac{-9 \text{ kg}}{\text{tonne}}$ **95.** Practical **96.** (Nine) 9 m

97. $3.42 \frac{\text{m}}{\text{S}^2}$ **98.** 10.5 m **99.** 3% **100.** Boundary

101. Monotony **102.** 100 cm 120 cm **103.** Inner **104.** 2.25 m **105.** Skid

SECTION 23

Tunneling

Definitions – 16

Multiple Choice Questions – 25

True & False – 12

Fill in the Blanks – 13

DEFINITIONS

American Method of Tunneling: In this method a drift driven into the top of the tunnel and supported by loggings, caps and two vertical poles. The sides of the drift are then widened and additional support is provided using timber planks and struts.

Army Method of Tunneling: Devised by the U.S. Army for construction of small tunnels of fairly shallow depth. The method mainly used for laying underground sewers.

Common Pattern of Cut Holes: Cut holes are drilled very near to each other as parallel as possible. The most commonly used pattern is the central wedge cut for square and rectangular sections. For circular sections sometime pyramid cut is also used.

Cut Hole: The blasting sequence in a tunnel or drift always starts from the **cut**, a pattern of holes at or close to the centre of the face, designed to provide ideal line of deformation. The placement, arrangement and drilling accuracy of the cut is crucial for successful blasting in tunneling.

Drift Method of Tunneling: A drift is a small tunnel measuring 3m × 3m, which is driven into the rocks and whose section is expanded in the later processes till it acquires the size of the tunnel.

Drilling and Blasting Method: A number of holes are drilled into the rock, which are then partially filled with explosives. Stemming, inert material, is packed into the holes to direct the explosive force into surrounding rock. Detonating the explosive causes the rock to collapse. Rubble is removed and new tunnel surface is reinforced.

Emergency Exit: In a tunnel emergency exits leading to open air are usually obligatory. Two standardised panels should be installed in front of the emergency exit, one of them visible from the direction of traffic and the other in opposite direction.

Forepoling Method in Tunneling: It has been frequently used to stabilize the ground around the cutting face, and to control the settlement of ground surface. Steel tubes (spiles) are installed in an umbrella shape and then injected to form a protective arch.

Needle Beam Method: It is a method which permits the roof of the tunnel section to stand without support for a few minutes. The needle beam consists of a stout timber beam or a composite flinched beam which form the main support during excavation. This method is suitable for soils where roof can withstand for a short time.

Road Tunnel: It is a subterranean passage designed for vehicular traffic, typically constructed beneath a city, mountain, body of water, or other physical obstructions. It enables faster and safe movement of traffic.

Steel Ribs: Also known as **steel arches** or **steel sets**, are structural elements used in tunneling and underground structures to provide temporary or permanent support. They are typically made of steel beams or plates and are installed to reinforce and stabilize the excavated tunnel or underground space.

The cut and Cover Method: It is the traditional form of tunneling that involves opening of the ground surface and excavating to the required depth. Once the construction is complete, the excavation is backfilled.

Trimmers in Tunneling: These are the **shot holes** drilled around the periphery of a shaft or tunnel that break or trim the sides of the excavation to the shape and size required.

Tunnel Boring Machine (TBM): It is a mechanical device used for digging tunnels through soil and rock. They are commonly employed in construction of road tunnels.

Tunnel Lining: Tunnels in loose rock and soft soils are liable to disintegrate and, therefore, a **lining** is provided to strengthen their sides and roofs so as to prevent them from collapsing.

Ventilation of Tunnel: It is the system installed in tunnels to ensure the circulation of air and remove vehicle exhaust flumes, smoke and heat. Proper ventilation is essential for safety, especially in longer tunnels.

MULTIPLE CHOICE QUESTIONS

1. Which of the following methods of tunneling is used for long tunnels at great depth.
   (a) Army method (b) Needle beam method
   (c) Austrian method (d) English method
2. Drifters can be used to drill.
   (a) Only up holes (b) Only down holes
   (c) Horizontal or up holes (d) Horizontal down or up holes
3. Which of the following lining material is useful for shield driven tunnels in sub-aqueous regions?
   (a) Stone masonry (b) Timber (c) Cast iron (d) Cement concrete
4. Ribs are used for strengthening and stiffening the liner plate for tunnels of diameter greater than
   (a) 2m (b) 3m (c) 4m (d) 5m
5. The most suitable soil for compressed air tunneling is
   (a) Silt (b) Sand (c) Clay (d) Gravel
6. Drift method of tunelling is used to construct tunnels in
   (a) Soft grounds (b) Rock (c) Broken grounds (d) Self supporting grounds
7. In 'full face' method of constructing tunnels, the first operation relates to
   (a) Removal of bottom portion (b) Excavation of one drift in the centre
   (c) Removal of top portion (d) Excavation being done along the periphery
8. Which one of the following tuneling methods is used for laying underground sewers?
   (a) Needle beam method (b) Army method
   (c) German method (d) Italian method
9. Which of the following is a component of a shield for tunneling?
   (a) Liner plate (b) Trench jack (c) Stiffener (d) Cutting edge
10. If 'D' is the diameter of tunnel in (m), then the thickness of lining in (mm) is recommended as
   (a) 82 D (b) 42 D (c) 104 D (d) 124 D

11. If (N) is the number of shafts used, then the total number of faces available for attacking the excavation and construction in tunnels are
   (a) 2 N (b) N + 2 (c) 2 N + 1 (d) 2 N + 2
12. Which of the following are percussion drills?
   (a) Shot and Diamond (b) Wagon and Churn
   (c) Shot and Churn (d) Diamond and Wagon
13. Circular section of tunnels is not suitable for
   (a) Carrying water (b) Non-Cohesive soils
   (c) Tunnels driven by shield method (d) Placement of concrete lining
14. The most commonly adopted pattern of cut holes is
   (a) Horizontal wedge cut (b) Pyramid cut
   (c) Fan cut (d) All these
15. To attain required shape of tunnel section, we use
   (a) Easers (b) Trimmers (c) Cut holes (d) Chisels
16. Which of the following methods of tunneling does not require the use of time bearing?
   (a) Linear Plate method (b) Shield method
   (c) Compressed air method (d) All these
17. The length of the needle beam used for tunneling is normally
   (a) 2 m to 4 m (b) 2.5 m to 6 m (c) 4 m to 7 m (d) 5 m to 6 m
18. For full face method, the excavation to be done is divided into sections
   (a) 3 (b) 2 (c) 4 (d) 5
19. For tunnels exceeding 300 m in length, the grade should be provided below (of ruling gradient)
   (a) 50 % (b) 60 % (c) 75 % (d) 80 %
20. Forepoling method is generally adopted for tunneling in
   (a) Soft ground (b) Firm ground (c) Running ground (d) None of these
21. For Broad Gauge single track railway, the height of tunnel above top rails should be
   (a) 5.5 m to 5.8 m (b) 5.8 m to 6.2 m (c) 6.7 m to 7.3 m (d) 7.3 m to 7.5 m
22. A good blast with good yield is obtained if the cut hole is
   (a) Normal to face (b) Inclined at 45° (c) Inclined at 15° (d) Inclined at 30°
23. American method of tunneling
   (i) is suitable for large size tunnels (ii) is not suitable for railway or highway tunnels
   (iii) Requires heavy timbers
   (a) Only (i) is correct (b) (i) and (iii) are correct
   (c) (ii) and (iii) are correct (d) (i) and (iii) are correct
24. In case of drift method of tunnelling, the drift may be excavated at
   (a) Centre (b) The top or bottom
   (c) The side (d) All these

25. Pick up the correct statement regarding drilling a tunnel from the following
    (a) Holes are drilled by pneumatically operated rock drills
    (b) One drill is required for each 4 to 5 m^2 force area
    (c) The diameter of the bore hole at the deepest point should be 6 mm more than the diameter of the cartridge
    (d) All these

TRUE OR FALSE QUESTIONS

1. In case of Railways tunneling is preferred over other options like detour or open cut. (T or F)
2. Railway tunnels are generally Rectangular or Circular. (T or F)
3. Shotcrete is commonly used as a temporary support in tunneling project. (T or F)
4. The tunnel boring machine (TBM) is only suitable for hard rock conditions and cannot be used in soft ground tunneling. (T or F)
5. The drill and blast method is rarely used in modern tunnel construction due to its high cost and low efficiency. (T or F)
6. Tunnel ventilation is not needed in tunnels shorter than 500 meters. (T or F)
7. Cut-and-cover tunneling is a method typically used for shallow tunnels. (T or F)
8. The use of explosives is not allowed in Urban tunneling due to safety concerns. (T or F)
9. A cross passage connecting two parallel tunnels often used as emergency escape route. (T or F)
10. Trimmers are used to attain required shape of tunnels. (T or F)
11. Tunnels are constructed for railways only. (T or F)
12. Ribs are used for strengthening and stiffening the liner plate for tunnels of diameter more than 3 m. (T or F)

FILL IN THE BLANKS

1. Circular section of tunnel is not suitable for placement of .... lining.
2. Drifters can be used to drill down or up and also .... holes.
3. If N is the number of shafts used then the number of faces available for attacking the excavation and construction in tunnels are ....
4. 6 m^3 per minute per unit of face area free air is provided in .... air tunneling.
5. For shield driven tunnels the most suitable material for lining is ....
6. For a tunnel of diameter D meters, thickness of lining is recommended as .... in mm.
7. Fore poling method is generally adopted for tunneling in .... ground.

8. Explosive used for tunneling in soft rocks is .... dynamite.
9. For highways, tunneling is preferred to if the open cut exceeds .... in depth.
10. Concrete lining is provided concurrently with the driving operation in case of .... terrain.
11. For a broad gauge single track, the section of the tunnel must have a width .... to ....
12. A good blast with good yield is obtained if the cut hole is inclined at .... degrees to the face.
13. For tunnels of more than 300 m long the grade should be below .... % of ruling gradient.

ANSWERS

Multiple Choice Questions

1. c **2.** d **3.** c **4.** b **5.** c **6.** b **7.** d **8.** b **9.** d
10. b **11.** d **12.** b **13.** d **14.** d **15.** b **16.** d **17.** c **18.** a
19. c **20.** b **21.** c **22.** b **23.** a **24.** d **25.** d

True & False

1. T **2.** F Polycentric **3.** T **4.** F **5.** F **6.** F **7.** F **8.** F
9. T **10.** T **11.** F **12.** T

Fill in the Blanks

1. Concrete **2.** Horizontal **3.** 2N + 2 **4.** Compressed
5. Cast Iron **6.** 82 D mm **7.** Soft **8.** Ammoniya
9. 15 m **10.** Rock **11.** 4.9 m to 5.5 m **12.** 45O **13.** 75 %